Tulane
School of
Architecture

what
is
affordable?

Tulane School of Architecture
Tulane University, May 2024

Editor
Andrea Bardon de Tena

Publication Assistants
Chelsea Kilgore
Grace Ostendorf
Olivia Roginson
Livi Bowers

Copy Editing
Olivia Roginson

Images
[cover] Ella Weeks & Jess Seidl
[inside cover] Akhil Singh - 100 faces of TuSA, 2023
[p 14] Riley Siltler
[p 92] Seth Laskin
[p 94] salazarsequeromedina
[p 138] Brianna Baldwin
[p 204] Preservation Program Visit
[p 306] Barbados Research Studio Field Trip
[p 308] Sofia Vladimir
[p 376] Zaynab Eltaib

Graphic Design
Actar –Tulane School of Architecture

Printing and Binding
Arlequin, SL. Spain

Indexing
ISBN PRINTED EDITION 978-1-63840-860-4
Library of Congress Control Number: 2024946610

Publication date:
September 2024

HOW TO USE THIS BOOK

THIS IS NOT A YEARBOOK

Indeed, like a yearbook, this book is composed of the materials produced by students and professors during two academic years, spanning six semesters: 2021-2022 and 2022-2023. However, its ambition goes beyond presenting the work being done at Tulane School of Architecture. It is designed to spark conversation about potential opportunities within the built environment.

You don't have to read it in order.

You can use it as a compilation of sheets, organized to explain the school's research in the context of real-world application. The topics are created to orient the work, categorizing them by variations of affordability.

You can use it to understand the physics of Tulane School of Architecture. Explore it in its entirety. Look at the work our professors do, the courses we develop, or the final theses that students present as a culmination of their journey here.

The following captions are guides to help you navigate this document.

KEY

Whose work

s>	**i>**	**UG**	**GR**	**UG + GR**
Student	instructor	undergraduate	graduate	both

When: year and semester

FA	**SP**	**SU**
Fall	Spring	Summer

Programs

arch	**desg**	**pres**	**sred / msred**
architecture	design	preservation	real estate

CONTENTS

008	Forewore	Dean, TuSA	Iñaki Alday
010	Introduction	The ReView Editor	Andrea Bardon de Tena

1 / ECONOMICALLY AFFORDABLE /

016	Article	Economically affordable	Casius Pealer
022	Faculty work	Pine Street	Emilie Taylor Welty
026	ARCH 2/6022	Inhabitation	Affordable development
036	ARCH 3031	Urban	Kansas city
048	ARCH 3/6032	Integrated	Poydras Offices
060	SRED 3080	Design Development Studio II	
064	MSRED 6740	Directed Research	
066	MSRED 6220	Sustainable Design and Planning	
070	Research studio	Town. Houses.	Jonathan Tate
078	Research studio	URBANBuild 17	Byron Mouton
086	Final Thesis	UG	Ethan Lewis
088	Final Thesis	GR	Daniel Tighe

2 / ENVIRONMENTALLY AFFORDABLE /

Materials life cycle

096	Article	Climate change & design	Sonsoles Vela
104	Faculty work	Outdoor room	Juan Medina Revilla
108	Faculty work	Entangled Matter	Cordula Roser Gray
112	ARCH 6012	Tech-Shelter!	
118	DESG 2005	Design Studio I	
124	DESG 3005	Textile Design	
128	Research studio	Resilience Reinforced	Kentaro Tsubaki + Charles Jones
134	Final Thesis	UG	Alex Langley + Sam Spencer
136	ARCH Thesis	UG	Seth Laskin

Natural resources

140	Faculty work	Circle Bar Ranch	Kenneth Schwartz
144	Faculty work	Shade-water Pavilion	Irene Keil + Judith Kinnard + Nick Jenisch
148	ARCH 2/6021	Jefferson Island	
168	Research studio	Gulf NAS	Margarita Jover + Liz Camuti
178	Research studio	Yamuna River Project	Iñaki Alday + Pankaj Vir Gupta + Andrea Bardon
186	Research studio	Addis Ababa	Ruben Garcia Rubio
194	ARCH Thesis	GR	Sean Fowler
196	ARCH Thesis	GR	Megan Spoor
200	Final Thesis	UG	Leah Bohatch + Camille Kreiser

3 / CULTURALLY AFFORDABLE /

206 Article US Canal Zone Research Carol Reese
210 Faculty work Luanco Center Ruben G Rubio + Sonsoles Vela
214 ARCH 1011 Discover
224 ARCH 1012 Space
244 ARCH 6011 From concepts and to programs
250 DESG 3010 Prototyping Solutions
256 Research studio Arch. Historic Contexts Ammar Eloueini
264 Research studio Ecuador Edson Cabalfin
276 TuSA Trip Ecuador
280 TuSA Trip Barbados
284 TuSA Trip Mexico
288 TuSA Event Re-New Orleans Enmanuel Osorno
292 TuSA Event Design Symposium
294 TuSA Event Design Showcase
298 DESG 2305 Visual Communication & Advocacy
302 Final Thesis UG Tiger Thepkanjana
304 Final Thesis GR Giuliana Vaccarino Gearty

4 / SOCIALLY AFFORDABLE /

Housing

310 Article Socially affordable Edson Cabalfin
316 Faculty Torre Baro Housing Iñaki Alday + Margarita Jover
320 ARCH 3/6032 Hostel
334 ARCH 2/6022 Inhabitation
346 TuSA Trip Barcelona Housing
350 TuSA Event Home sweet home Omar Ali + Nimet Anwar
354 Research studio URBANBuild 18 Byron Mouton
362 Research studio Competitions Juan Medina Revilla
372 Final Thesis UG Alyssa Barber + Olivia Georgakopoulos

Public infrastructure

378 Faculty work Lines of Flight Adam Marcus
382 ARCH 3031 Urban: Los Angeles
394 Research studio STEM Library Emilie T Welty + Jose Cotto
402 Research studio Sugar Roots Emilie T Welty + Jose Cotto
410 Research studio Public Space & Scrutiny Tiffany Lin + Emilie T Welty
418 Final Thesis UG Mitchel Hubbell
420 Final Thesis UG Andreea Dan + Tess Temple
422 Final Thesis UG Merrie Afseth + Connor Little

428 TuSA Events & Lectures
432 TuSA Faculty & Staff

2024 FOREWORD

Iñaki Alday
Dean of Tulane School of Architecture

In a world of interlocking systems, we know that the main ones – environmental, social, cultural, and economic – do not re-balance themselves (Purdy). *"What's Affordable"* asks about what we can afford as a society in the built environment. At a time of crisis in integrative thinking, in which political discourse has shifted towards questions of identity, perhaps it is time to recover the humanist framework formulated in 1932 (originated almost 100 years earlier) based on economics (oikonomos, the management of the house, humanities common home) and the aspiration to equity. Systems do not find balance by themselves and inequality is growing -exacerbated by the environmental crisis- among the most economically, socially, racially or geographically vulnerable populations. New Orleans is one of the grounds zero in the United States in terms of environmental and social vulnerability, dramatically increased by climate change. Under this circumstances, what can we afford?

In the school's work on affordability, you won't find too many pro-formas with figures and quantitative arguments. However, the pedagogies of all programs are imbued with social, ecological, and economic (in its current and etymological meanings) responsibility designed to educate leaders who will transform the built environment. The research projects also aim to change policies, to have an economic impact, and to improve living conditions in New Orleans, Louisiana, and elsewhere around the world. Not only do URBANbuild homes provide low-cost housing, but they have changed city regulations by doubling the number of units per plot, with the potential of doubling density using the same infrastructure and reducing the impact of land cost. Gulf Futures explores the urgent energy transition and the restoration of the ecological and social fabrics of the Gulf South, preparing scenarios that can be translated into policies and concrete actions. Learning about domestic space, human scale, horizontal and vertical aggregations, and city-building is disciplinary core. And it's similarly core to do it with a vision, creating intentional communities and systems adaptable to each inhabitant.

Sustainable Real Estate uses finance not as an end in itself but as one more component of the spatial, environmental and social impact that Development has on the transformation of cities and territories. The discourse of Preservation goes beyond conservation and material studies: How much more can we tear down and rebuild? Isn't this the time to preserve, adapt and reuse everything built around us? The Design program applies this principle to the scale of the matter and problem solving, starting from the remains of noble and humble materials to design backpacks, lamps or clothing.

"What's Affordable" questions how far we can go with our usual practices and how to change them through pedagogy, research and advocacy. But it's also literal. All actions in the built environment have an economic cost, with a column of benefits and another of losses always assigned to the same social groups. "Winners take all..." Pollution, resource exploitation, displacement of communities, demolition of structures and the blight of our neighborhoods can and should be accounted for in the balance sheet, corrected and reallocated to those with responsibility and resources. But, above all, rethought with a new socio-ecological vision, socially just and designed to sustain the planet and life on it.

INTRODUCTION

Andrea Bardon de Tena
The ReView editor

Affordable can be an ambiguous term. In architecture, it is used so commonly that it often loses its clarity and rigor. Critically, we examine the term in the context of contemporary challenges: climate change, social inequalities, cultural heritage connections and the unbalanced global economic situation. This book, entitled "What is Affordable?", explores the different facets of affordability in the world of architecture through the multidisciplinary work of the Tulane School of Architecture.

Before addressing specific topics within these pages, it is imperative to explain how we understand "affordability" within contemporary reality. What is affordable now must also be sustainable over time. Therefore, we understand affordability to be directly connected to the economic possibilities, environmental circumstances, cultural complexities, and social balances of the different contexts. Affordable and sustainable architecture cannot be detached from the commitment to be a long-term response and the multidisciplinary collaborative work.

This book brings together the work of the different programs (Architecture, Design, Real Estate, Historic Preservation, Landscape, and Social Innovation) that coexist and intersect within the Tulane School of Architecture, New Orleans, USA. This volume includes a sample of academic coursework (between 2021 and 2023), faculty professional practices, students' final theses, the school research, and activities and events held in the school. Among them, we highlight the written contributions of professors Casius Pealer, Edson Cabalfin, Carol Reese, and Sonsoles Vela. The university presents a great diversity of topics and approaches contributing to a common desire to address problems of the contemporary world.

To unravel the enigmas of affordability, we divide this broad term into four dimensions that organize the chapters of this book: social, cultural, economic, and environmental. The student's and faculty's contributions present a diversity of contexts, dilemmas, and multi-scalar proposals. Due to the multidisciplinary vision developed by the university, many works respond to more than one dimension of affordability. We invite the reader to identify these different connections between works and themes.

Environmentally

Architecture's relationship with natural resources, especially water, is a fundamental lens for reimagining connections between the built environment and the natural world. Our research about local environments and our architectural proposals appear interconnected, whereas we, architects, respond to the challenges and opportunities of each site. On the other hand, supported by the life cycles of materials, we think of construction, of any scale, with the logic of time and the use that its components provide us with.

Economically

Through the prism of economic viability, we unveil innovative strategies to optimize the use or reuse of resources and maximize the multi-function of our spatial proposals. In classes throughout the school, particularly in the Real Estate program, we seek to promote long-term economic resilience. Locally, through the URBANbuild program and other research initiatives, we explore the possibilities for sustainable housing construction for the city of New Orleans.

Culturally

The city of New Orleans, as well as many spotlighted locations in Barbados, Ecuador, India, etc., is a sample of the diversity of cultural contexts where students and instructors develop their research and proposals. The dialogue with cultural heritage is not only a goal of the Historic Preservation program but a common focus across the entire school. The school also seeks to explore different ways of communicating and connecting contemporary ways of life and culture, both in design and architecture.

Socially

Lastly, working towards social affordability, the work explores two major topics: access to public housing and the provision of public infrastructure. Both point in the direction of working towards contexts where social equity prevails. In this chapter, we can also find the work of The Albert and Tina Small Center for Collaborative Design, which empowers marginalized local voices, seeks social cohesion, and fosters a sense of belonging through empathetic design interventions and participatory processes.

In an era of unprecedented global challenges, the Tulane School of Architecture feels an obligation to link its efforts to the sustainability of the built environment. Collective work, both theoretical and propositional, must serve as a response to the sustainability of ecosystems and must consider the term affordability as a paramount characteristic. Immersed in the multidisciplinary complexity, we invite the reader to join in this commitment to collectively reimagine the contours of affordability in architecture and transcend conventional terms. And, through this collaboration, we can work together towards a path where economic, social, cultural, and environmental sustainability lead to a more inclusive, resilient, and regenerative future.

Barbados
Studio + Trip
Cultural
Reese
Article
Panama Canal
Eloueini
Historic Contexts
G.Rubio/Vela
Luanco
ARCH 1011
Discover
Osorno
Re-New Orleans
ARCH 6011
Culture
ARCH 1012
Space
Ecuador
Studio + Trip
DESG 3005
Textile
Vaccarino
Sicily
Design
Symposium
Alday/VirGupta/Bardon
Yamuna River Project
Design
Showcase
Langley & Spencer
Immaterials
DESG 2005
Lamps
DESG 3010
Prototyping
Laskin
Antarctica
G. Rubio
Addis Ababa
Tsubaki/Jones
Resilience Reinforce
Jover/Camuti
Gulf Coast
Fowler
New Orleans
Environmental
Spoor
Welsh Coastline
Taylor Welty
Pine St
Roser Gray
Entangled Matter
Vela
Climate change and design
Schwartz
Circle Bar Ranch
ARCH 2021
Site

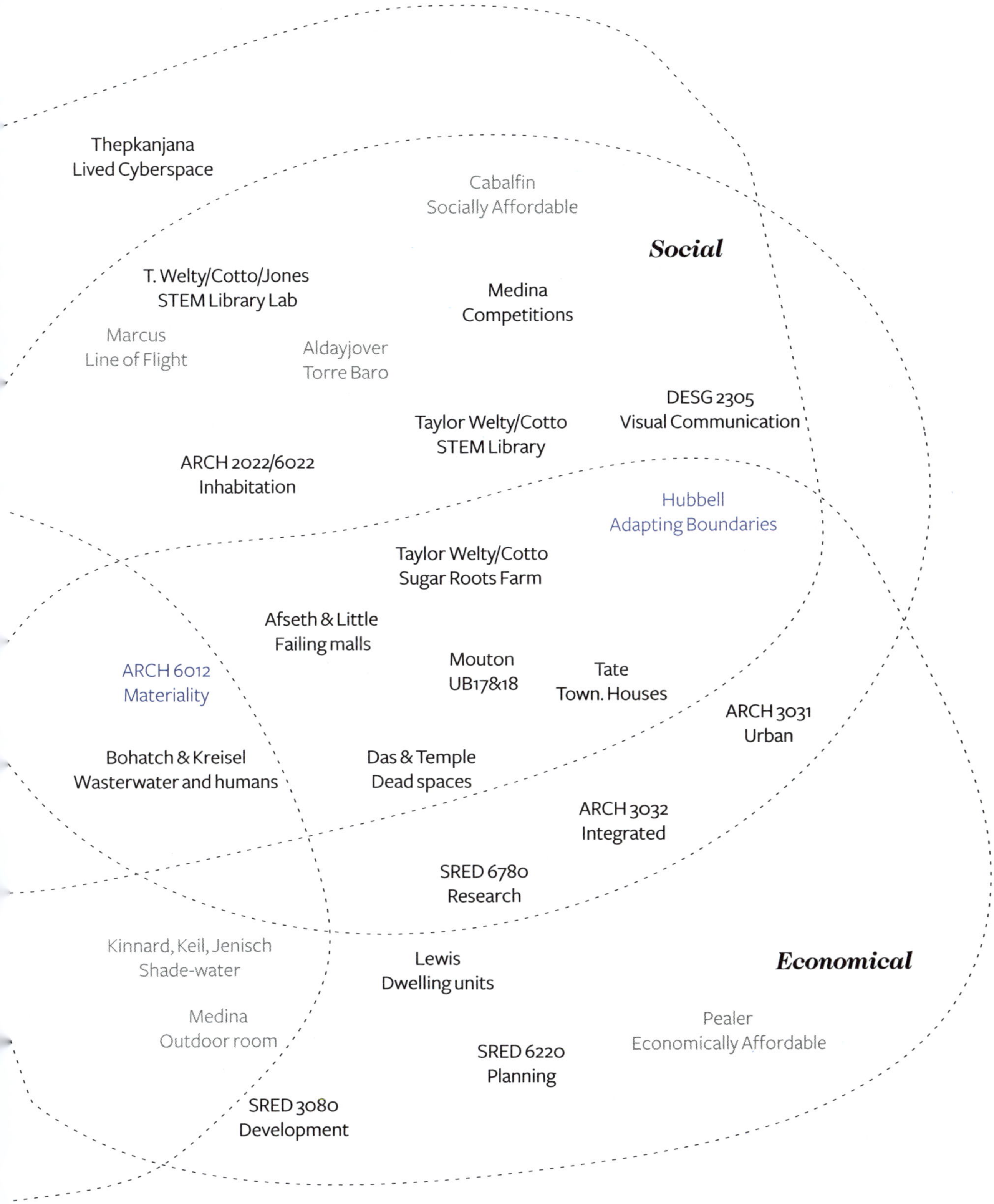
Thepkanjana
Lived Cyberspace
Cabalfin
Socially Affordable
Social
T. Welty/Cotto/Jones
STEM Library Lab
Medina
Competitions
Marcus
Line of Flight
Aldayjover
Torre Baro
DESG 2305
Visual Communication
Taylor Welty/Cotto
STEM Library
ARCH 2022/6022
Inhabitation
Hubbell
Adapting Boundaries
Taylor Welty/Cotto
Sugar Roots Farm
Afseth & Little
Failing malls
Mouton
UB17&18
Tate
Town. Houses
ARCH 6012
Materiality
ARCH 3031
Urban
Bohatch & Kreisel
Wasterwater and humans
Das & Temple
Dead spaces
ARCH 3032
Integrated
SRED 6780
Research
Kinnard, Keil, Jenisch
Shade-water
Lewis
Dwelling units
Economical
Medina
Outdoor room
Pealer
Economically Affordable
SRED 6220
Planning
SRED 3080
Development

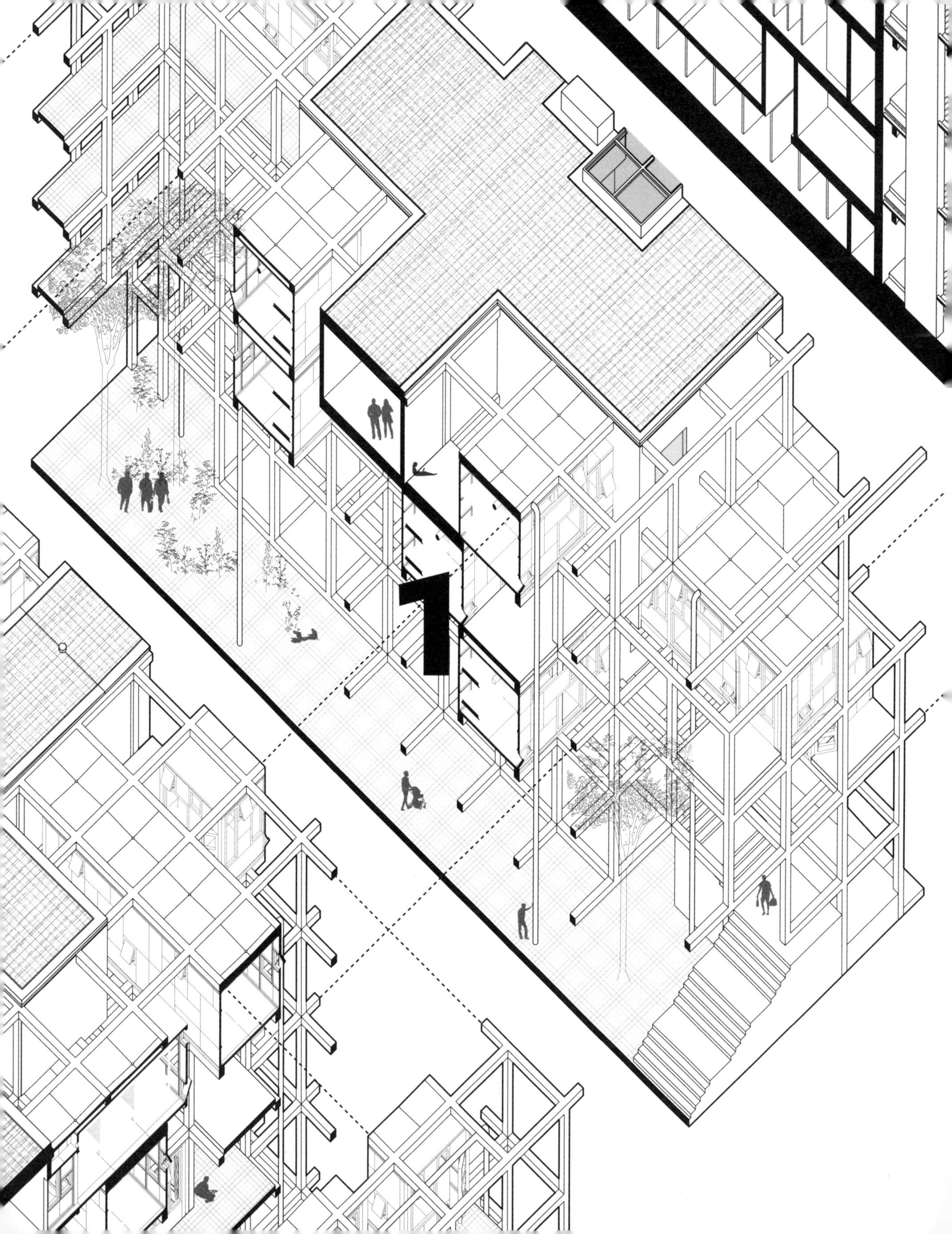
1

ECONOMICALLY
AFFORDABLE

CASIUS PEALER
Shane Professor of Practice
Director of Real Estate Development Program

WHAT IS ECONOMICALLY AFFORDABLE IN THE BUILT ENVIRONMENT?

"Don't tell me what you value—show me your budget, and I'll tell you what you value."

Economics, broadly, is just a means of inquiry into how individuals and societies allocate scarce resources. Those resources can be financial, but they can also be environmental, social, or even cultural. In that sense, economics is not always or certainly not only about money, but rather about efficiency. Economics doesn't determine affordability—rather it helps to identify the trade-offs necessary in order to afford something we value.

For example, if my daughter says, "Dad, I want a pony," I might respond truthfully that "honey, we can't afford a pony." But my focus on affordability here purposely avoids any discussion about values and goals. It suggests that I agree it would be great to have a pony, but that we can't do it. And because our 6-year-old doesn't understand finance, the conversation will mostly end there.

But instead, if my wife turns to me and says, "I want a pony"—well, then we'll have a much longer conversation. That conversation will start with values and goals ("why exactly do you want a pony?"), and then identify various assumptions, trade-offs, and timeframes. It will include money, certainly. But we'll also do additional research ("do we need a zoning variance?"), consider alternatives ("can we rent a pony?"), and in the end quite possibly develop a plan to, in fact, afford a pony.

In 2011, Tulane's School of Architecture developed our first real estate finance courses for the graduate program in sustainable real estate development. In 2015, we expanded our real estate finance courses at both the graduate and undergraduate levels. These courses generally produce the most anxiety for students, in part because there is more obviously a right or wrong numerical answer to be entered into a specific cell in a complex spreadsheet. In the best cases, this anxiety gives way to a sense of confidence when the numbers "work."

Our long-time finance professor, Richelle Allen, likes to say that "each cell in your spreadsheet represents a person." That person may be a lender or investor, or it may be a contractor, insurance broker, permitting officer, utility provider, maintenance team, leasing staff, etc.—but it helps students to see and understand the relationships among the numbers. Relatedly, these courses are often as much about language and fluency as they are about numbers or math. The real importance for students across disciplines to understand project finance is not to be able to calculate a cap rate or after-tax cash flow, but to be able to stay in the room when decisions are made and to effectively lead a larger conversation about values and goals. Because otherwise "the project can't afford it" will be the end of the conversation rather than the beginning.

"If you torture the data long enough, it will confess to anything."

The most widespread measure of financial affordability in real estate is the federal government's metric for housing affordability: housing costs (including utilities, taxes, and insurance) should be less than 30% of a household's gross income. By this metric, a family of four earning 80% of the Area Median Income (AMI) in New Orleans ($65,000 in 2023) is able to afford $1,625/month in rent and utilities.[1] However, according to

1 *New Orleans HUD Median Family Income*, City of New Orleans, Office of Community Development, 2023 (available at https://nola.gov/nola/media/Community-Development/Direct%20Homebuyers%20Assistance/2023-New-Orleans-HUD-Income-Limits.pdf).

U.S. Census data, over 50% of renters and nearly 30% of homeowners in New Orleans were cost-burdened in 2022, meaning they were paying more than 30% of their income in housing costs. And fully 32% of renters and 16% of homeowners were severely cost-burdened, defined as paying more than 50% of their income towards housing costs.[2]

Economically, this level of unaffordability is very much afforded by individual households and by the country as a whole. Nationally, just ¼ of households that qualify for housing subsidies actually receive them due to funding limitations at the federal level.[3] But even this measure of affordability is just a widely shared assumption. When Congress first established a payment threshold for housing in 1969, the assumption was that just 25% of income was affordable. This initial metric was based on a Depression-era rule of thumb of "a week's wage for a month's rent."[4] When Congress passed legislation in 1983 raising the measure of affordability from 25% to 30% of household income, a lot of newly "affordable" units were created overnight. But nothing about economics or income or actual costs changed—only the underlying assumptions did.

More recently, researchers have argued for expanding this core housing affordability metric to include transportation costs, since those expenses are usually closely tied. The thinking is that housing costs far from transit or other community resources might look affordable in isolation, but come with hidden transportation costs (typically a family's second-largest expense) and thus not ultimately be affordable.[5] The point isn't that they are right or wrong, but simply that starting with different assumptions and having different goals can yield quite different results in what might otherwise seem like just a math problem.

---

2 *Who Lives in New Orleans and Metro Parishes Now?*, The Data Center, October 6, 2023, analysis of U.S. Census Bureau data from American Community Survey 2004 and 2022 (available at https://www.datacenterresearch.org/data-resources/who-lives-in-new-orleans-now).

3 *Funding Limitations Create Widespread Unmet Need for Rental Assistance*, Center ofr Budget Policies and Priorities, February 15, 2022 (available at https://www.cbpp.org/research/housing/funding-limitations-create-widespread-unmet-need-for-rental-assistance).

4 Pelletiere, Danilo, *Getting to the Heart of Housing's Fundamental Question: How Much Can a Family Afford?*, NLIHC 2008, available at https://nlihc.org/sites/default/files/affordabilityresearchnote2-19-08.pdf

5 The Center for Neighborhood Technology produced an interactive H+T Affordability Index to visualize the difference in affordability generated by this different assumption, available at https://htaindex.cnt.org. An example from Metro New Orleans of the difference in affordability generated by this assumption is shown here in Figs. 1 and 2.

At a project level, similar changes in background assumptions and conditions can impact whether a project is affordable or feasible. Many project teams have experienced this recently with a rapid increase in what were previously historically low interest rates, making a lot of conventional

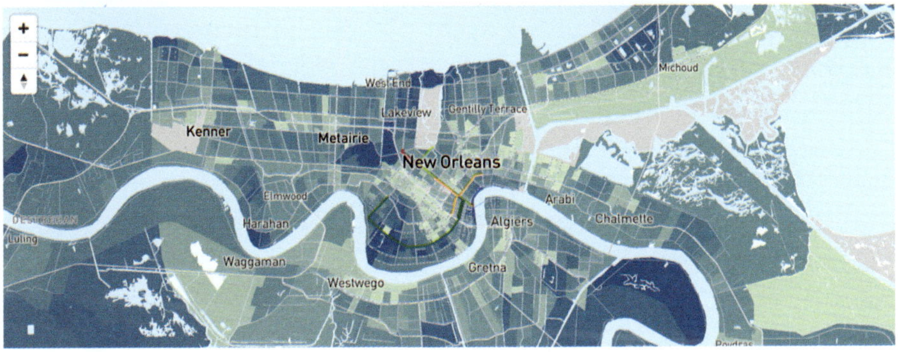

Fig.1 & 2
Darker areas in the maps highlight census tracts that are unaffordable, and lighter areas suggest affordability.

Figure 1 shows housing costs alone (30%), indicating greater affordability in suburban New Orleans.

Figure 2 shows housing + transportation costs (45%), indicating much more widespread unaffordability.

projects suddenly no longer affordable. Of course, this isn't a surprise and is affecting all projects—but even a fairly simple development budget includes possibly hundreds of assumptions. Typically, these assumptions are based on long experience in a particular market; but often they include extrinsic factors and a risk analysis that can be changed outside of the project analysis itself. The result is that a project where "the numbers don't work" can often become one where the numbers do work simply by changing factors outside of the project itself.

A good deal of a lender's or an investor's underwriting process is designed to ensure that their assumptions and level of risk aligns with the project budget and pro forma presented by a developer or owner. This is a level of detail that few outside parties will have access to engage in, but overly conservative assumptions can sink a deal or force significant concessions elsewhere. Conversely, overly optimistic or aggressive assumptions can make a project "pencil out" but be a highly risky investment. Of course, a developer who is willing and able to put up a large down payment or a personal guarantee can greatly reduce the risk and make the financing work on a project that might not work for another developer.

"It is difficult to make predictions, especially about the future."

A key goal to achieve affordability then is to reduce risk. One variable central to risk is time. In the early 1900s, most Americans could not afford to own their own home. Typical loans required a 50% down payment, with the full amount due or fully refinanced in 5-10 years. These terms were designed primarily to reduce the risk of loss for lenders, and just over 40% of American families owned homes of any kind, often meeting only minimal fire and building codes.

Following the Great Depression, however, the federal government structured and capitalized a system of housing finance based on a standard 30-year self-amortizing mortgage, typically with a 20% down payment. These terms were decidedly not driven by the private market, but rather by government allocations of capital and loan purchases that served to greatly reduce the risk to private lenders willing to adopt these terms. As a result, homes were suddenly more affordable and accessible to American families, even when incomes and home prices did not change. [6]

On commercial developments however, standard repayment periods vary widely but are typically much shorter, largely to reduce risk for investors and lenders. The result is that a project where "the numbers don't work" can often be made to work with different financing on different terms—terms or products not always available to every developer. In particular, this shorter repayment period can have a significant impact on the affordability of energy efficiency measures that are otherwise projected to pay for themselves well before the end of useful life of the equipment—say, 7-8 years. But if the investors require repayment in five years, then these cost savings alone might not be enough to make any additional initial investment pay off. Of course there may be other reasons to make the investment, and the owner might believe she can capture the future value in a sale or refinancing, etc. But a seemingly straightforward discussion about financial affordability just became a lot more complex.

---

6 *Historical Census of Housing Tables: Homeownership*, U.S. Census, 2000, revised October 8, 2021 (available at https://www.census.gov/data/tables/time-series/dec/coh-owner.html).

Relatedly, one reason the federal government and many community institutions have led the way in green building efforts is that they often have much longer timeframes for projected ownership than private sector or commercial enterprises. A major challenge for green building and energy efficiency efforts remains retrofitting existing multifamily buildings. But local public housing authorities were among the early leaders in this work, in part because they had access to longer term (20-year) financing paid for by utility savings. [7]

Questioning assumptions, creatively addressing risks, and understanding the value of time.

In the end, there is real strength in a design and development professional who is fluent in the language of finance and who can lead a discussion centered on advancing a shared vision. Economic affordability is crucial, and it can only be achieved by questioning assumptions, creatively addressing risks, and understanding the value of time. Building confidence in these skills is a goal of the finance coursework and multidisciplinary offerings at the Tulane School of Architecture.

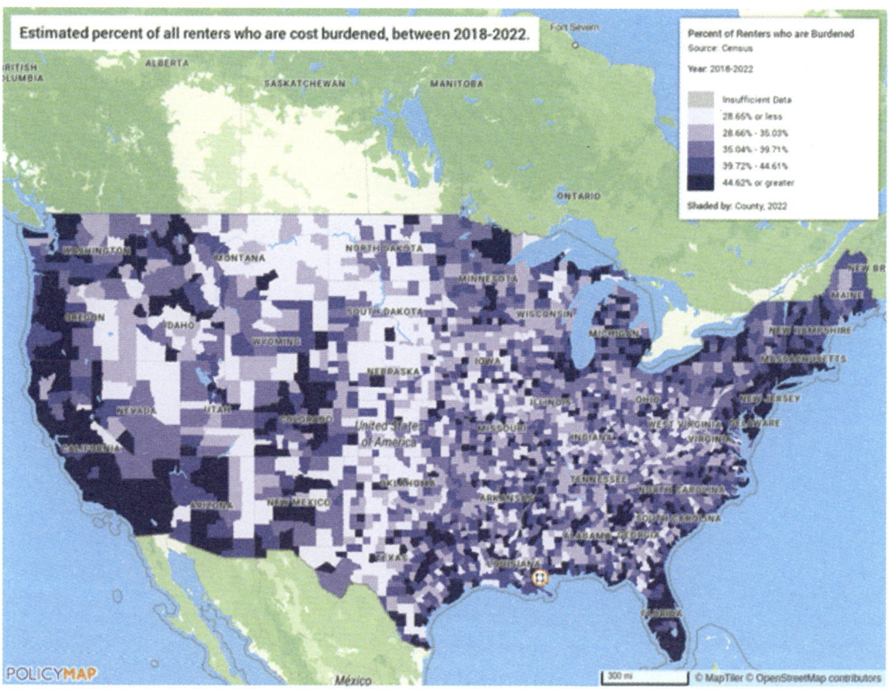

Fig.3
Estimated percent of all renters who are cost burdened, 2018-2022. PolicyMap, https://plcy.mp/271bsfKs (based on data from the U.S. Census; Accessed Feb 19, 2024).

---

7 *Review of Energy Performance Contracts in Public Housing,* U.S. Department of Housing and Urban Development, Office of Policy Development & Research, March 20, 2020 (available at https://www.huduser.gov/portal/publications/epc-evaluation.html).

EMILIE TAYLOR WELTY
& SETH WELTY

PINE STREET

RESIDENCE
New Orleans, 2022

Credits
Colectivo (architecture + photography).
Nick Licausi (drone shots).

This home started with the question, How can we rethink an urban infill project to build higher density on high ground and cultivate more life on site? This home stacks three units into a thin L-plan to capture light, natural ventilation, and views while saving ground space to create lush native plant gardens, stormwater gardens, and covered outdoor spaces for crawfish boils and gatherings.

Additionally, the material choices nod to context while reimaging common things in uncommon ways. The white ribbed metal of the upper main unit is in conversation with the industrial sites on the working riverfront nearby. Meanwhile the base is grounded in a terracotta rain-screen salvaged and repurposed from a nearby house - the earthen material brings a rich, tactile, and human scale to the home while also hardening the base in our flood-prone city.

The design team worked to create a site plan that captures rainwater on site and creates native habitat for the bees the owners keep in the yard. What was once a lot housing a low-slung single-family home spread out across the buildable footprint of the site, is now a lot teaming with rain gardens, pollinator plants, compost, a treehouse, and a dozen fruiting trees along the street front. This home provides an example of ways that we as architects can increase density while also repairing the ecological damage our profession has done in the urban landscape.

In sum, the Pinecone house is a test case in building sustainably and adaptively in a disaster-prone context. The house is designed to weather extreme conditions but also nurture backyard ecosystems and day-to-day connections with the outdoors.

Fig.1
Backyard
Photo. Collectivo.

Fig. 2 Level 1 plan
Access, Main floor

Fig.3 Stairs
Photo. Collectivo.

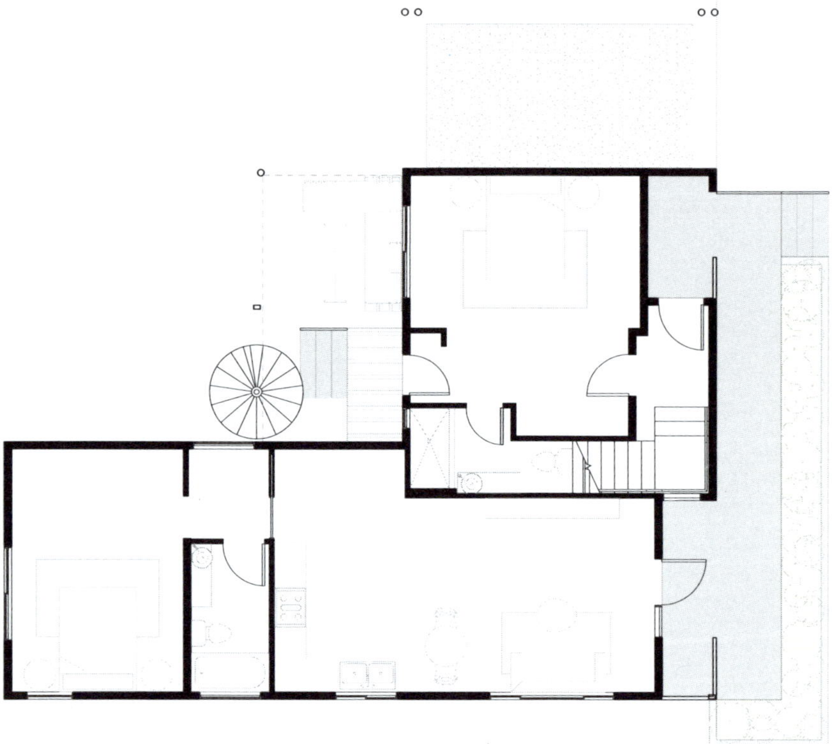

Fig. 4
Dining space
Photo. Collectivo.

COLLECTIVE HOUSING

Margarita Jover [C]
Cynthia Dubberley
Juan Medina
Sonsoles Vela
Jesús Utrillas
Emmanual Osorno
Pia Mendaro

ARCH 2022-6022 Core studio is organized into two parts. The first 7 weeks are dedicated to designing 4 small quick projects: horizontal, vertical, sectional homes and the inhabited stair. The second part of the course is dedicated to the design of the collective housing project that combines the previous designs.

The project is the design of collective housing with shared amenities and collective open spaces- it occurs densifying an existing urban block of 300 feet by 300 feet (100meters). The cultural challenge is to keep the 'suburban' lifestyle but with mixed-use, mixed-income and ecologically minded strategies. To that end the notion of the meta-home is put forward for exploration.

This core studio works with the hypothesis that the site has 21 owners -all except 2 agree to redevelop together. One of them decides to participate in the redevelopment and relocate its business (Frankie and Johnny's). Owners are determined to redevelop themselves self-organizing as an 'intentional community'. They hire an architect to study the transformation from a group of homes to a meta-home that must generate collective usable space, hypothetical profits, and ecological benefits.

The three spheres of knowledge students get in this course are:

1. Visualization, that includes the consolidation of the universal language of 'orthogonal projection'.
2. The architectural theme of collective housing with its set of constraints; the methodology of the project as the main field for architectural thinking.
3. Critical thinking in relation to the way we live today and can live tomorrow if we were to do it in a more sustainable way.

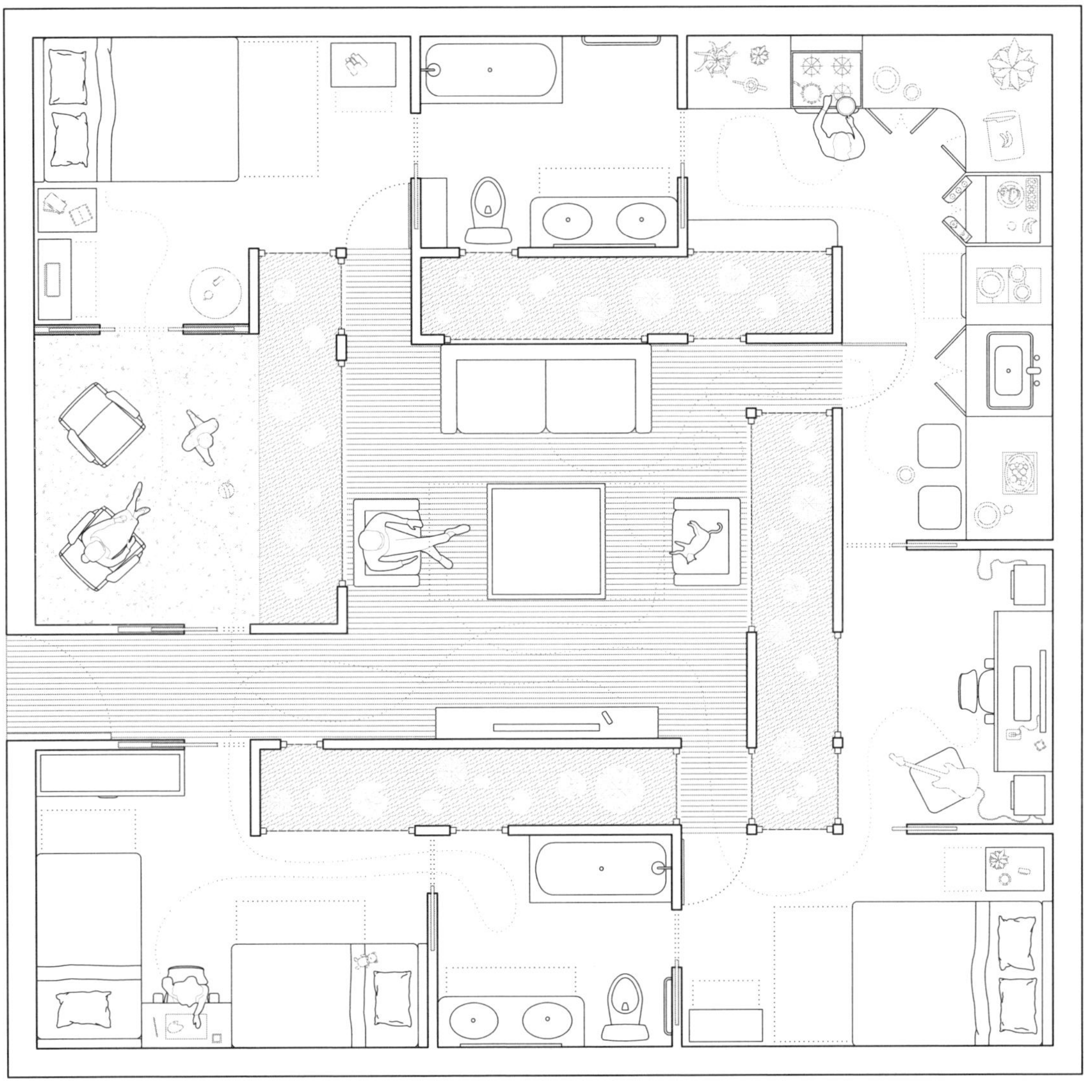

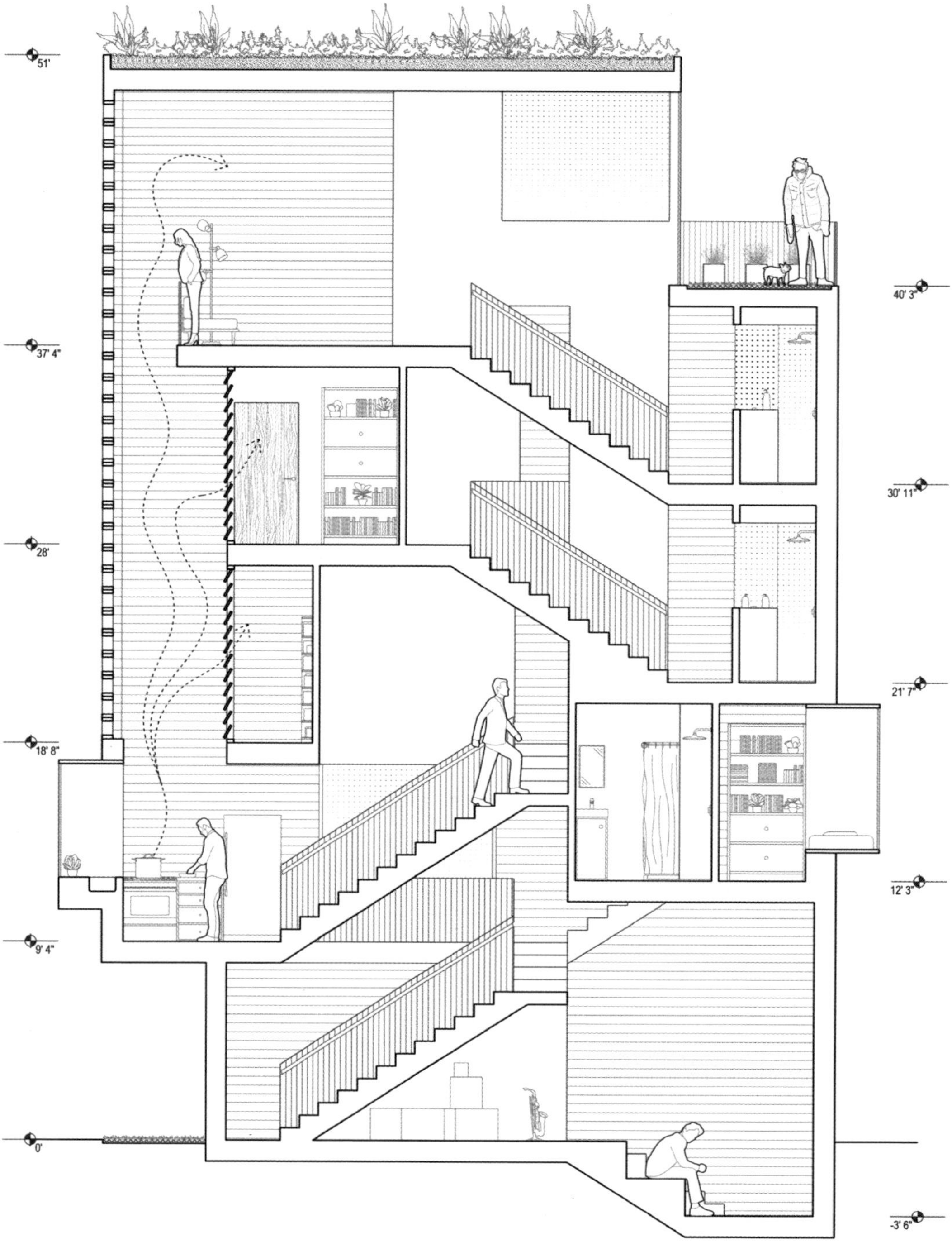
51'
37' 4"
28'
18' 8"
9' 4"
0'
40' 3"
30' 11"
21' 7"
12' 3"
-3' 6"

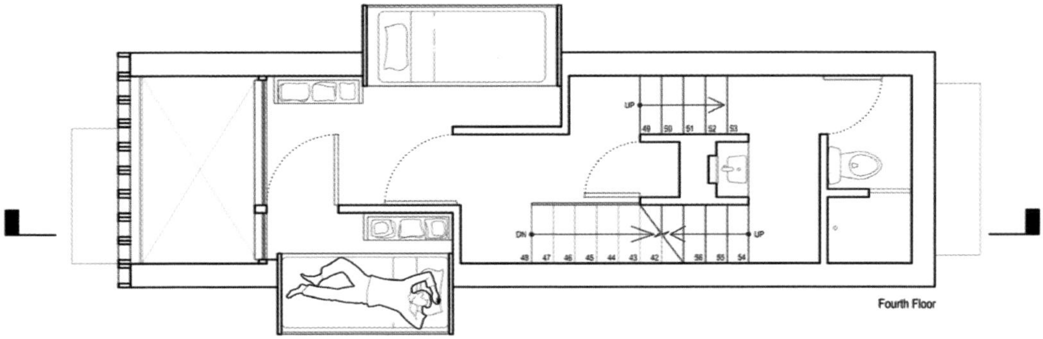

Fourth Floor

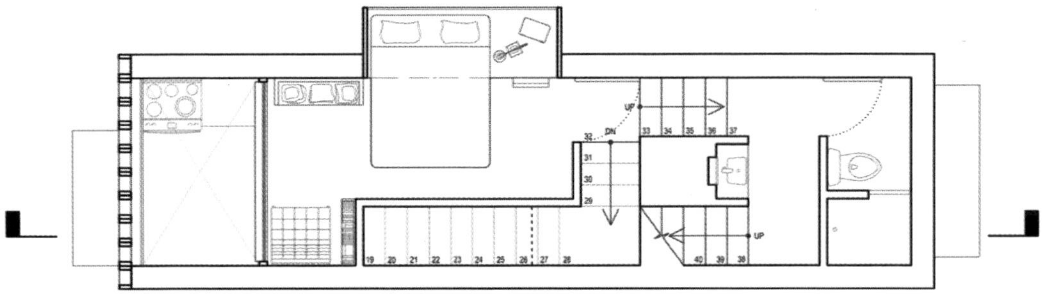

Third Floor

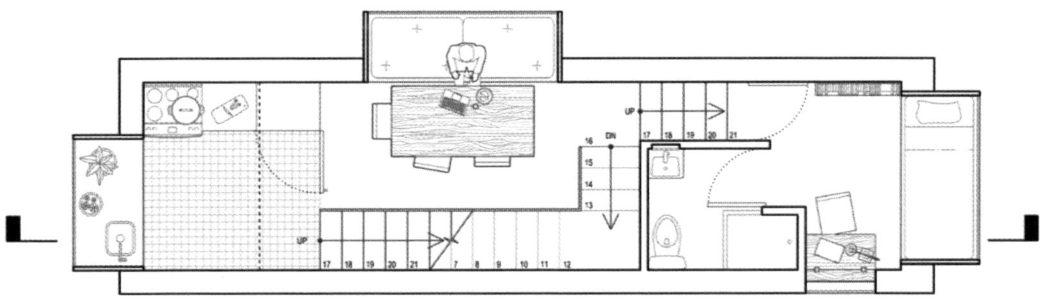

Second Floor

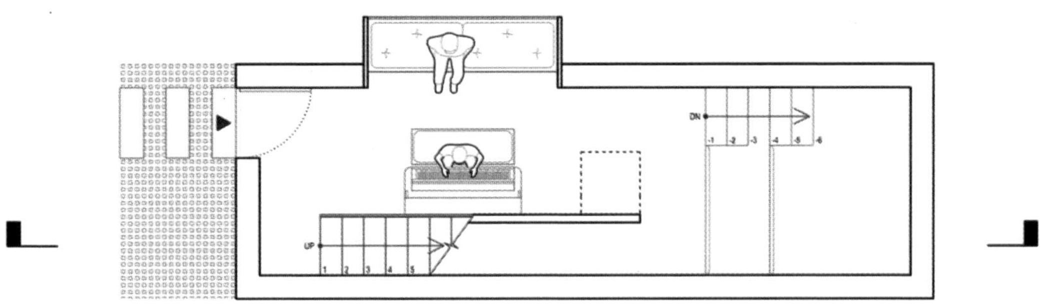

First Floor

 what is **economically** affordable

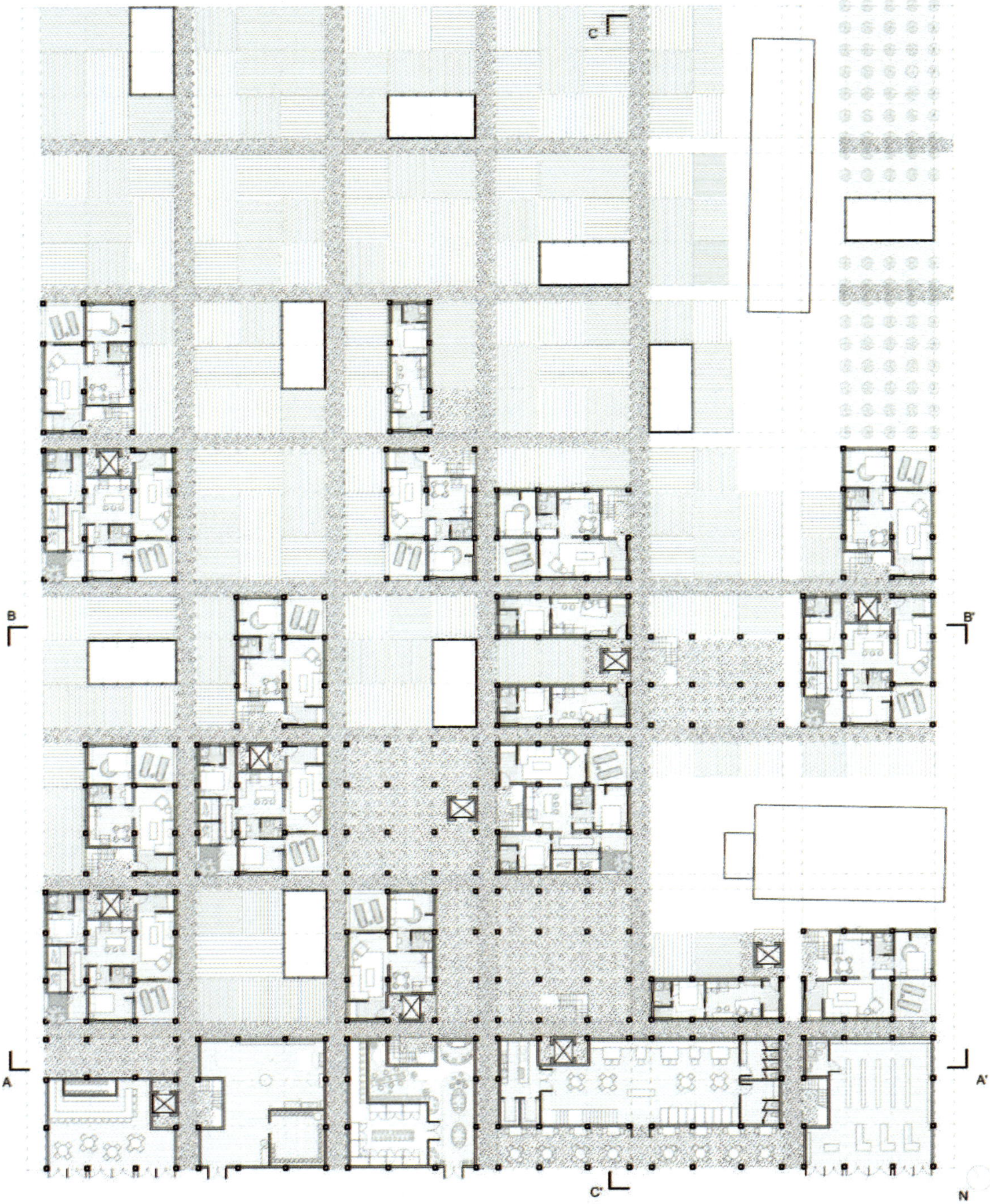

 what is **economically** affordable

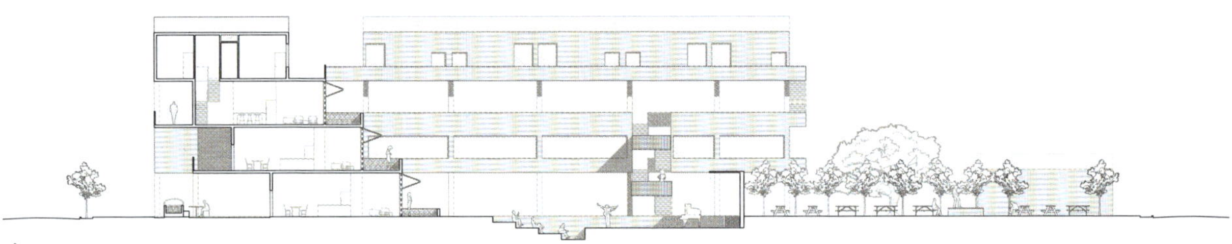

A

s> Ben Cornett   **i>** Emmanual Osorno

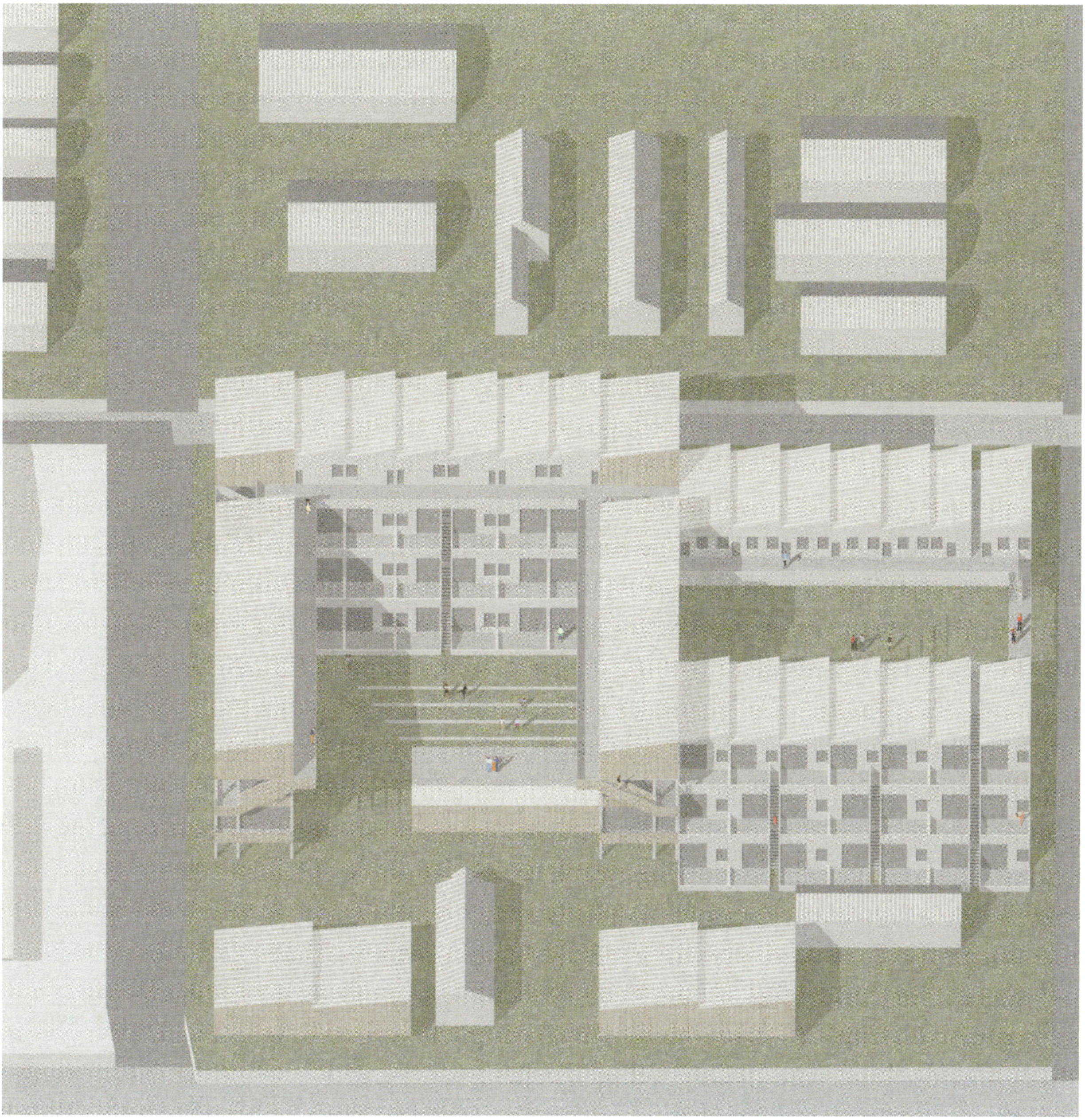

KANSAS CITY: SCENARIOS

Irene Keil [C]
Omar Ali
Sonsoles Vela

The core 3031 architecture design studio introduces students to scale in the context of a city. It reinforces and expands lessons on building organization and program circulation by looking at the relationship between a city block and its surrounding areas. The chosen site to design is a series of mostly vacant blocks within the freeway loop of downtown Kansas City.

The studio is structured in two phases. In Phase I, Urban Scale, students work in teams to establish a conceptual scenario for the development of the site. They begin by learning the vocabulary and instruments of urban design and by researching precedents. Together they investigate, analyze, and map the urban context at various scales. Within the given site, which intentionally emphasizes sustainable density, resilience, and mobility, students speculate on the development of a series of parcels with respect to context, program, zoning parameters, and site opportunities.

In Phase II, Building Scale, students individually design a building in more detail with the established urban scenario as a basis. Using the context of the Kansas City site, students are encouraged to make connections to neighborhoods cut off by the freeway layout.

s> Chelsea Kilgore

i> Omar Ali

UG

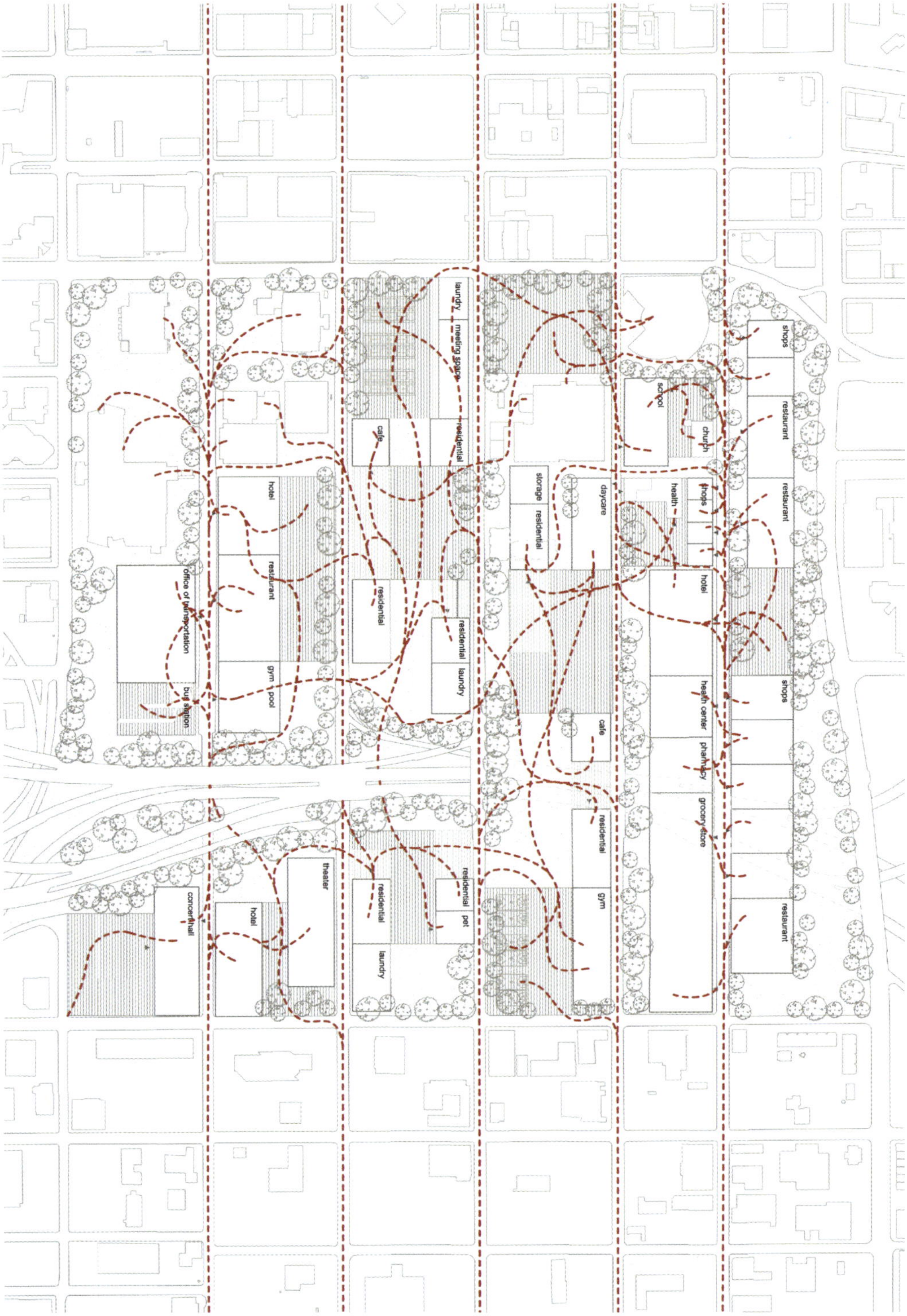

what is **economically** affordable

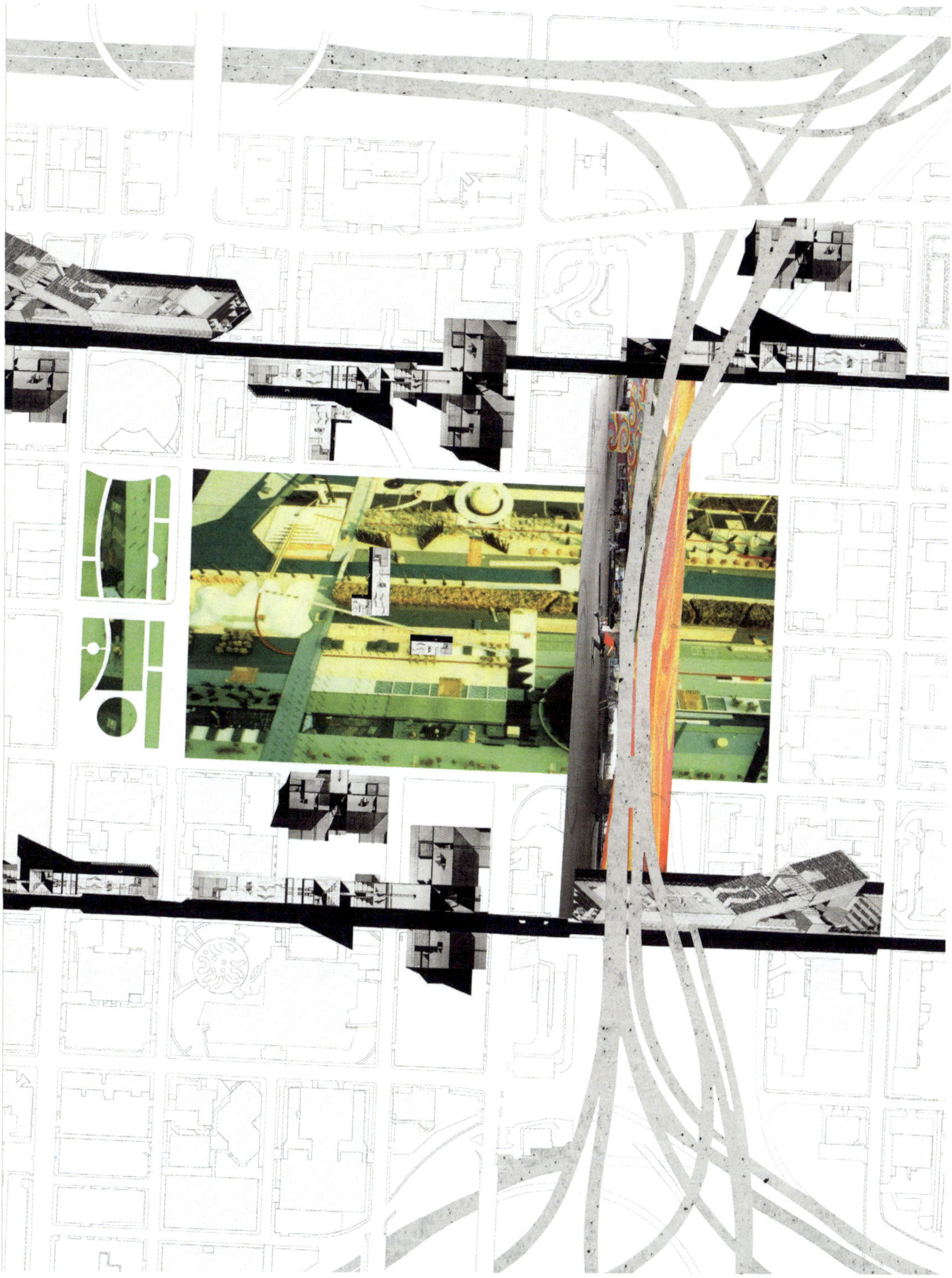

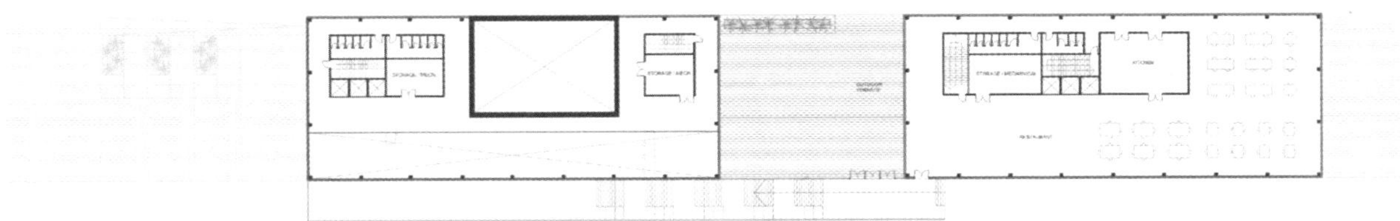

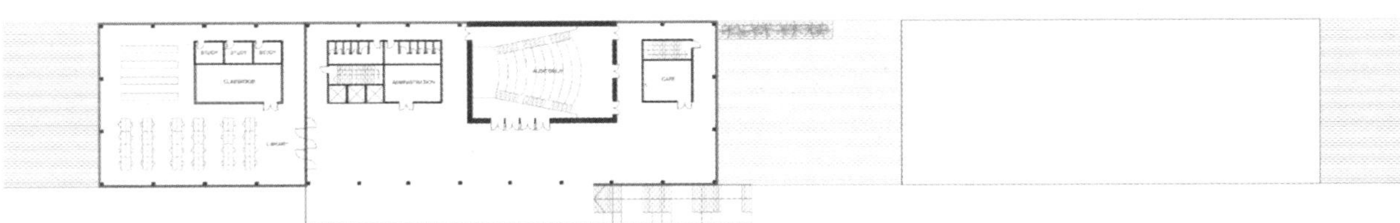

what is **economically** affordable

s> Ella Weeks

i> Irene Keil

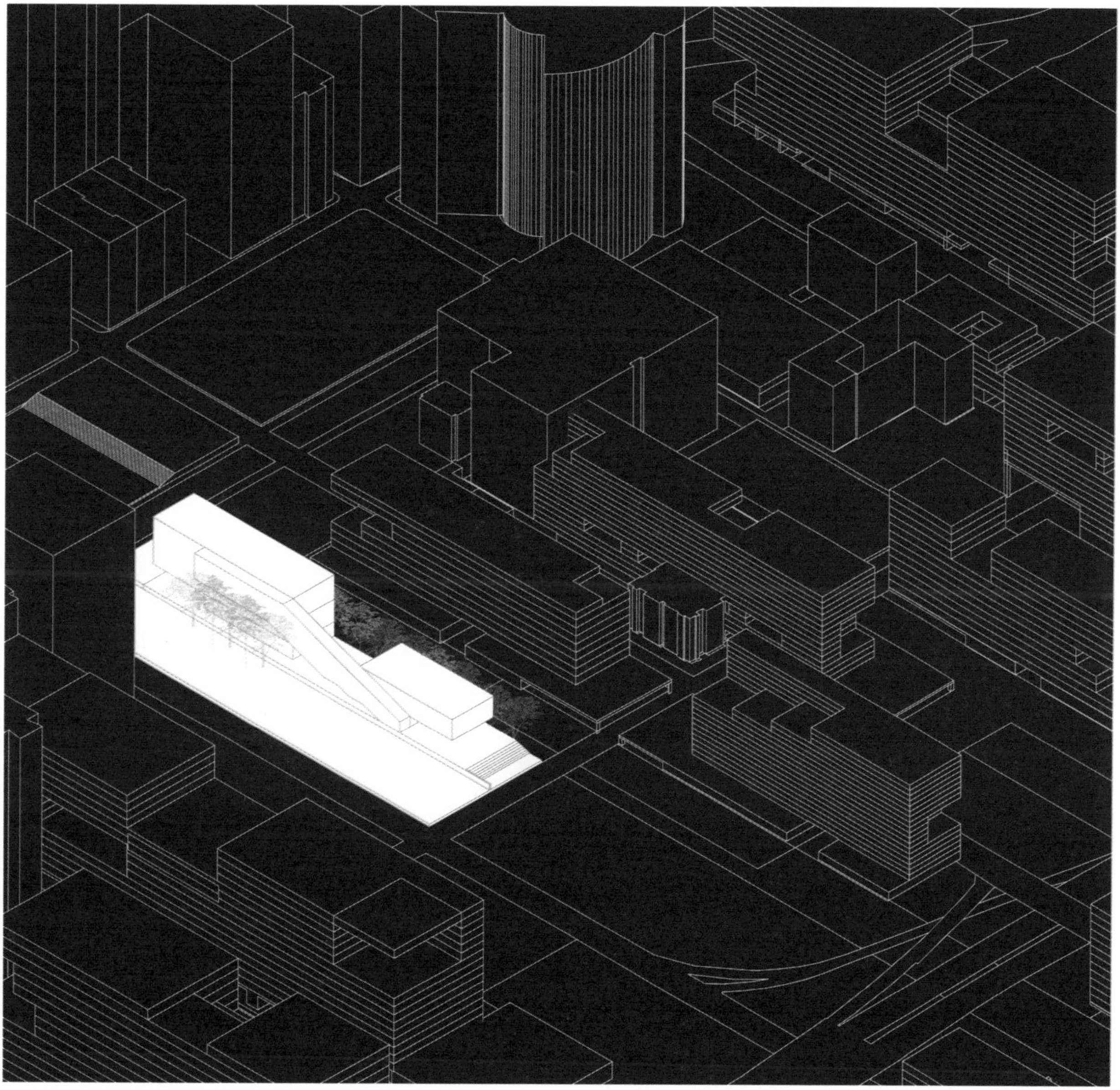

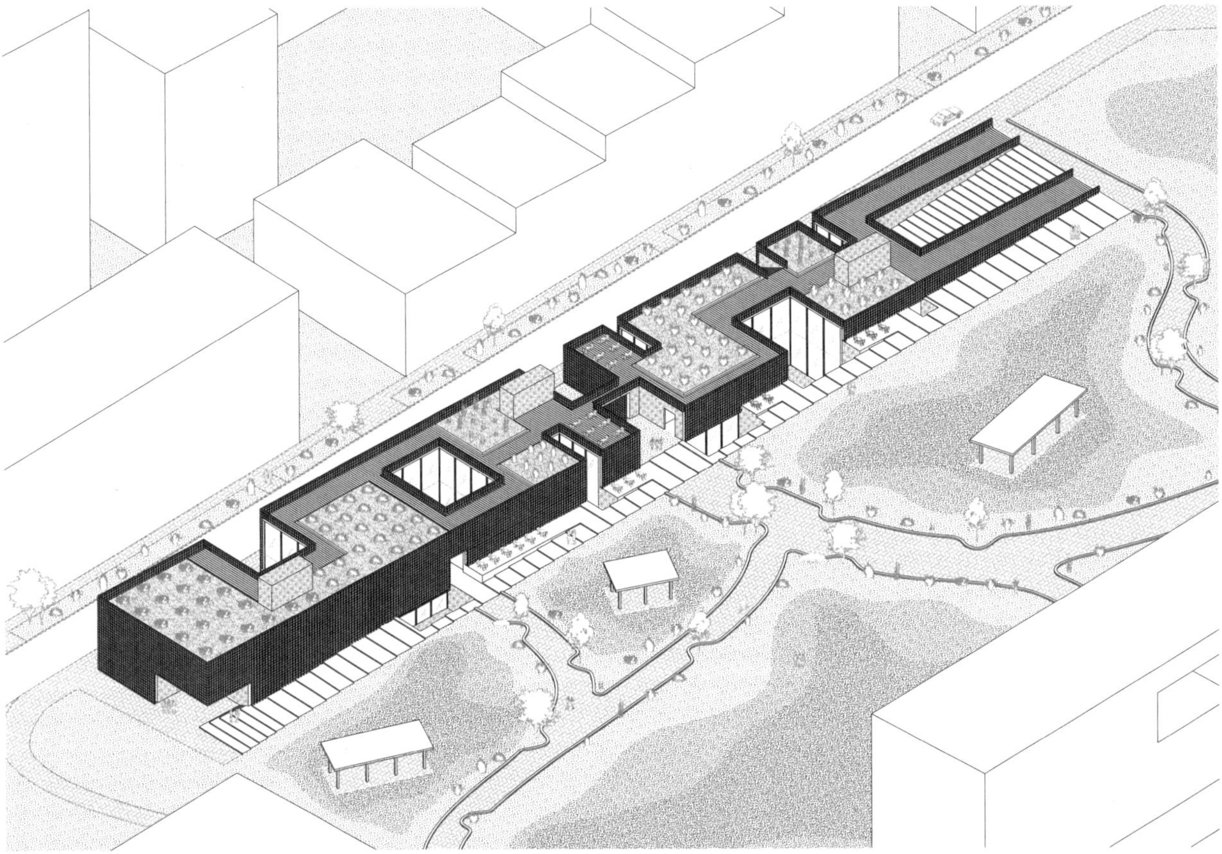

what is **economically** affordable

s> Frank Taylor

i> Omar Ali

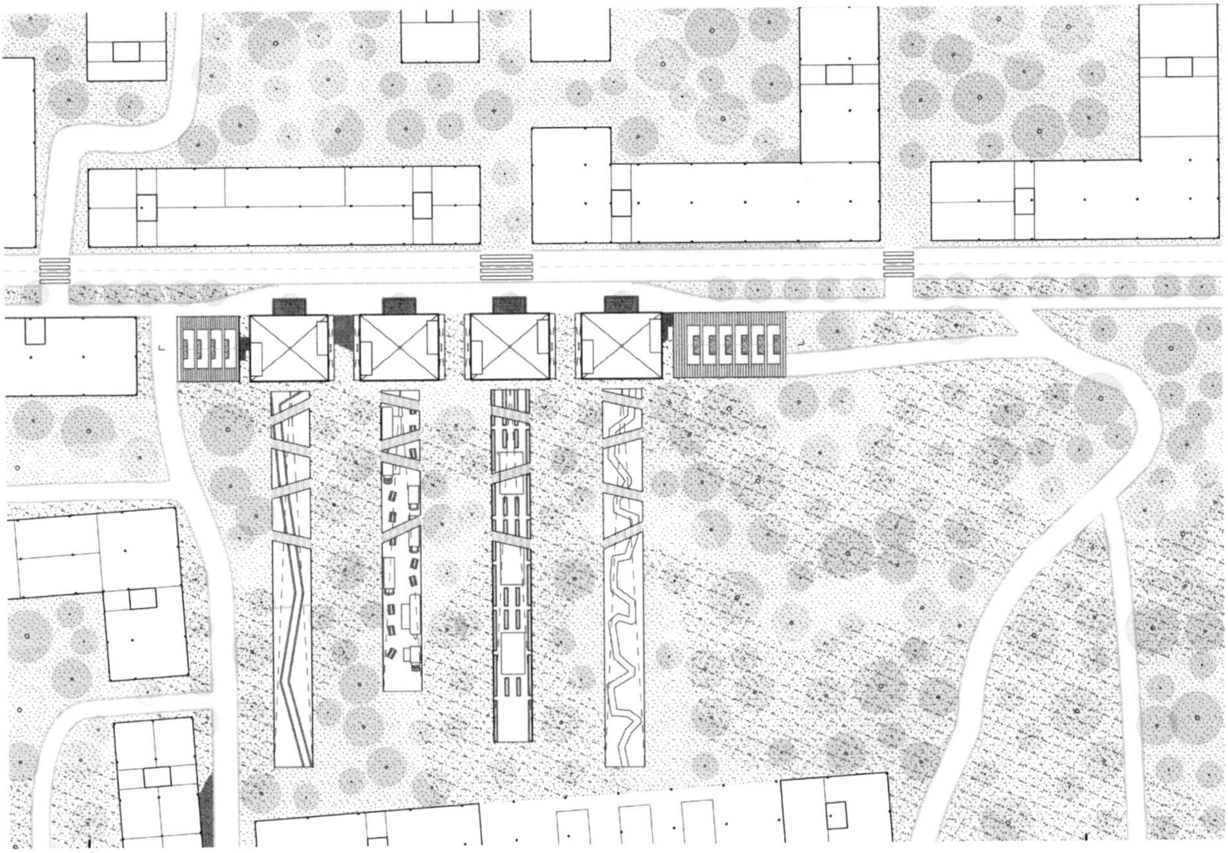

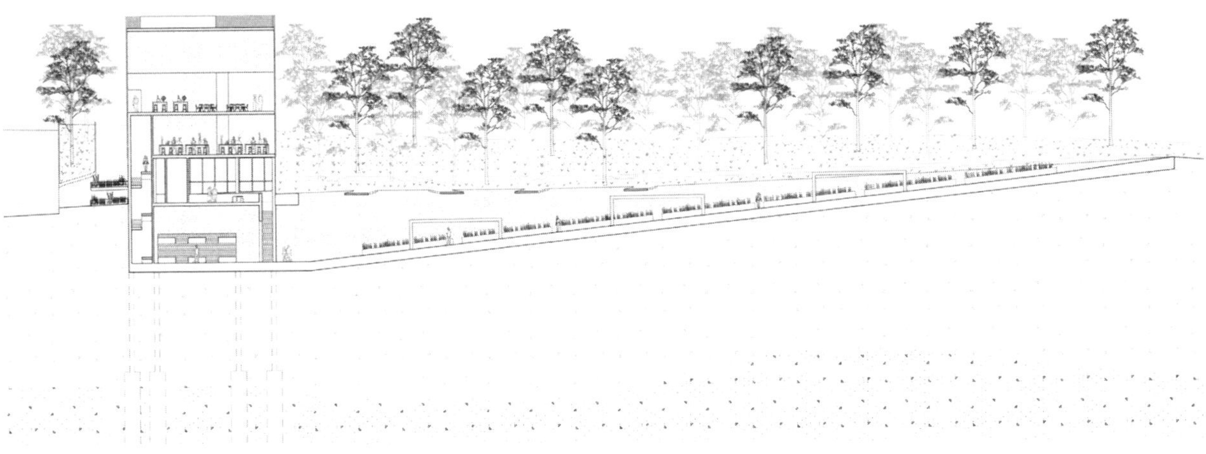

what is **economically** affordable

s> Jose Varela Castillo

i> Omar Ali

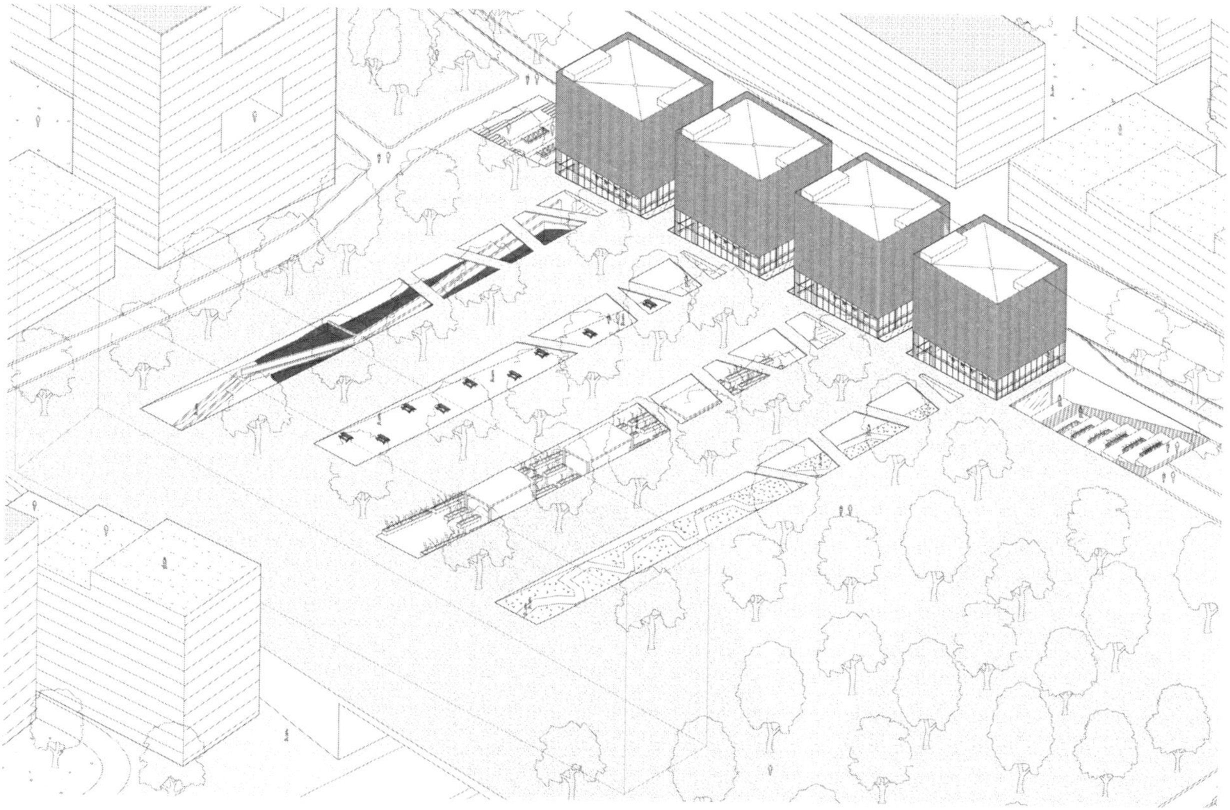

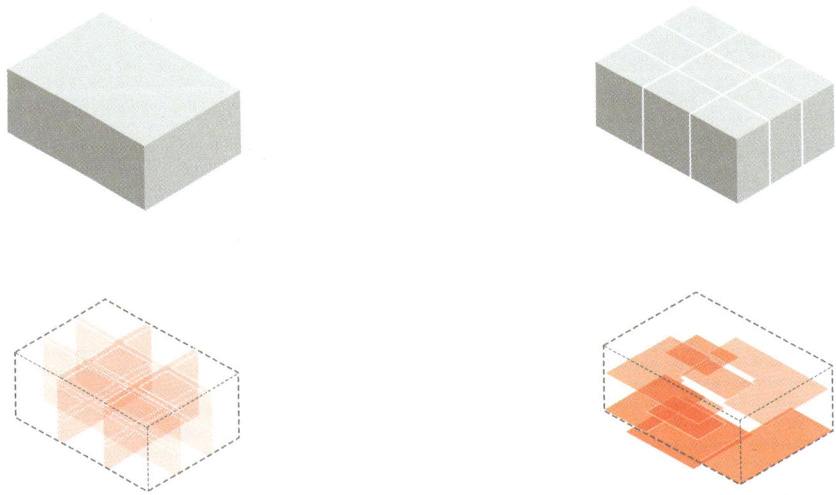

what is **economically** affordable

s> Olivia Vercruysse

i> Omar Ali

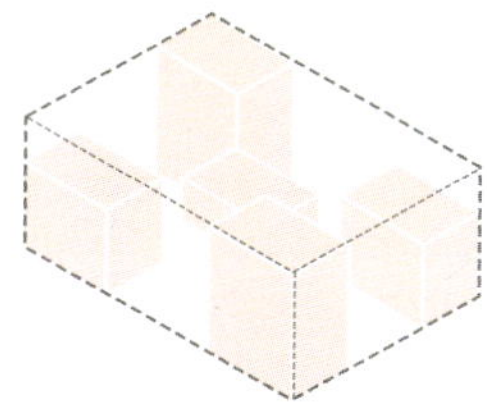

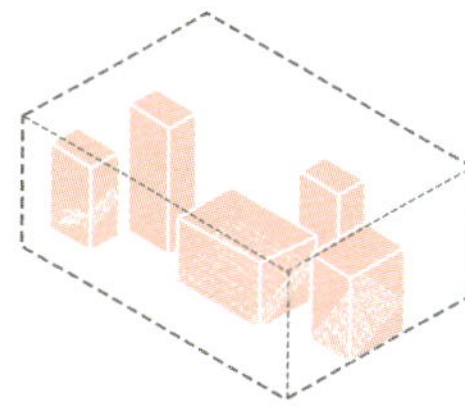

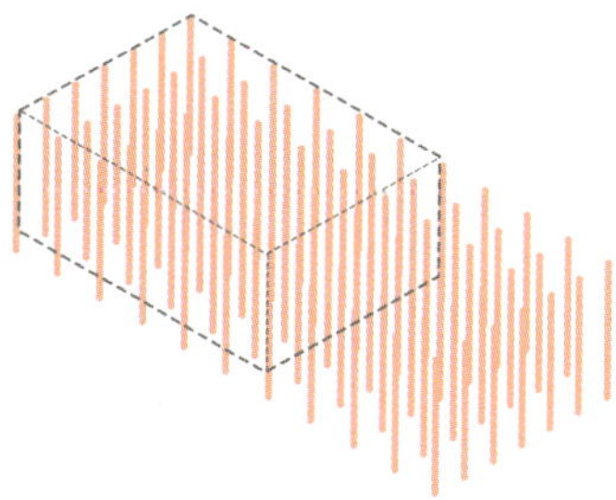

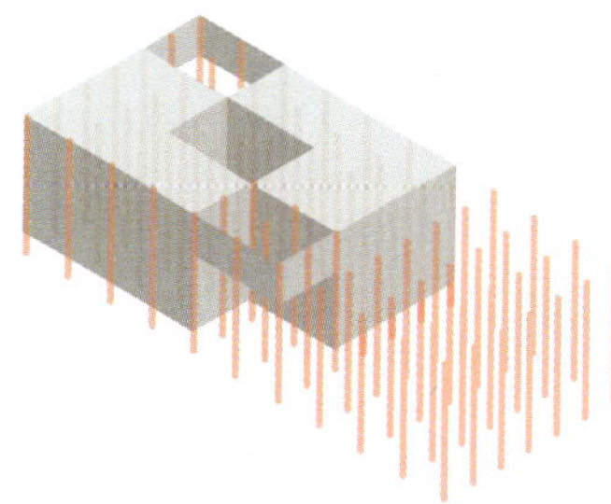

REPETITION & SPECIFICITY

Irene Keil [C]
Andrea Bardon de Tena
Matt DeCotiis
David Schneider
Ruben Garcia Rubio
Nick Marshall

The Core 3032 Integrated studio focuses on the development of an architectural project with regards to site relationships, historic context, structure, systems, materiality, and building codes, meeting the NAAB accreditation requirements of the integrated studio. "Integration" encompasses a coordination of building systems during the project's development, and reciprocal feedback between intent and material manifestation. The project demands a careful calibration of intent, form, system, material, and technology where relationships have inherent potentials and unanticipated possibilities.

The pedagogical objective of this studio is to lead the student through a design process where architectural ideas, spatial planning, and building technologies are integrated. Students learn to comply with all applicable building code regulations, such as life safety and egress. The content addresses and explores the implications of construction techniques, building technologies, and systems.

The typical project of an integrated studio addresses the design and resolution of a normative building program and building type - a mid-scale building in an urban context. The project focuses on a mixed-use building on an infill site. Students work individually on one design project over the semester, integrating group constellations for components of the design process.

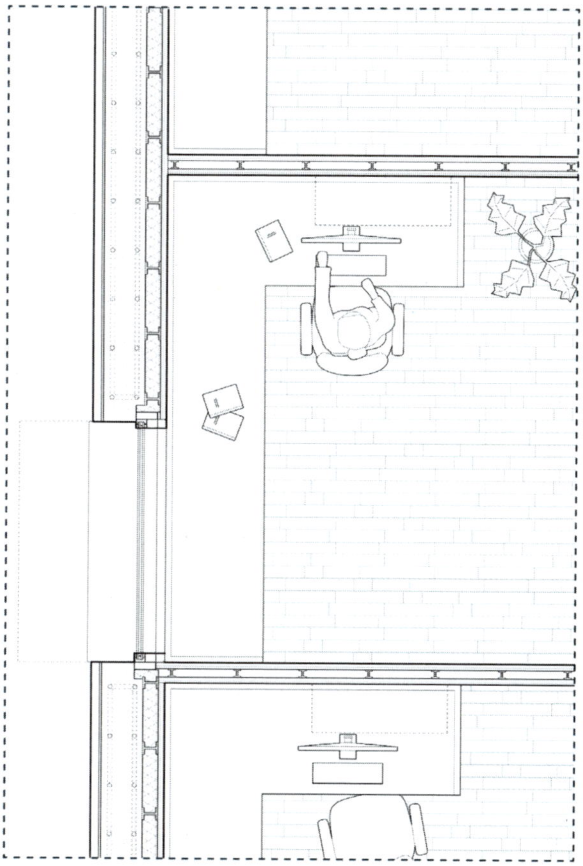

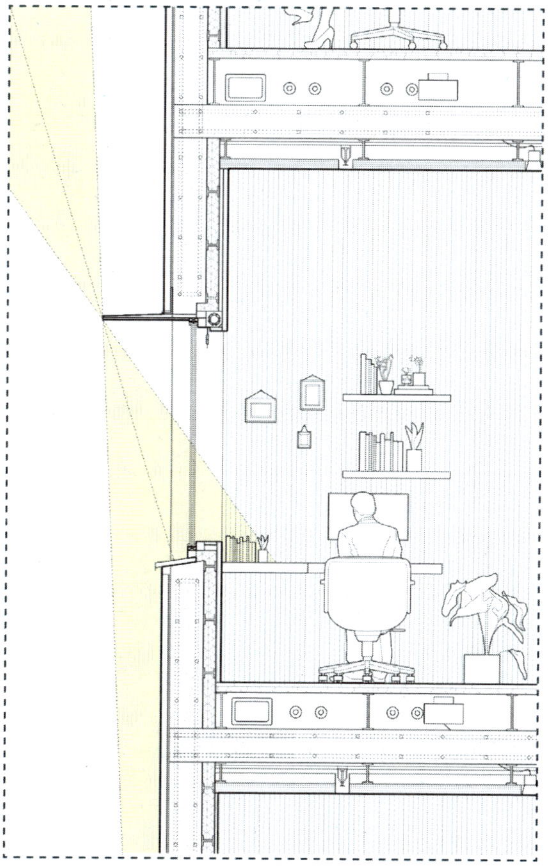

1
2
3
4
5
6
7
27′
24′
A
B
C
D
E
F
G

s> Anna Katherine Becker

i> Andrea Bardon de Tena

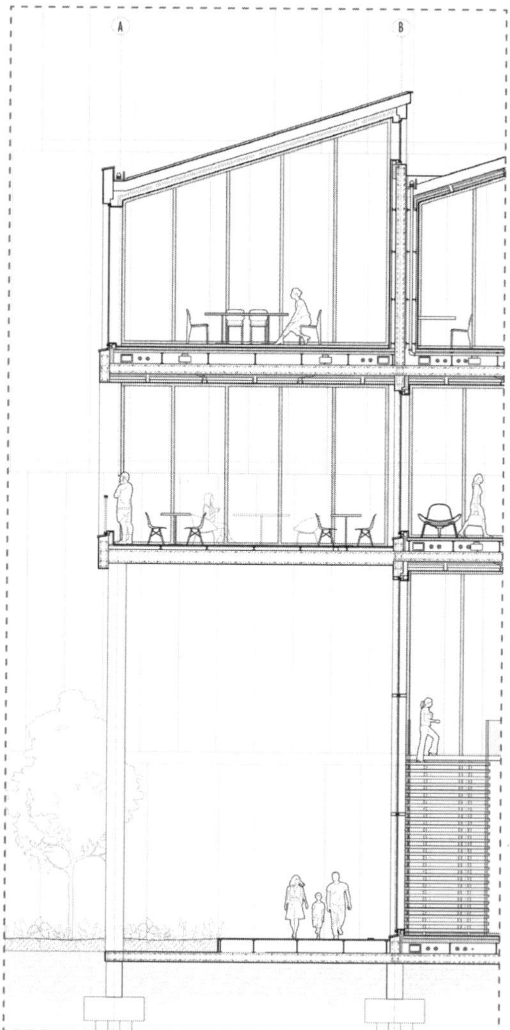

SECTION

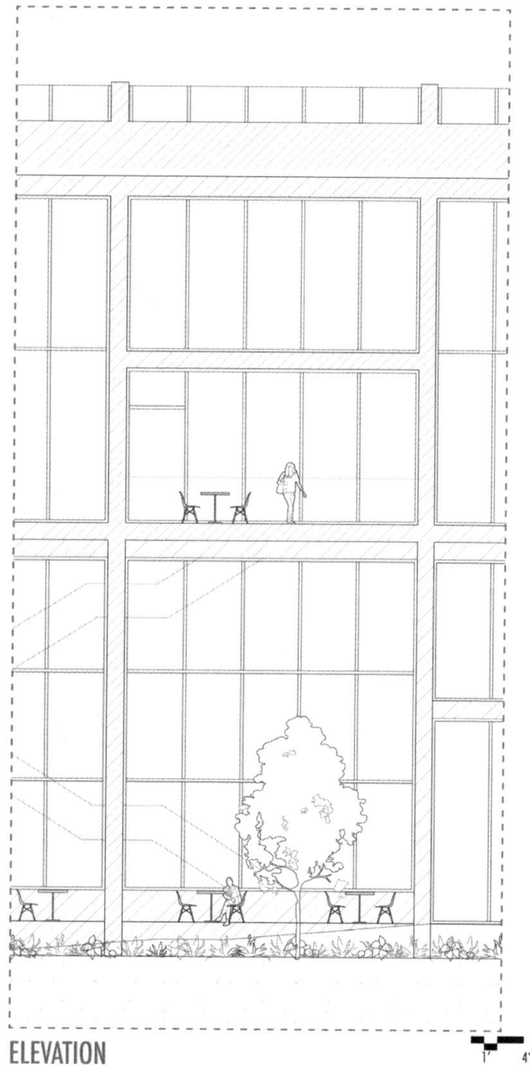

ELEVATION

what is **economically** affordable

what is **economically** affordable

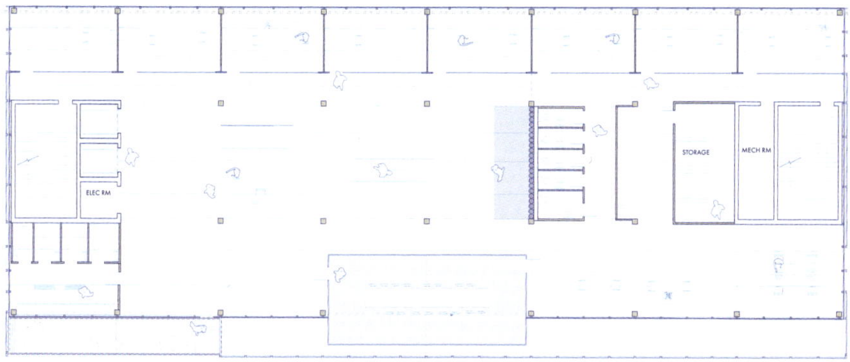
ELEC RM
STORAGE
MECH RM

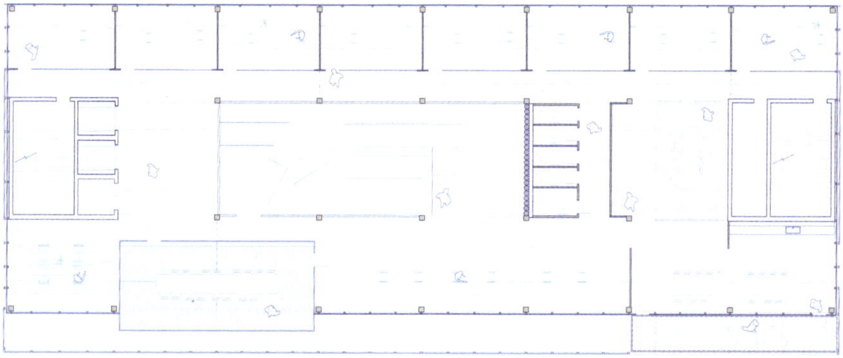

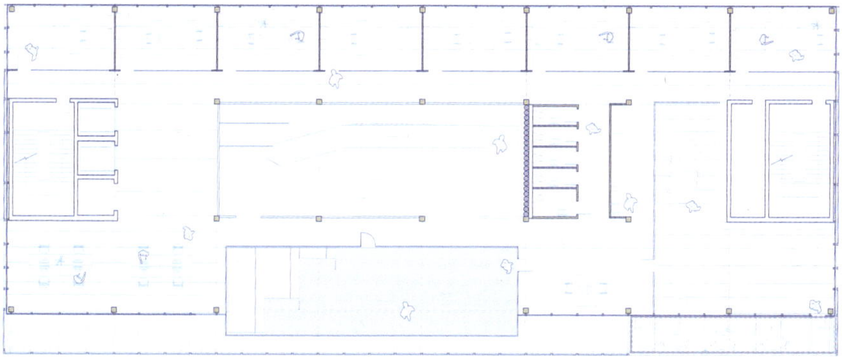

s> Sarie Keller

i> Andrea Bardon de Tena

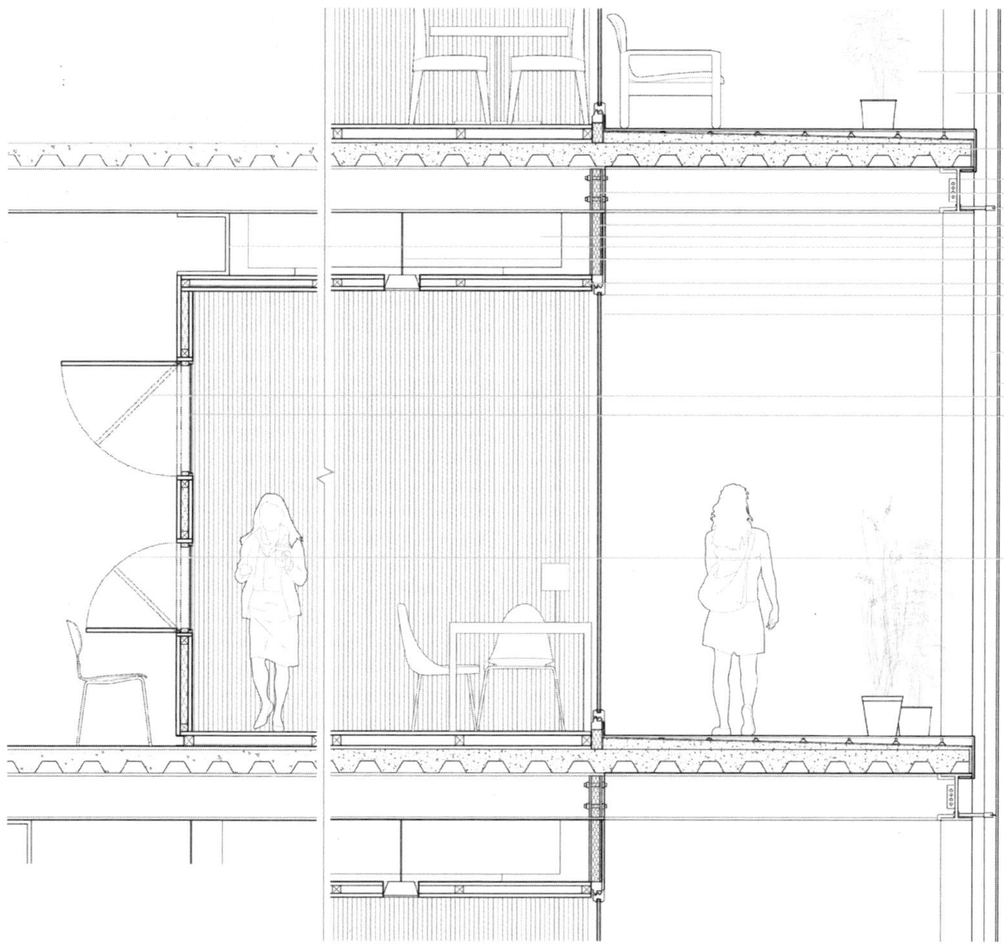

1. BALCONY
2. STEEL COLUMN
3. COMPOSITE STEEL DECKING
4. STEEL BEAM
5. SCHÖCK ISOKORB THERMAL BREAK
6. CONTINUOUS INSULATION
7. AIR SUPPLY DUCT
8. RECESSED CAN LIGHTING
9. WOOD CONSTRUCTION
10. WOOD MULLION
11. GLASS ASSEMBLY
12. FACADE SUBSTRUCTURE
13. ETFE SCREEN FACADE
14. FACADE STRUCTURAL ARM
15. "C" CHANNEL STEEL DETAIL
16. OPERABLE INTERIOR WINDOW
17. OPERABLE FOLD-DOWN TABLE AND WINDOW

GALLERY AND BOX

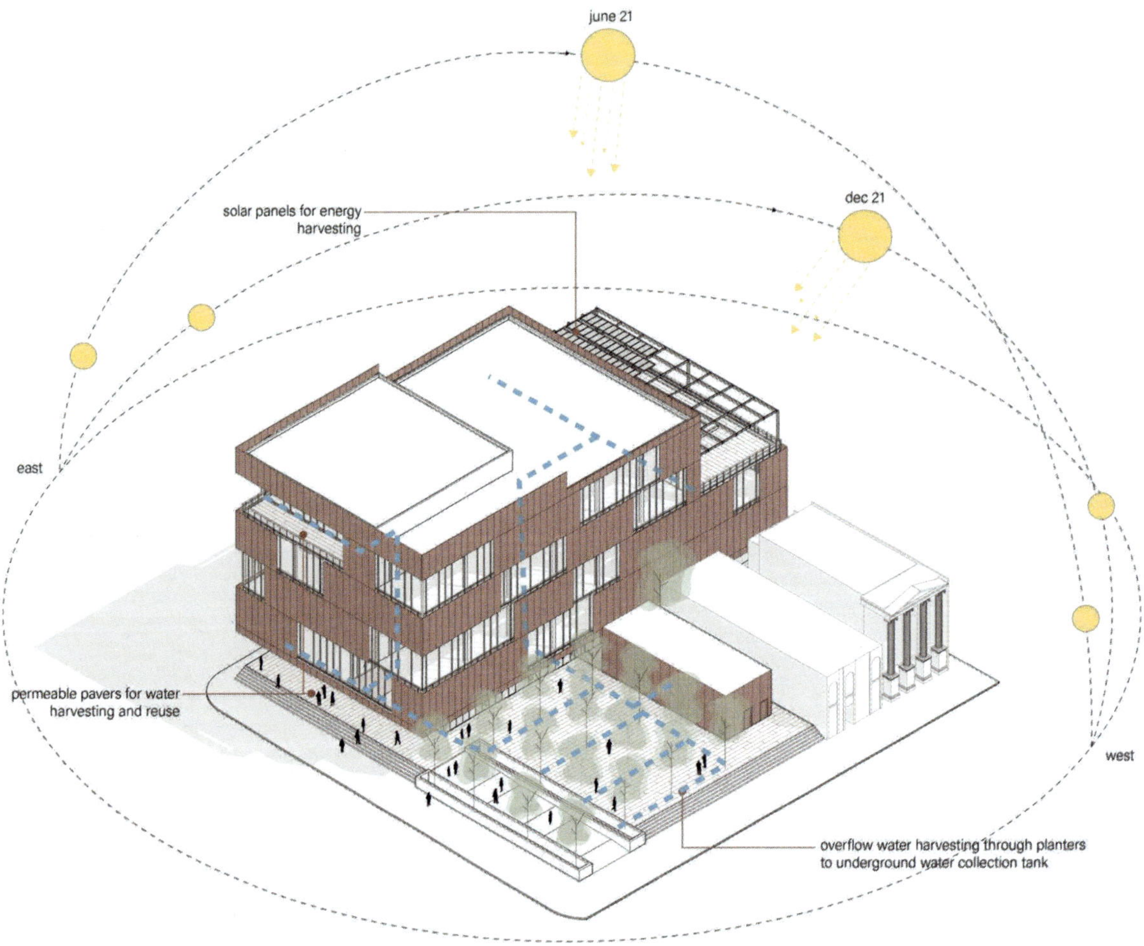
june 21
dec 21
solar panels for energy harvesting
east
west
permeable pavers for water harvesting and reuse
overflow water harvesting through planters to underground water collection tank

s> Brianna Baldwin i> Ruben Garcia Rubio

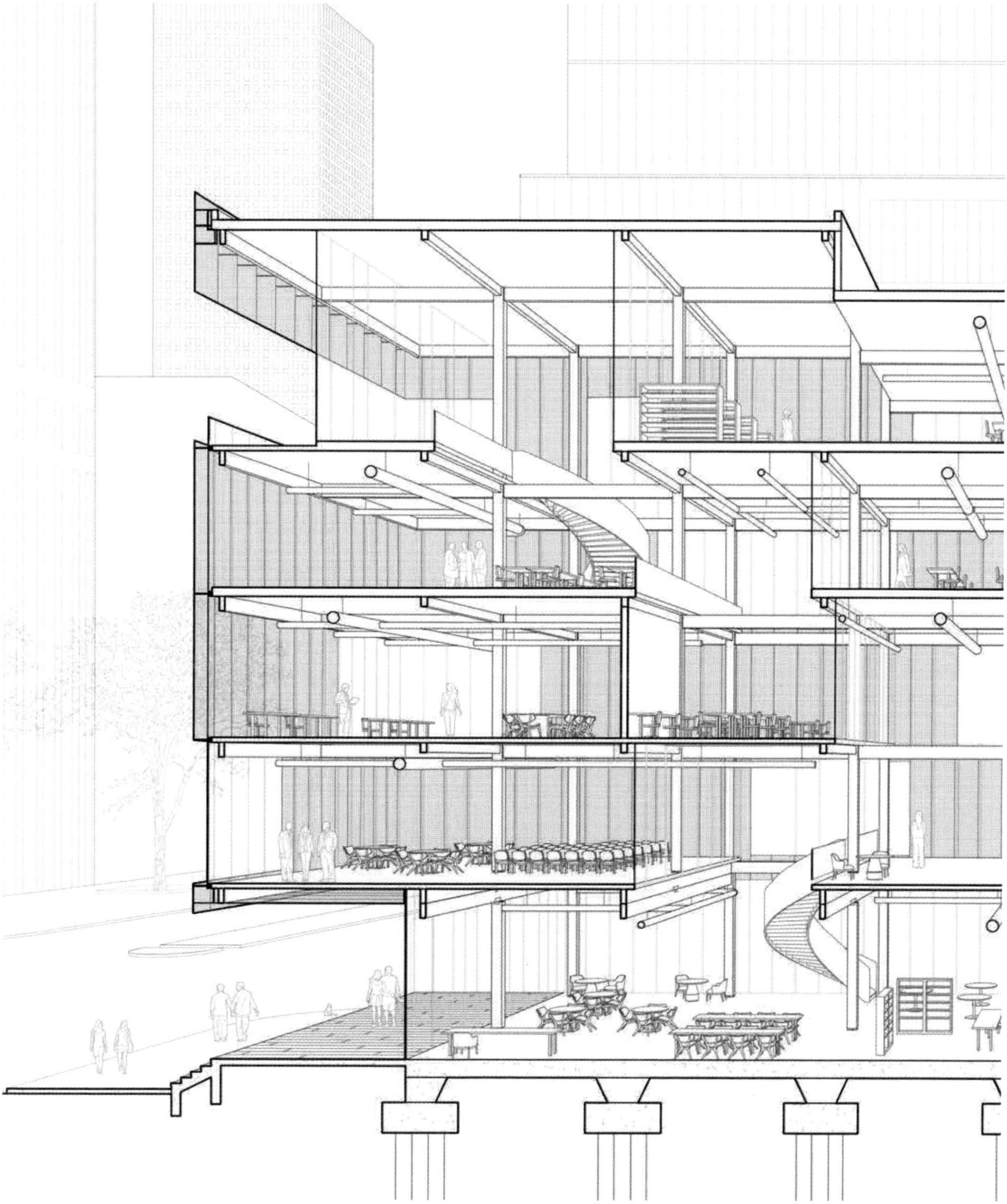

DESIGN DEVELOPMENT STUDIO II

Gina LaMacchia
William Bradshaw

The goal of this course is to introduce students to the design and financial process of real estate development in tandem. Real estate development is a creative process in many ways, and students are encouraged to be innovative in their thinking (financial, programmatic, etc.) and design work.

As a result of this course, students should be able to:

1. Effectively work in groups to design a real estateproject which is realistic, financeable, is contextual toits surroundings, and meets public (zoning andbuilding codes) and private (potential userexpectations) requirements.
2. Communicate to members of your team (architect,lenders, appraisers)
3. Understand how financial constraints impactdesign and vice versa
4. Comfortably utilize financial (Excel) and design (Adobe, SketchUp) programs regardless ofprevious skill.
5. Observe market conditions and understand howthey impact the development costs, revenues, andoperating expenses of an intervention in the builtenvironment.

s> Kate Vannini + Sidney Seale + Joe Connelly + Nick Pappas

i> Gina LaMacchia + Will Bradshaw

NEIGHBORHOOD MARKET DATA

Population

1,794

1 **Predominant Race and Ethnicity:**

50-70% Black

2 **Homeownership Rate:**

20.33%

3 **Median Age:**

31

4 **Median Household Income:**

$20,688

5 **Per Capita Income:**

$18,010

Timeline

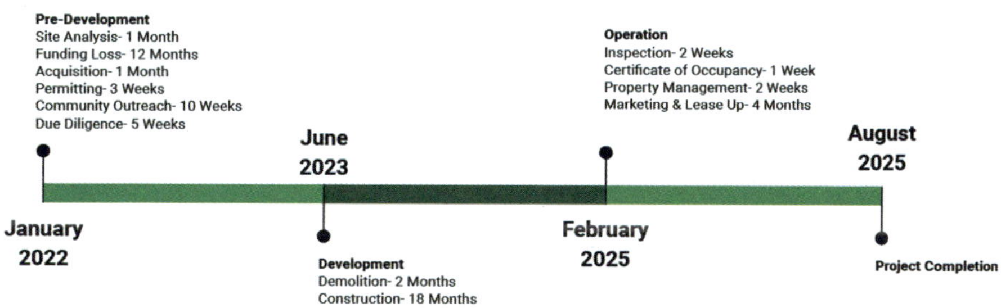

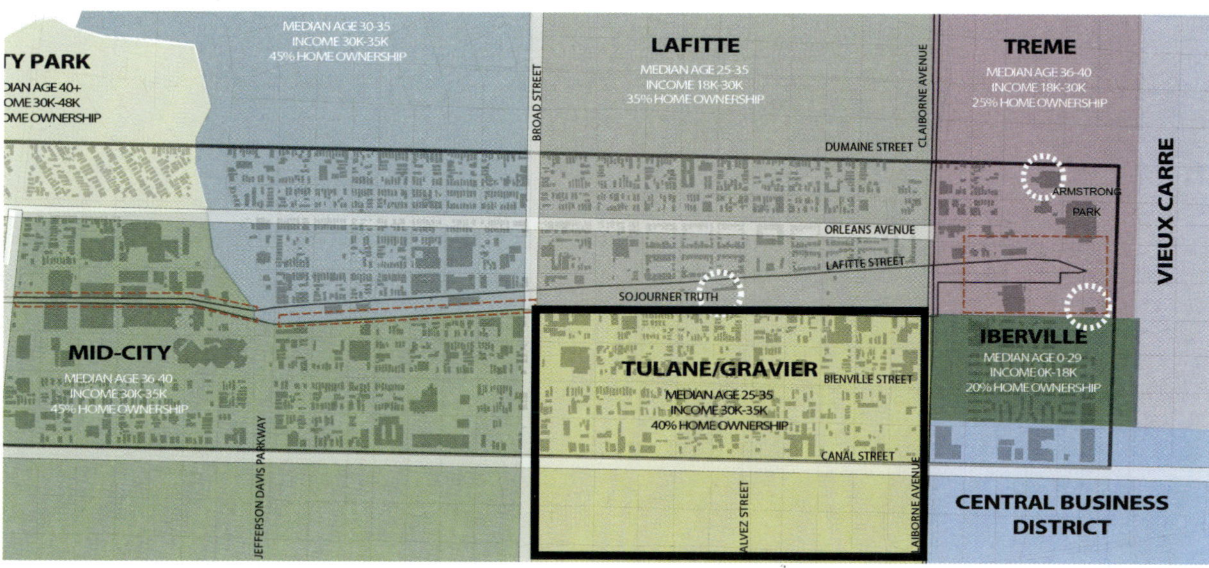

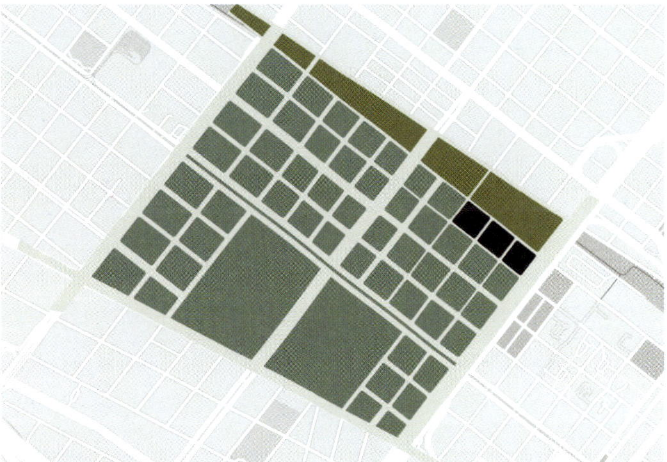

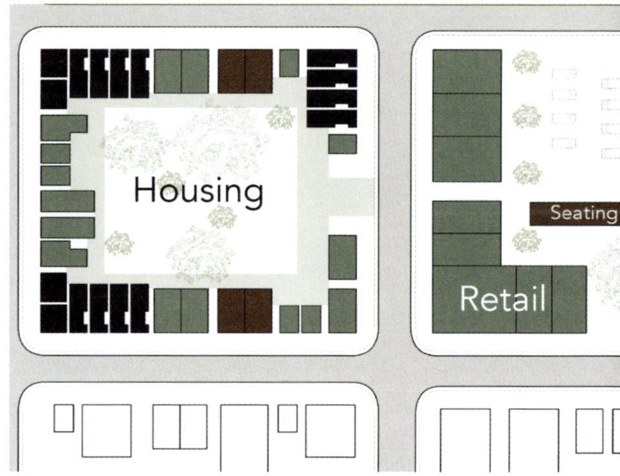
Housing
Seating
Retail

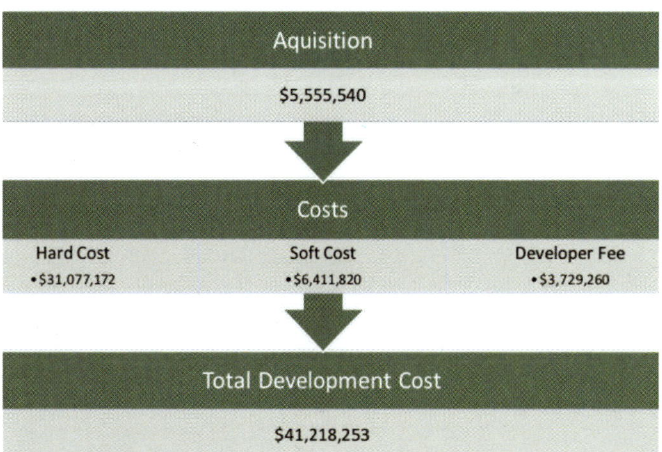
Aquisition
$5,555,540
Costs
Hard Cost
•$31,077,172
Soft Cost
•$6,411,820
Developer Fee
•$3,729,260
Total Development Cost
$41,218,253

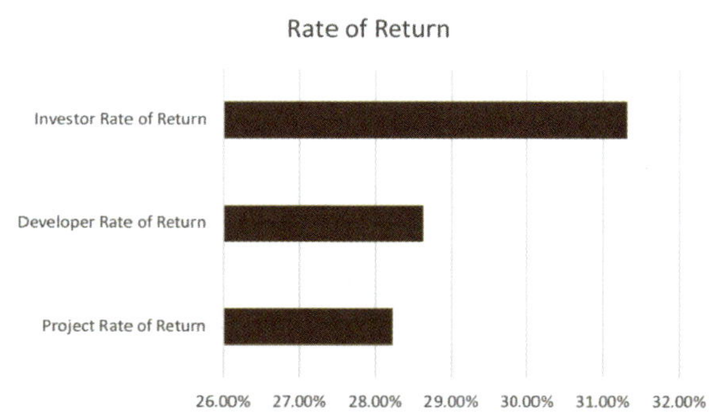
Rate of Return
Investor Rate of Return
Developer Rate of Return
Project Rate of Return
26.00%
27.00%
28.00%
29.00%
30.00%
31.00%
32.00%

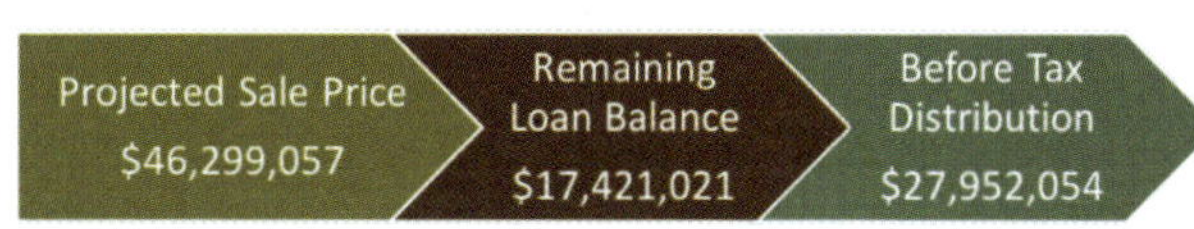
Projected Sale Price
$46,299,057
Remaining Loan Balance
$17,421,021
Before Tax Distribution
$27,952,054

s> Kate Vannini + Sidney Seale + Joe Connelly + Nick Pappas

i> Gina LaMacchia + Will Bradshaw

130,950 SF Built Space	**27%** Black Equity Ownership	**92,231 SF** Open Space
3.8M Local Tax Revenue	**36** Affordable Housing Units	**41.2M** Total Investment
18% Women Equity Ownership	**250** Permanent Jobs	**750** Construction Jobs

DIRECTED RESEARCH

In this course students conduct an original investigation in order to acquire new knowledge within a framework set by a client in practice. The primary goal of this research class is to deepen their personal understanding of a particular topic or issue in real estate development. A key secondary goal is to build your skills and confidence to make succinct public presentations of complex material. The students assess both on their final work product (written and oral) as well as their professional client communication along the way.

Students identified a particular client and general topic to focus on prior to the start of the semester to ground their research, most of your research will consisted of secondary research to be summarized and analyzed this to apply it to the particular situation or future project.

Casius Pealer

Students described their research question, conducted a literature review and collected relevant data, interviewed relevant professionals and others with knowledge of the topic, and ultimately analyzed and interpreted all this information in order to compile a final report.

In addition, they developed a Statement of Work, including key deliverables and milestones along the way, the standard document to start any professional or consulting contract to help ensure agreement among the parties.

Housing Cost Burdened

Percentage of Cost Burdened Households per Census Tract in New Orleans

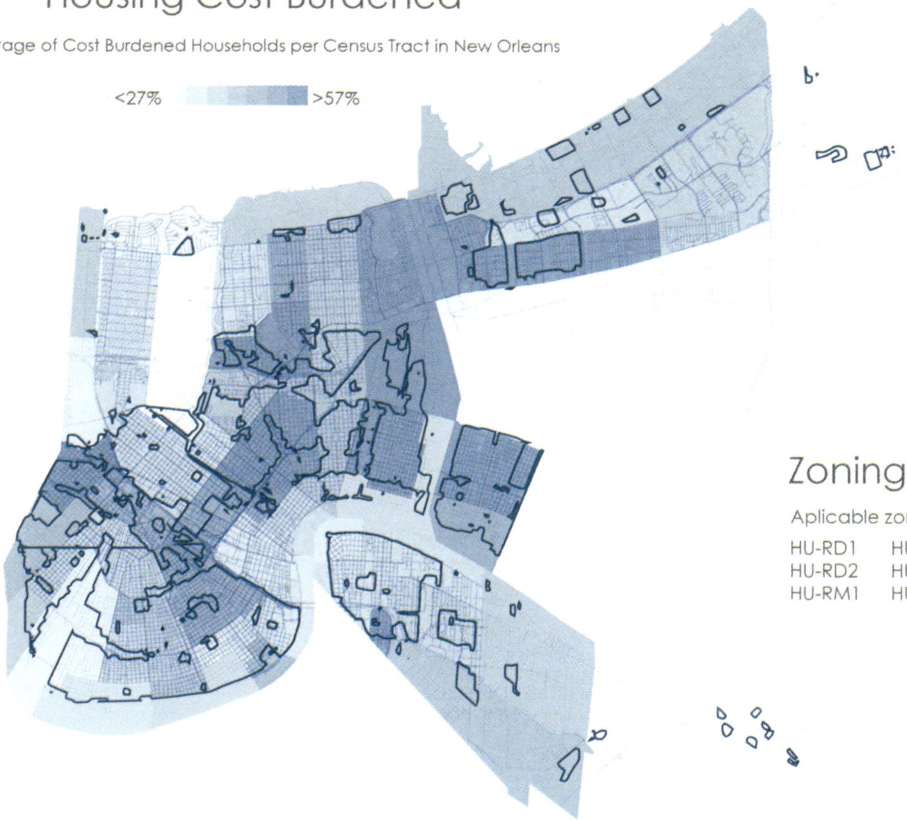

Zoning

Aplicable zones:

HU-RD1 HU-B1A HU-RM2
HU-RD2 HU-B1 MU-1
HU-RM1 HU-MU MU-2

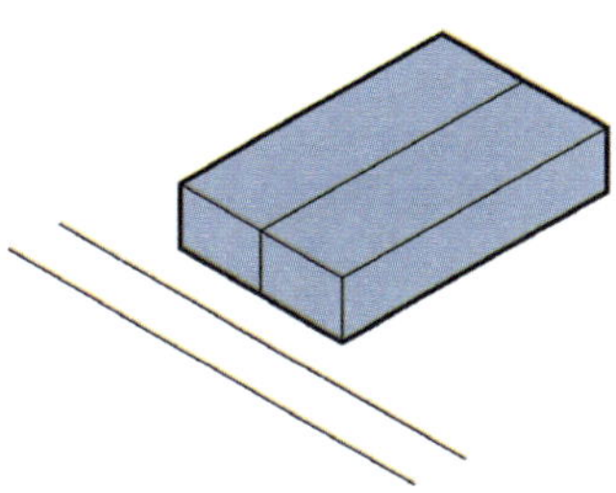

Before: Two units per lot maximum

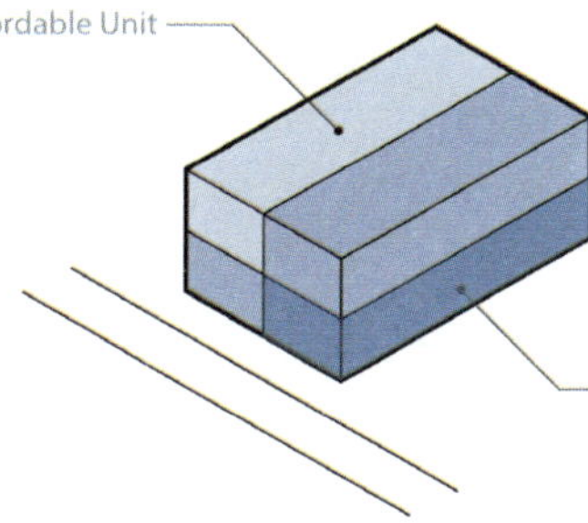

Now: up to four units provided one is rented at affordable rates

*Affordable unit must be rented at rate affordable to a household making 70% AMI to a Household making up to 80% AMI

SUSTAINABLE DESIGN AND PLANNING

Fred Neal Jr.

The goal of the course is to explore the intersections of plan-ning and sustainable development. The course balances his-tory, theory and practice with a focus on the nuts and bolts of planning and design processes.

Students will gain a conceptual understanding of how deci-sions are made that shape the built environment. The course provides applied skills and experience in synthesizing large development projects. Additionally, the course examines both planning and sustainable design from a lens of community engagement with equitable outcomes. The course builds upon students' introductory coursework in real estate finance and sustainable urbanism.

As prospective developers, students should have a firsthand knowledge of the complexities of development including the purpose and goals of sustainability; providing/connecting to infrastructure in the built environment; supporting community goals and equity in outcomes; and synthesizing constraints into an implementable physical development project. This course is rooted in the creative process as real estate devel-opment is inherently a creative problem-solving profession.

s> Meredith Clark + Andrew Mann + Andrew Marymount

i> Neal Jr.

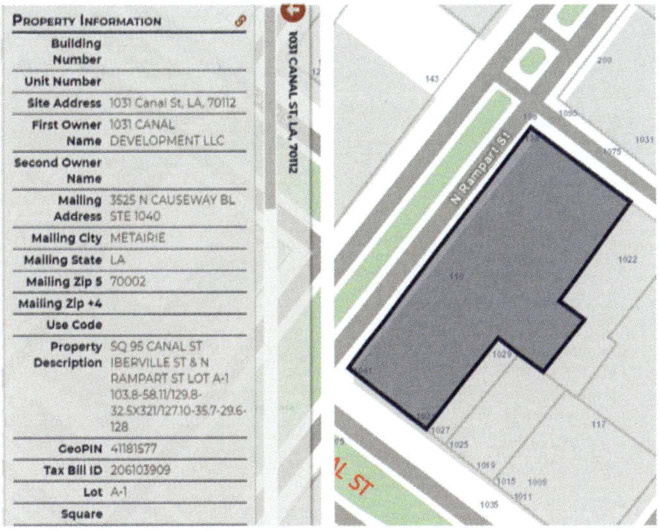

Building A (Canal St.)	Use
Floor 1	Restaurant + Commercial
Floor 2	Restaurant and Hotel
Floors 3-6	Hotel
Roof	Green Roof

Building B (Rampart St.)	Use
Floors 1-4	Parking
Floor 5-12	Multifamily
Roof	Green Roof

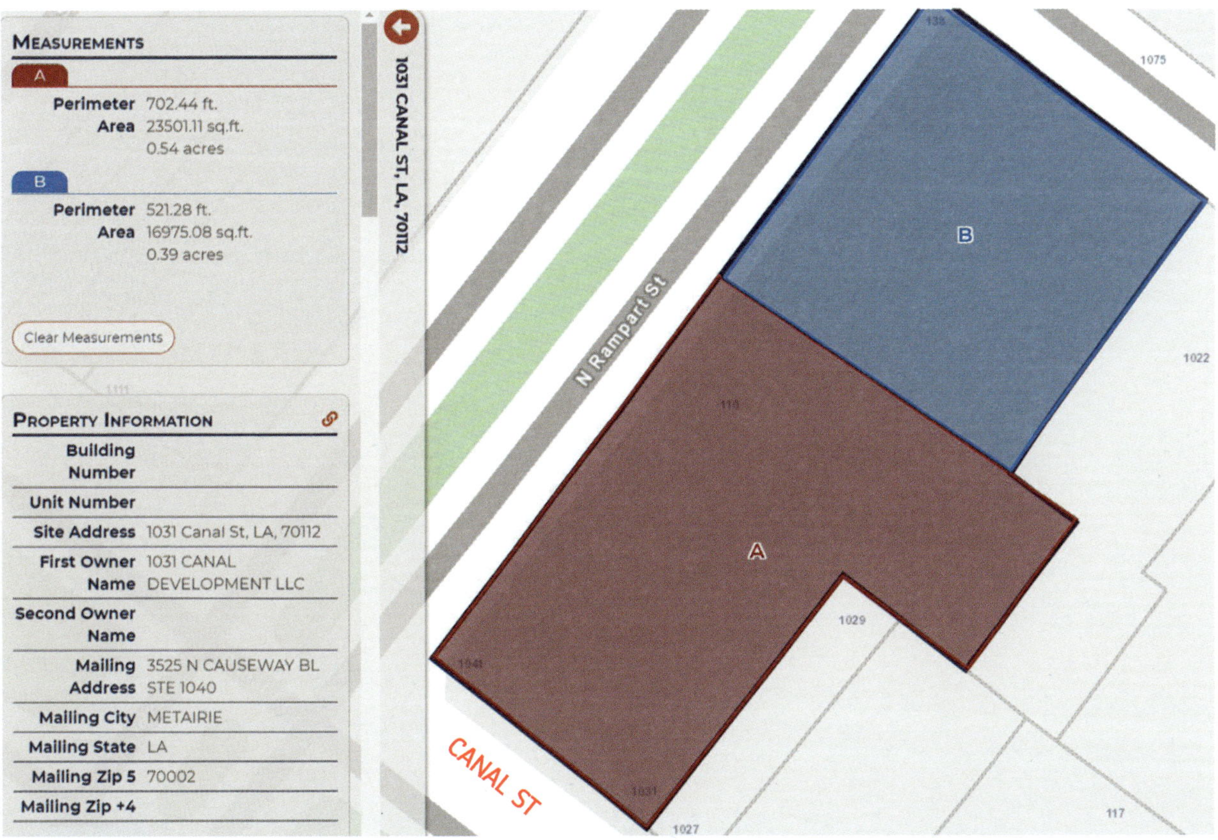

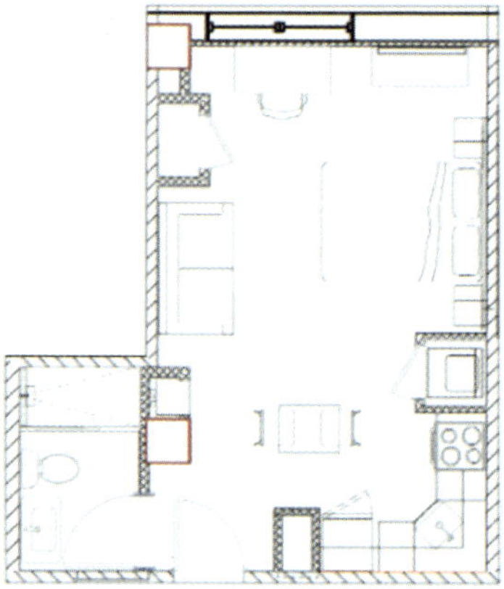

Micro Studio | 360 sf

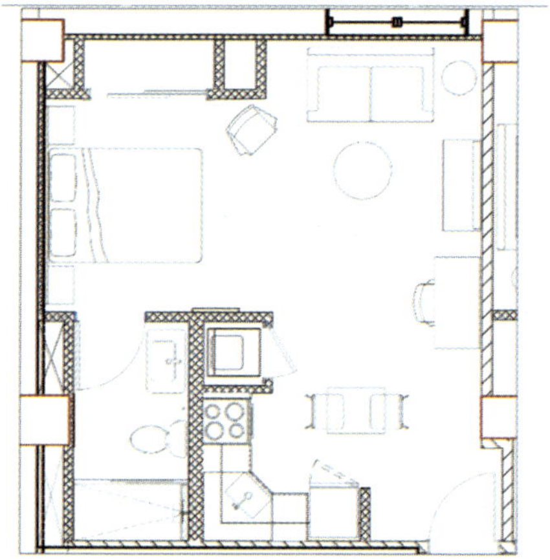

1-bed | 600 sf

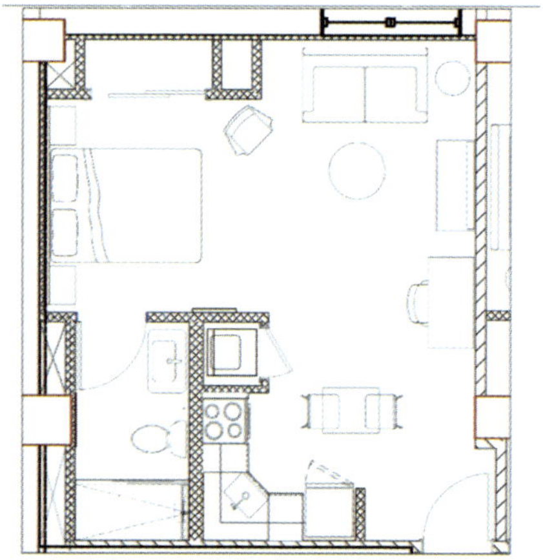

Studio | 500 sf

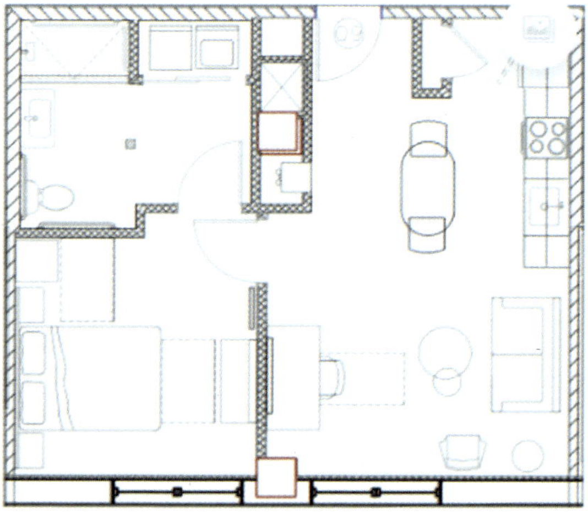

2-bed | 800 sf

s> Megan Fitzgerald + Kosta Sevic + Jiaqi Luo

i> Neal Jr.

Housing Occupancy

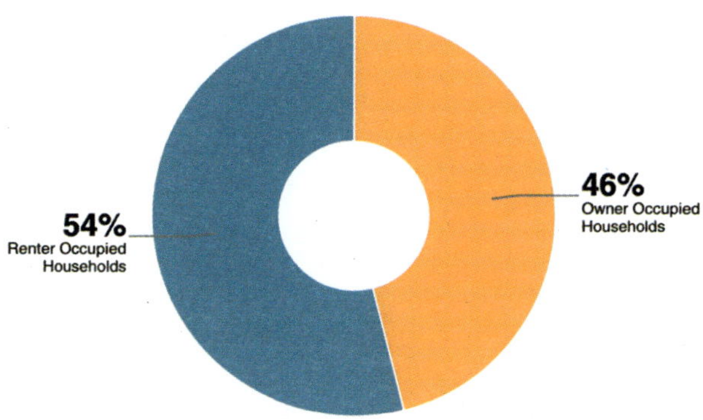

Housing Type

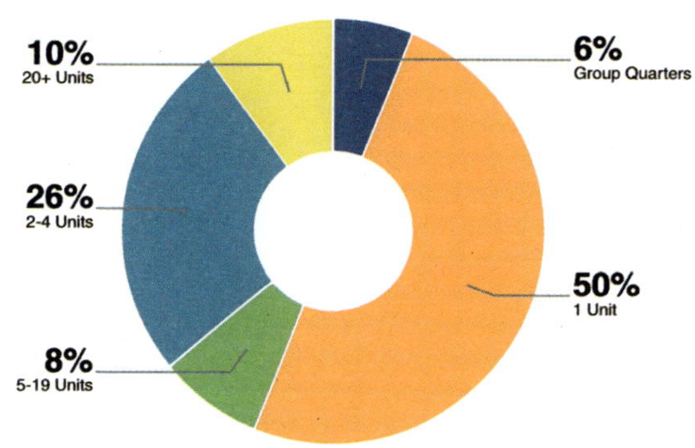

TOWN. HOUSES.

Jonathan Tate [C]

Town \'taûn\ n, 3 : A place that is a population and business center and is so recognized geographically and politically as 1). a compactly settled area of any size as distinguished from surrounding rural territory; 2) a compactly settled area usually larger than a village but smaller than a city in population and usually incorporated in and given definite boundaries and powers by law: a small municipality.

Webster's Third New International Dictionary of the English Language, Unabridged.

The Town. Houses. research studio is intended to formulate, as an introductory phase, the ideation of Town. This is a zone of particular interest given its notable absence in contemporary urban and architectural discourse. The studio begins by cataloging what has been said about a town, and comparing this to what should be said from an architectural and urban planning perspective. The research phase culminates in the "taxonomy of town."

The second part of the studio focuses on housing within these locations. Unlike larger urban centers, towns do not have the conventional suite of financial incentives, policy enticements, or market pressures. In absence of these normative forces, housing is created leveraging the conditions present and available, requiring a certain resourcefulness and imagination. TOWN. HOUSES. Johnathan Tate.

With both Town and House, the objective is to use context and a chosen scale to realize architectural possibilities and adaptive urban engagement.

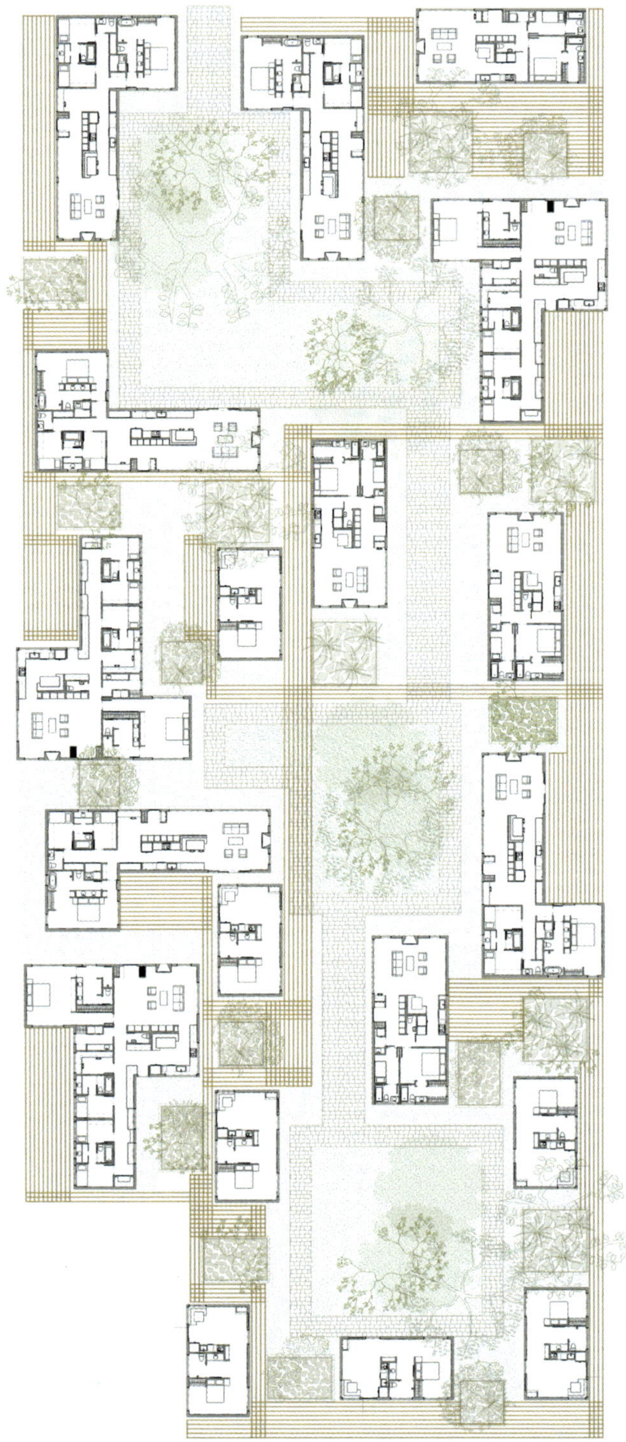

what is **economically** affordable

what is **economically** affordable

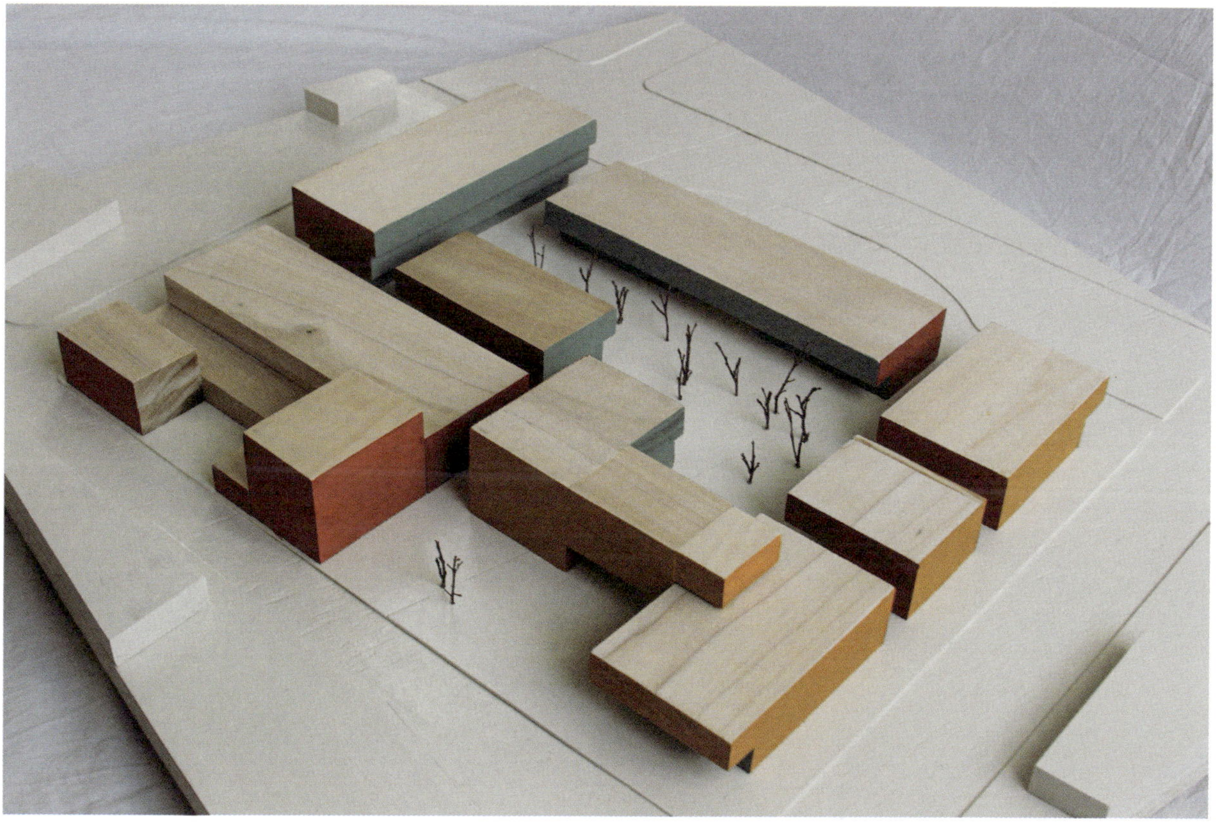

 what is **economically** affordable

s > Mitchell Hubbell

i > Jonathan Tate

URBANbuild 17

Byron Mouton [C]
Sara Allen Harper

As a continuation of efforts initiated by the Fall URBANbuildstudio, the Spring studio will further develop and fabricate a dwelling to be sited in the Central City Neighborhood of New Orleans.

Last year, as the fabrication of URBANbuild16 was underway, a relationship was established with members of the neighboring Bethlehem Lutheran Church – the oldest historically black Evangelical Lutheran Church in the continental United States (https://www.blcnola.org/history). The size of the congregation has waned since hurricane Katrina, but the current pastor, Ben Groth, is working to increase membership and has spearheaded a very successful community outreach initiative known as Community Table (https://www.blcnola.org/communitytable). In addition, Pastor Groth has recently defended his Graduate Thesis at Tulane's Department of History, and that research focused on the relationship between white supremacy and Christianity. Therefore, in addition to URBANbuild's focus on the topic of housing, his involvement with this studio and the resulting conversations will aim to broaden each students' awareness of community needs through their direct engagement on site and with the outreach programs of Bethlehem.

Debate continues with regard to what should be done with the many historically marginalized areas of largely low-income underprivileged communities of this city. During the fall semester, Master planning options were proposed for Bethlehem's primary church site, and several housing strategies were developed for another property of theirs just across the street at 1822 Washington Avenue.

One has been selected for construction, and groundbreaking will occur mid-January. Students considering this studio are urged to visit the URBANbuild website and acquaint themselves with the programs agenda and pedagogy. WWW.URBANbuild.tulane.edu.

The spring studio group will work at the scale of dwelling and fabrication - focusing on material issues and the development of fabrication details through the realization of a built project. Sixteen substantial projects have already been designed and constructed by the School's URBANbuild program, and students will continue to build upon the lessons offered by those accomplishments.

Students:

Asher Behar
Le'Bryant Bell
Hayden Boyce
Rebecca Dunn
Olivia Georgakopoulos
Elisabeth Keigler
Quinn Kelsey
Ethan Lewis
Sam Lindley
Abigail Lovins
Mitchell Lyons
Lauren McGarth
Matthew Meloy
Jenna Norton
Bhumika Shirole
Sean Tichnor
Daniel Tighe
Andres E Perez Yepez

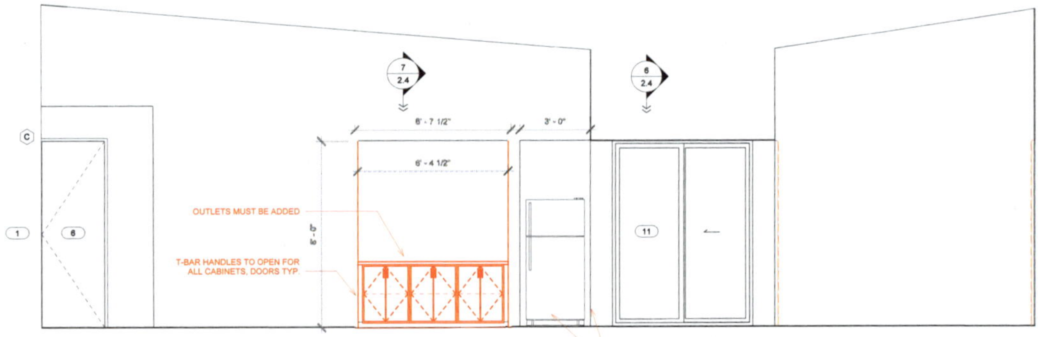

1 ELEVATION - INTERIOR KITCHEN RIVERSIDE REVISED
2.1 | 2.4 3/8" = 1'-0"

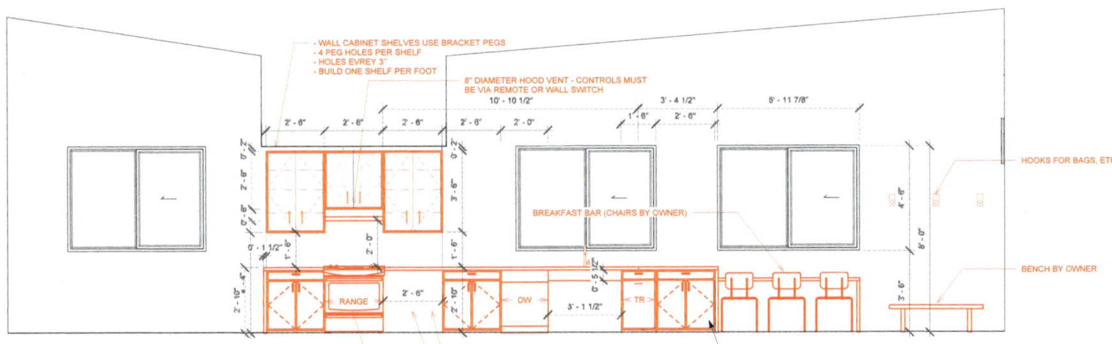

2 ELEVATION - INTERIOR KITCHEN LAKESIDE REVISED
2.4 | 2.4 3/8" = 1'-0"

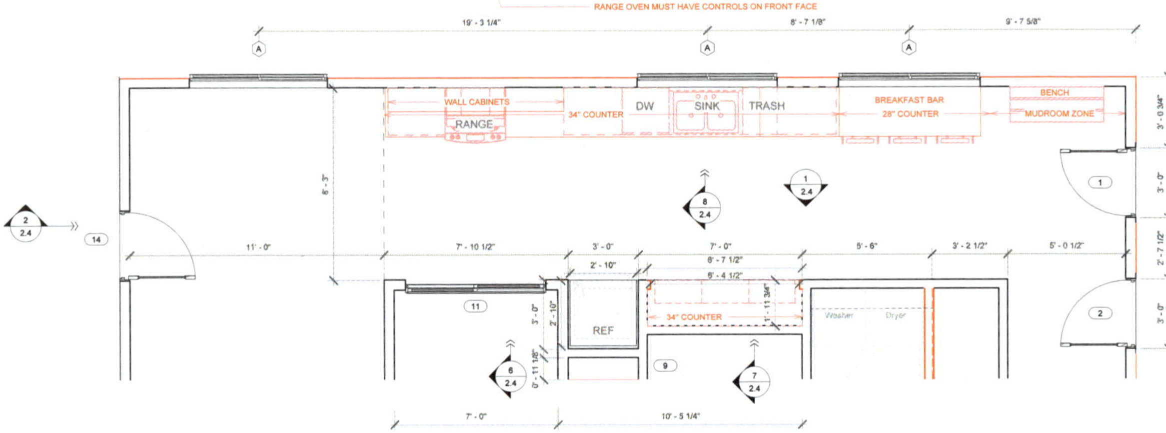

3 KITCHEN PLAN REVISED
2.3 | 2.4 3/8" = 1'-0"

4 | 2.4 KITCHEN VIEW 1 REVISED

5 | 2.4 KITCHEN VIEW 2 REVISED

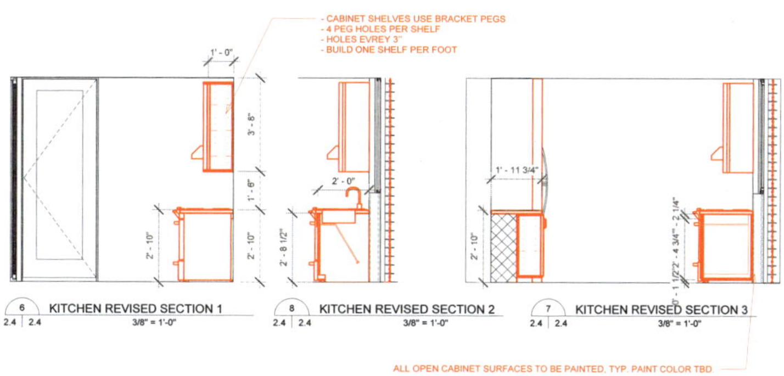

6 | 2.4 | 2.4 KITCHEN REVISED SECTION 1 3/8" = 1'-0"

8 | 2.4 | 2.4 KITCHEN REVISED SECTION 2 3/8" = 1'-0"

7 | 2.4 | 2.4 KITCHEN REVISED SECTION 3 3/8" = 1'-0"

ETHAN LEWIS

BACKYARD BUILDING

Rethinking New Orleans Single-Family Neighborhoods Through Prefabricated Accessory Dwelling Units

ARCH 5990 Instructors
Cordula Roser Gray & Ammar Elouini

This thesis comprises a real estate development investigation and architectural proposal focusing on revitalizing stagnating single-family neighborhoods through legalizing accessory dwelling units (ADUs).

America is in the midst of a housing crisis. Home costs and rents have risen far past the rate of income. Americans need cheap housing and fast. However, solutions are possible and not unknown. Time and time again, analysis has shown that increasing the housing supply leads to a decrease in the price of homes.

The housing crisis in America requires urgent solutions. The current system fails to provide affordable housing as costs surpass incomes. Evidently, single-family neighborhoods, prevalent nationwide, contribute to rising prices and limited affordability. Reviving ADUs can help address this issue without generating displacement and serve a greater good. ADUs can promote sustainability, affordability, and foster community development. New Orleans is no exception and an ideal city to investigate the history of ADUs and their promising future.

This system aims to promote sustainable living, facilitate wealth generation, and enhance neighborhood resiliency across multiple scales, including zoning, block, lot, and domestic levels. These chapters emphasize the significant potential of underutilized spaces on homeowners' lots to drive positive change for homeowners and the broader community

Fig.1
2060 ADU Benefit

Fig.2
Street Lot Axon

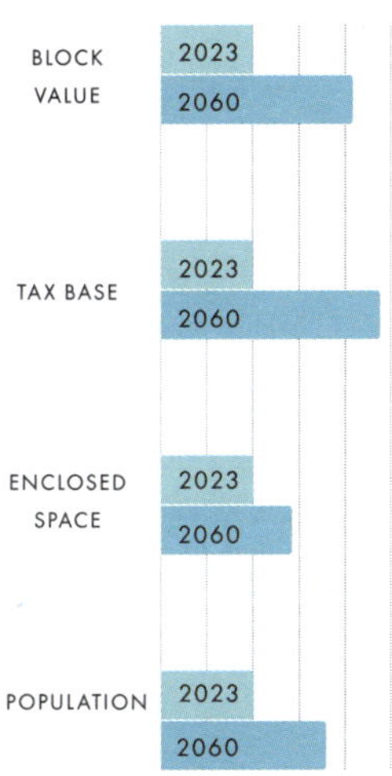

DANIEL TIGHE

SMALL, MULTIFAMILY, AFFORDABLE

Design and Development of Affordable Fourplexes in New Orleans

ARCH 6990 Advisor
Byron Mouton

s> Daniel Tighe **i>** Byron Mouton

Local advocacy group Housing Nola estimates the need for affordable housing to be somewhere north of 40,000 new units. Meeting this need will be a massive undertaking and will require strategic partnerships between many different organizations. One potential resource in the affordable housing market is faith-based organizations and mission-driven non-profits. While faith-based organizations, such as churches, mosques, and synagogues, have not traditionally been major providers of affordable housing, they may be uniquely situated to create hundreds, if not thousands, of affordable housing units throughout the city.

The biggest advantage these organizations is that in many cases, they already own land suitable for affordable housing development. In a study of Central City New Orleans, mission-driven non-profits were found to own a total of 214 vacant lots, 111 of which were owned by churches. Despite the great advantages faith-based and mission-driven non-profits have when it comes to the production of affordable housing units, they do not have the expertise, or capital, to develop complex housing projects.

What if they did? What if the city, or some other agency, created a subsidy for these institutions to build affordable housing on the vacant land they already own? This would create a huge opportunity and a challenge for architects to design new, four-unit, affordable housing complexes throughout the city. The goal is to create a compelling vision for how small-scale, scatter-site affordable housing projects can be built throughout the city, creating equitable neighborhoods for all.

Fig.1
Unit Experience

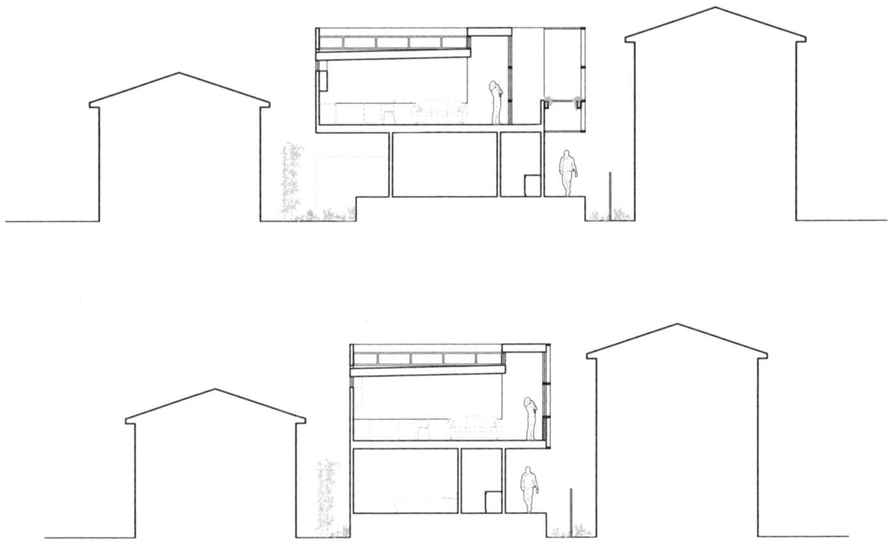

Fig. 2
Unit Sections

Fig.3
Unit Aggregation Options

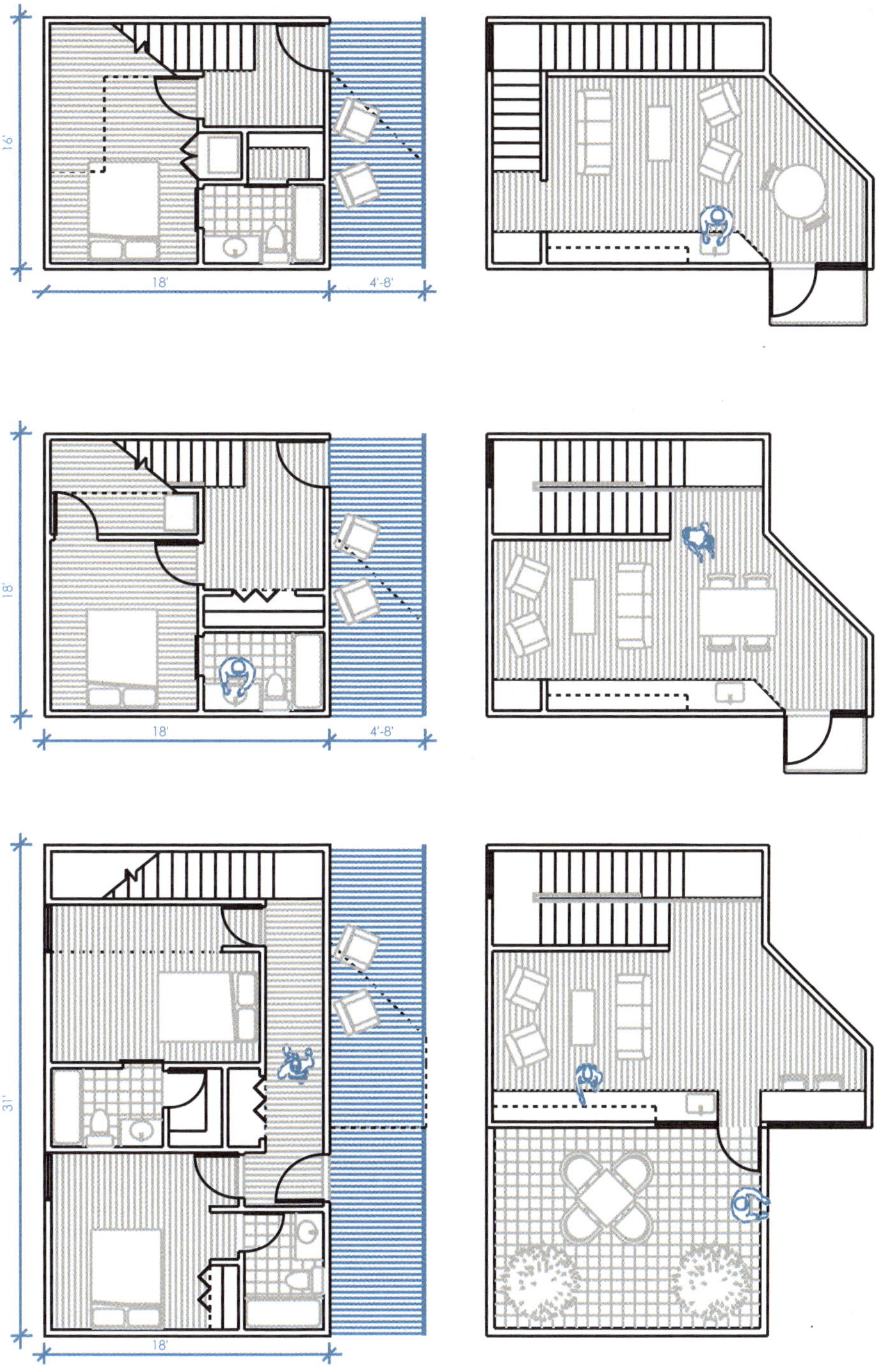

Fig. 4
Unit Variations

2

ENVIRONMENTALLY AFFORDABLE

ENVIRONMENTALLY AFFORDABLE

Materials' life cycle

SONSOLES VELA NAVARRO
Assistant Professor of Architecture

EMBEDDING CLIMATE CHANGE IN THE EARLY STAGES OF DESIGN PROCESSES

There is a global consensus that reducing our energy consumption and dependence on carbon-emitting fossil fuels is essential to prevent irreversible harm to our planet. By recognizing the significant contribution of buildings to the climate crisis, we can take steps towards mitigating their impact.

The global buildings sector consumes an estimated 30 percent of global energy and is responsible for around 27 percent of global operational-related CO2 emissions. Combined, CO2 emissions from building operations and the materials used in the construction of buildings are estimated to account for around 37 percent of global energy and process-related emissions.[1]

---

1 *2022 Global Status Report for Buildings and Construction: Towards a Zero-emission, Efficient and Resilient Buildings and Construction Sector,* DTI/2482/PA. Nairobi (Kenia): United Nations Environment Programme, 2022. https://www.unep.org/resources/publication/2022-global-status-report-buildings-and-construction

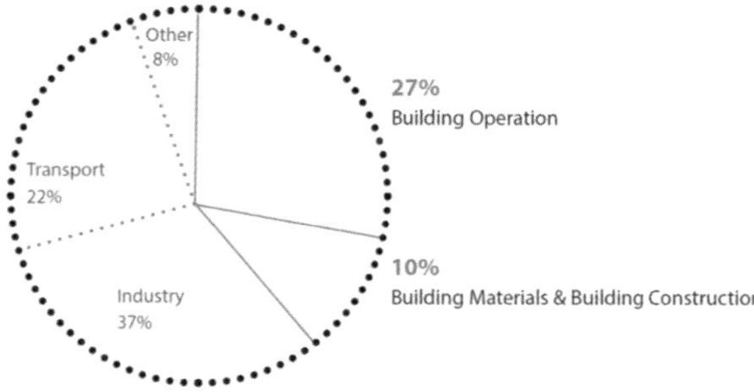

Fig.1
Annual energy consumption

Fig.2
Annual global CO2 emmissions

Source:
2022 Global Status Report for Buildings and Construction

According to the United Nations Global Status Report 2017, the projected construction plans for the next two decades indicate that the building floor area will double by 2060. The current projections indicate that between now and 2050, nearly half of new construction emissions will come from embodied carbon. Consequently, embedding climate change in the early stages of the design process is paramount for our future.[2]

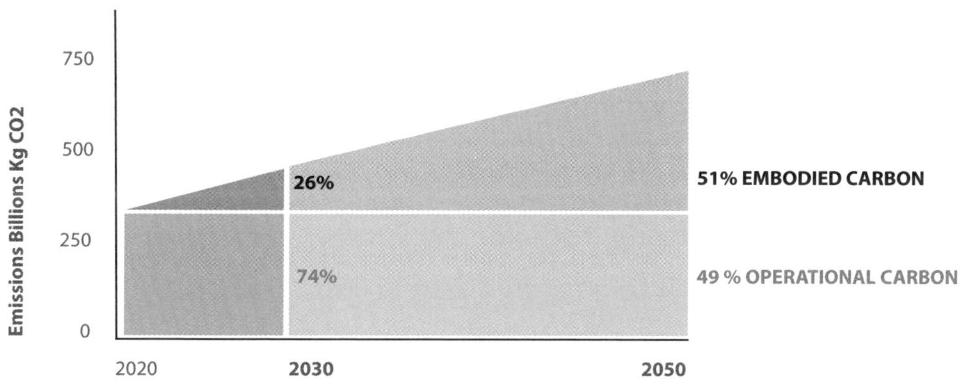

Fig.3
Total Carbon Emissions of Global New Construction from 2020-2050.

Source:
Architecture 2030.

Embody and In-use Carbon: how to tackle it?

The 2015 Paris Agreement on climate change established the necessity of capping global temperature rise to well below 2° Celsius, setting the target at 1.5°. In order to achieve that target, according to The Architecture 2030 Challenge, we must eliminate all CO_2 emissions from the built environment by 2040 to meet those 1.5° climate targets.

2 *Towards a zero-emission, efficient, and resilient buildings and construction sector. Global Status Report 2017.* UN Environment and International Energy Agency, 2017. https://worldgbc.org/article/global-status-report-2017/

A complete life cycle energy analysis of a building includes two types of energy: embodied energy and operational energy. Embodied energy is consumed during the raw materials extraction, manufacturing, transportation to the construction site, assembly-installation, replacement and maintenance, and demolition phases. Operational energy, on the other hand, is the amount of energy consumed over a building's lifespan during its use -phase for heating, cooling, hot water, electrical appliances and equipment, ventilation, lighting, and cooking.

Eliminating emissions from the built environment requires recognizing the critical role of embodied carbon. Unlike operational carbon emissions, which can be reduced over time through energy-efficient upgrades and the use of renewable energy, emissions from embodied carbon are inherently fixed once a building is constructed. As such, it is paramount that we take proactive steps to address embodied carbon now if we are to achieve our goal of zero emissions by 2040.

Adopting newer methods to reduce operational energy is a positive step towards more sustainable buildings. However, it is important to remember that these procedures can increase the embodied energy of a building. To ensure that new buildings have lower energy demand and are built sustainably, it is crucial to evaluate the energy performance of existing buildings and set meaningful targets for the construction industry to follow.

The building sector is the largest contributor to the world's energy consumption and greenhouse gas emissions. At the individual building level, design processes are decisive in the fight against climate change, determining energy performance and overall ecological footprint. Therefore, building life cycle energy evaluation plays a key role in achieving this goal, not only by assessing the energy efficiency of existing buildings but also by promoting sustainable practices in the construction industry.

Sustainable strategies in the architectural curriculum

Nowadays, the architectural curriculum requires a comprehensive evaluation of a building's impact on its environment. Sustainable building strategies, such as energy consumption, construction components, materials, environmental impacts of building subsystems, integrated renewable energy systems, and electrical and thermal systems, are essential topics for every school that promotes affordable and eco-friendly construction. Investing in quality education empowers future architects to tackle 21st-century challenges and achieve global sustainability goals.

The core course Building Comfort and Climate (BCC) at Tulane School of Architecture aims to bridge the gap between building performance and the architectural design process.

Methodology
The course methodology consists of learning by doing and testing back and forth, which is a continuous experimental process that involves both theoretical and practical components.

Lectures are used to provide students with foundational, structured knowledge. Laboratory sessions employ visual simulations to facilitate experimental learning in the field of building people and climate interactions. While pre-lab assignments are the seamless transition of establishing a connection between the theoretical concepts covered in a lecture and the practical aspects of the lab to start hands-on analysis performance. This is a continuous methodology that entails a high level of interconnection that is developed through every phase mapped below.

Phase 1: Environmental boundary conditions. Passive strategies
In this phase, passive strategies are integrated into a building's design, utilizing natural elements on site, such as sun and wind patterns, to provide natural heating and cooling throughout different seasons.

Environmental plugins and software can provide valuable insights into the environmental conditions that affect our built environment. By analyzing these conditions, we can better understand how to modify them, as well as identify opportunities for leveraging nature's resources to reduce our reliance on purchased energy. Climate analysis is critical for making informed decisions across a range of industries, including land use, agriculture, construction, and energy production. Considering environmental factors, we can minimize negative impacts and promote sustainable practices that benefit both people and the planet.

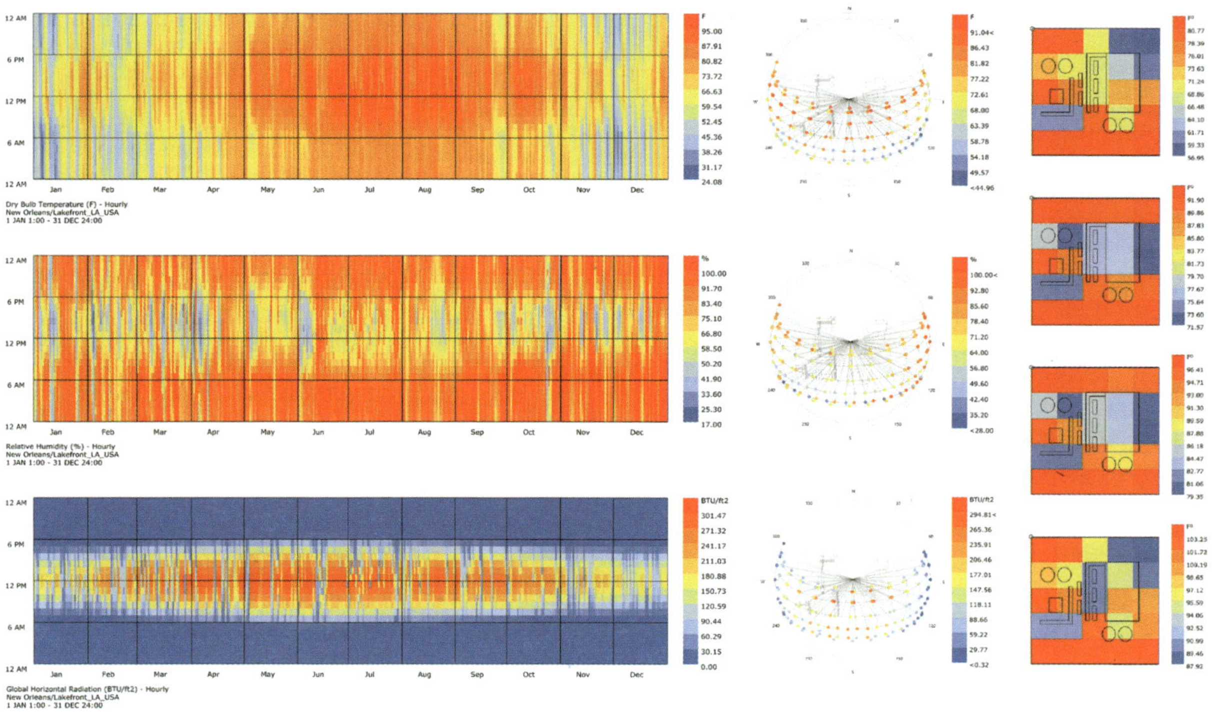

Fig.4
BCC Climate Analysis Parameters. Dry Bulb Temperature, Relative Humidity, Global Radiation.

Outdoor Comfort Conditions at noon March - June - September - December.

Phase 2: Building systems impact on design and ecological processes: technology and tools for "high comfort spaces" with "minimum environmental impact." Active strategies

This phase is based on active design strategies, including systems that use energy and systems that generate energy, such as solar electric and solar thermal panels, wind turbines, and geothermal energy exchangers.

The construction of buildings can have a significant impact on both design and ecological processes. As such, it is imperative to utilize technology and tools to achieve "high comfort spaces" with "minimal environmental impact." By incorporating a range of innovative technologies, such as energy-efficient HVAC systems and sustainable building materials, it is possible to create structures that meet the needs of occupants while minimizing their carbon footprint. Achieving this balance between comfort and sustainability requires a multifaceted approach that leverages cutting-edge research and design principles to create buildings that are both efficient and aesthetically pleasing.

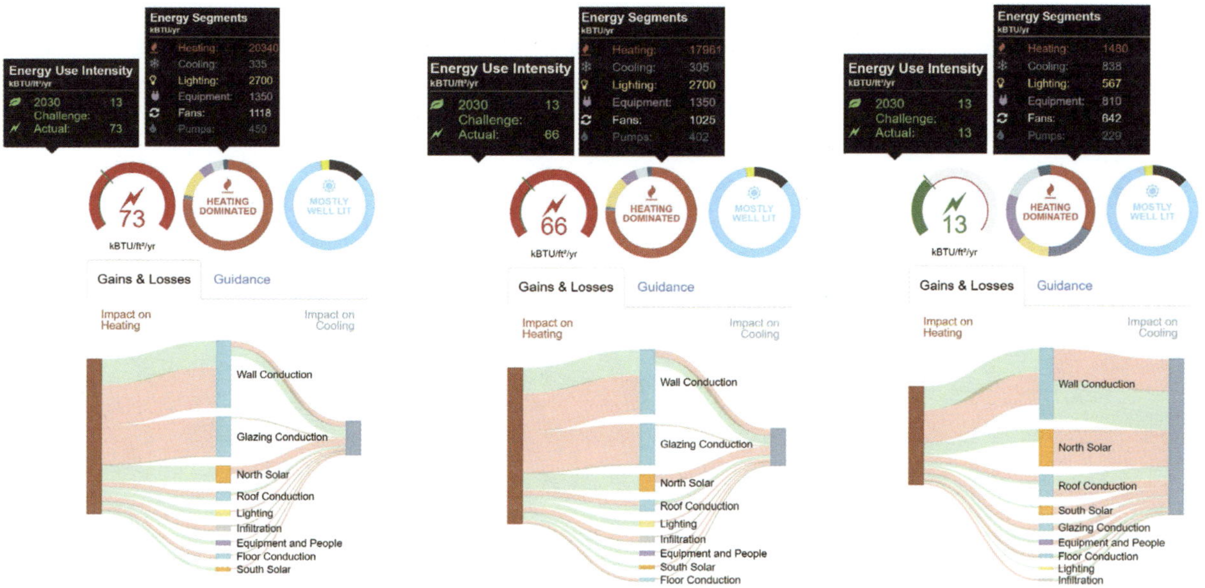

Fig.5
BCC Energy Use Intensity Analysis: ASHRAE 90.1-2013 baseline (left); Custom baseline, energy budget (center);
Challenge 2030 goal improvements (right).

Phase 3: Embodied and operational carbon targets

This last phase adopts a life cycle analysis approach to address both embodied and operational carbon emissions. This involves implementing three main strategies: "avoid," "shift," and "improve," all of which aim to achieve greater "adaptability." These strategies encompass a range of measures, such as building less, using fewer materials, incorporating low-carbon materials, implementing circular approaches, and designing buildings with a longer lifespan and lower operational emissions.

The estimated carbon emissions of each project during the design phase of the project will be measured using energy modeling and embodied carbon analysis. This data will be used to optimize the designs further. It will be taking into account the renewable energy or carbon offsets needed to meet the carbon budget established for each case.

These three phases aim to explore how integrating building performance early in the design process can contribute to the well-being of the planet.

It is important to understand how technology and design can be blended to prioritize energy efficiency in some areas and concepts in others. Energy modeling and environmental software can help make informed decisions while staying within the overall project's energy target and carbon budget.

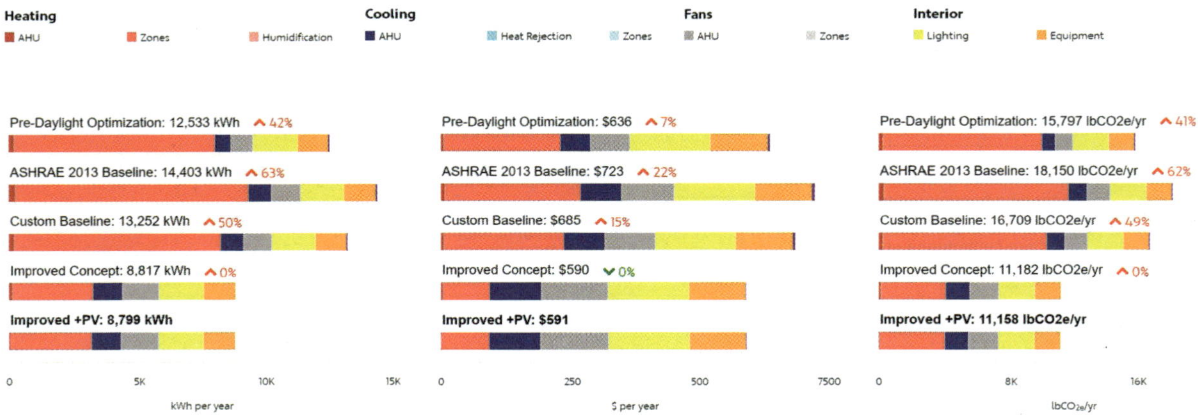

Fig.6
BCC Final Optimized Design Summary: Annual energy use comparison (left); Annual energy cost (middle); Annual CO_2 emissions (right).

Energy and carbon targets

An energy budget is a specific target for Energy Use Intensity (EUI) that should be considered. EUI is an annual measure of the total energy consumed in a building. A carbon budget is a maximum amount of cumulative net global carbon dioxide (CO_2) emissions that would, with some probability, limit global warming to a given level.

Prospective architects must understand that every design contains a concept, energy target, and carbon budget assigned that is all part of the project. By keeping these factors in mind, architects can create spaces that are both functional and efficient while contributing to a more sustainable future.

Benchmarking

Each phase of the process aims to measure metric values such as total CO_2 emissions, EUI, cost, renewable production, daylight efficiency, and thermal comfort to determine the impact of enhancing the building envelope, site conditions, integration of renewables, natural ventilation, and shading strategies. These improvements will be evaluated based on both current code requirements and beyond-code standards.

The starting point for benchmarking will be establishing a default baseline of parameters that comply with the current code on the specific location. The following step process will be defining an energy and carbon budget or target, The 2030 Challenge can be an option or establishing a % reduction related to the baseline diagnosis.

Displaying energy information on an architectural model helps with informed decisions regarding energy, indoor-outdoor thermal comfort, daylight, Co2 emissions, radiation, and sun hours.

Conclusion

The more we use energy modeling and environmental software, the better we understand their potential to design for climate action. Energy and Carbon design must become an essential part of the design process; the design strategy focuses on prioritizing the concept in specific areas while building energy modeling helps maximize efficiency opportunities in areas where design flexibility exists.

The built environment is responsible for a significant amount of energy consumption, but it also has the potential to generate energy. Individual buildings must be designed both to limit energy consumption and carbon emissions to produce and recycle energy and to increase reliance and robustness in the face of climate change.

JUAN MEDINA, LAURA SALAZAR,
PABLO SEQUERO & FRANK BARKOW

THE OUTDOOR ROOM

PAVILION
Seoul, Korea, 2023

Credits
salazarsequeromedina | Laura Salazar, Pablo Sequero, and Juan Medina | and Frank Barkow

THE OUTDOOR ROOM is a pavilion for the 2023 Seoul Biennale of Architecture and Urbanism, built in Songhyeon Green Plaza. It is an architectural device to sense land and city.

The simplicity of the square structure aims to capture the complexity of the environmental and cultural history which endows great significance to the opening of its site to the public. Once an ancient pine forest attached to a royal palace, in the second half of the 20th century Songhyong-dong was a representative site of American military occupation. The site was a void, enclosed by walls and isolated from the rapidly growing urban environment around it. As the walls come down and the park opens to the public, becoming available to the citizens of Seoul, it remains enclosed by the density of the city surrounding it. Therefore, The Outdoor Room is a void within a void which frames the history of change for land and city.

Its construction assembles materials which have been repurposed from another structure designed by Diagonal Thoughts. The lifecycle of these elements transcends the structure itself as they will carry on after the biennial closes and become adapted to a new site. Within the limits of its current form, visitors will discover a garden of native plants and a chimney which reverberates with the wind. The Outdoor Room will feel domestic and familiar, though strangely abstract. It will be a place of arrival, and a place to stay for a while and sense the sky, the mountains, and the tops of buildings.

Fig.1
Model

Fig. 2 Section.
Roofing.

Fig.3 Section
Greenhouse.

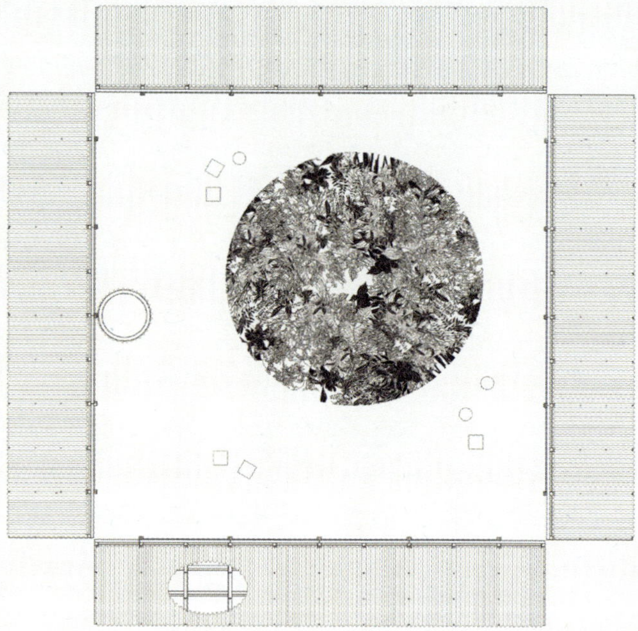

Fig. 4
Outside view
Photo.

ENTANGLED MATTER

INSTALLATION

Credits
X-Topia + CRGarchitecture
Marcella Del Signore, Cordula Roser Gray, Tatiana Teixeira (Project Leads.)
Marcella Del Signore, Cordula Roser Gray, Tatiana Teixeira, Marcelo Chaves, Lucas Coelho Neto (Design and Research.)
LaMaquina by Noumena (3D Printing & Engineering)

The drastic increase of environmental imbalances caused by resource extraction, mass production, and the insatiable need for consumption inevitably leads to the destruction of our environment. This creates the need for continued technological advances that, motivated by nature, visualize, and productively utilize the necessary convergence with existing ecological processes. To symbiotically and productively merge natural cycles with human habitats, Entangled Matter is a modular system for dynamic matter conversion. The project incorporates concepts from natural ecosystems to remediate climate change-induced stresses through new self-sufficient ecologies.

Aware that open systems promote adaptation, we propose a combinatorial network of scalable entanglements, where hygroscopic materials are combined with organic life forms to filter the air around us and cool our surroundings through natural processes to re-establish healthy microclimatic conditions. Allowing for a symbiotic relationship between human and non-human species, Entangled Matter replaces adverse interactions and consciously allows for direct access to make natural processes visible and accessible through closed-loop systems.

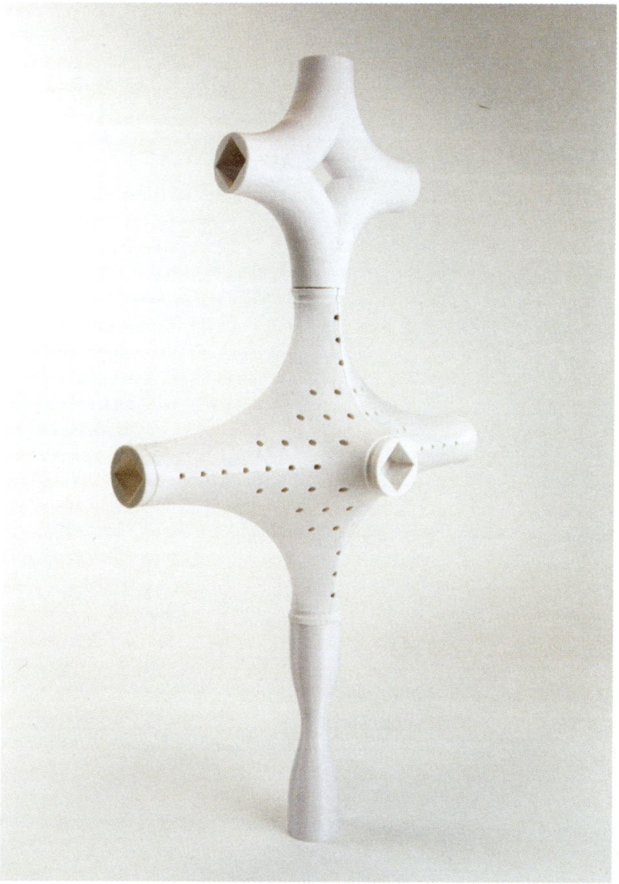

Fig.1
Piece of installation.
Photo.

Fig. 2 Elevation.
Drawing.

Fig.3 Axon.
Drawing.

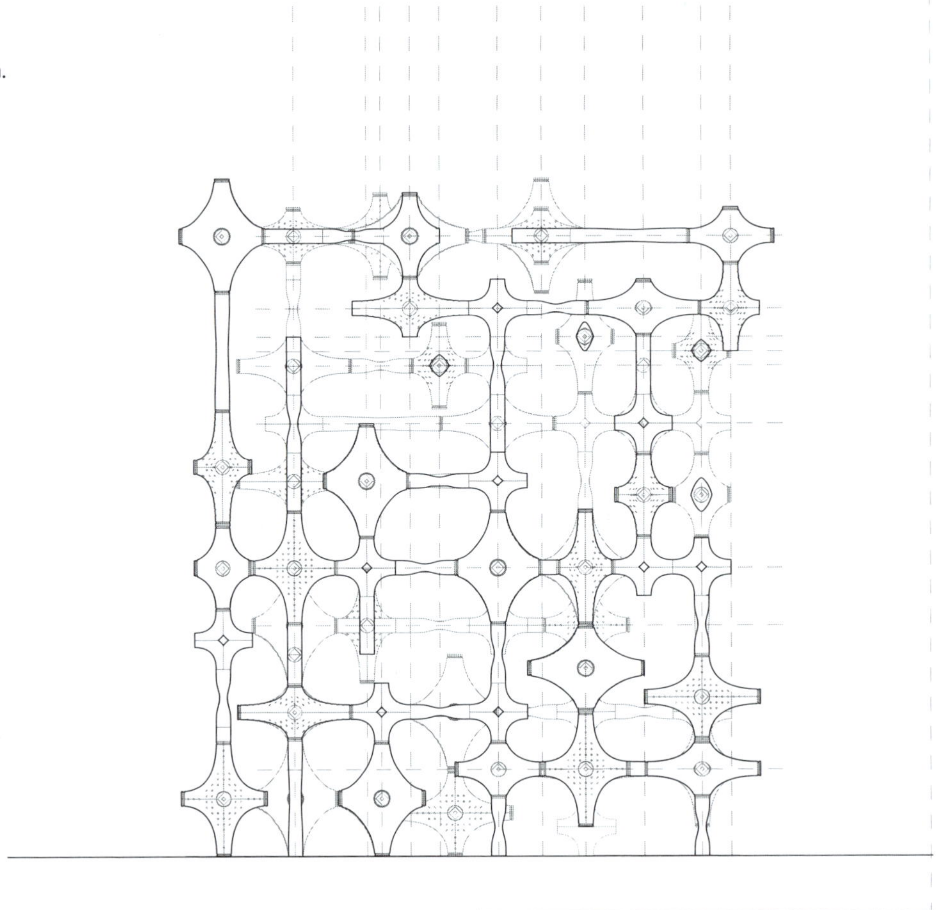

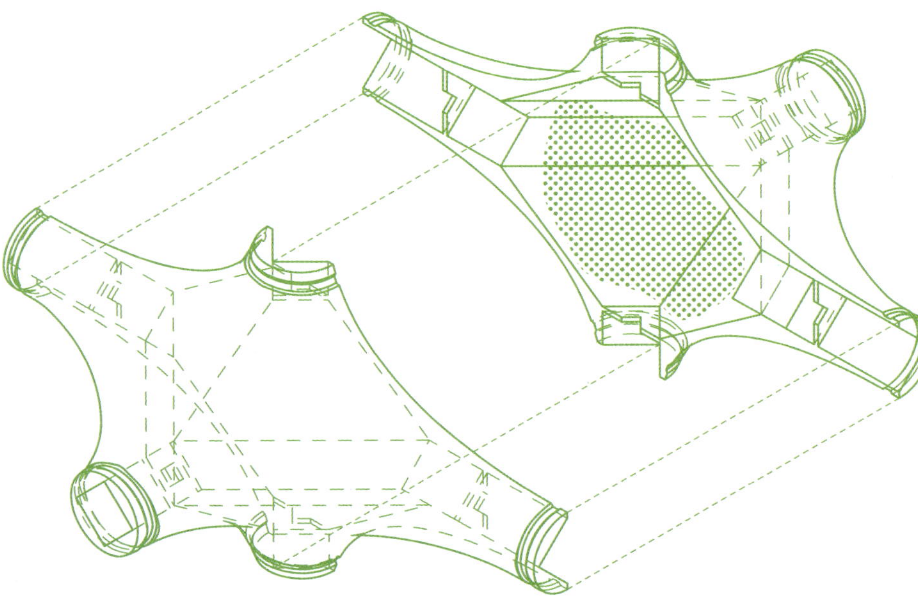

Fig. 4
Installation close up.
Photo.

TECH-SHELTER!

Andrea Bardon de Tena [C]

The main goal of this course is to get familiarized with the construction systems, material properties, and scale of the structures. The main objective for the students is to become "visual thinkers". During this course, the students explore different representational scales: the territory scale, to analyze and understand the problems and challenges of the place-ment through mapping, and the small scale to explore the structural systems and construction details.

How can we think and design a space with materials and con-struction systems as a starting point?

The course takes the history of the Eames House (Case Study House No. 8, Los Angeles, 1949) as a possible exercise hy-pothesis. Charles and Ray Eames, having the materials of their house already in place, decided to rethink their project to configure a different space with the same construction pieces. The students, over the next four weeks, found themselves in a similar situation.

In this case, we imagine that each student "lands" in a specific territory [a national park in the United States] with a load of materials from a building [a 20th-century case study]. The starting point is twofold. On the one hand, they will study the assigned case study, specifically, its most significant materials and construction systems to understand how they work and use its material in their project proposals: a shelter in nature. This is a new challenge. The use of smaller scales (1:50, 1:40, 1:20) both in drawings and models to get closer to working with detail, with materiality.

On the other hand, each student must study the National Park that corresponds to him/her. Each case will have very different physical conditions (climate, topography, relationship with water, etc.) so they will have to analyze in detail the challeng-es and opportunities latent in the site. To do so, they should use mapping as an analytical tool. After this study, they will have to choose the most suitable locations for their shelters.

s> Tracy Jones

i> Andrea Bardon de Tena

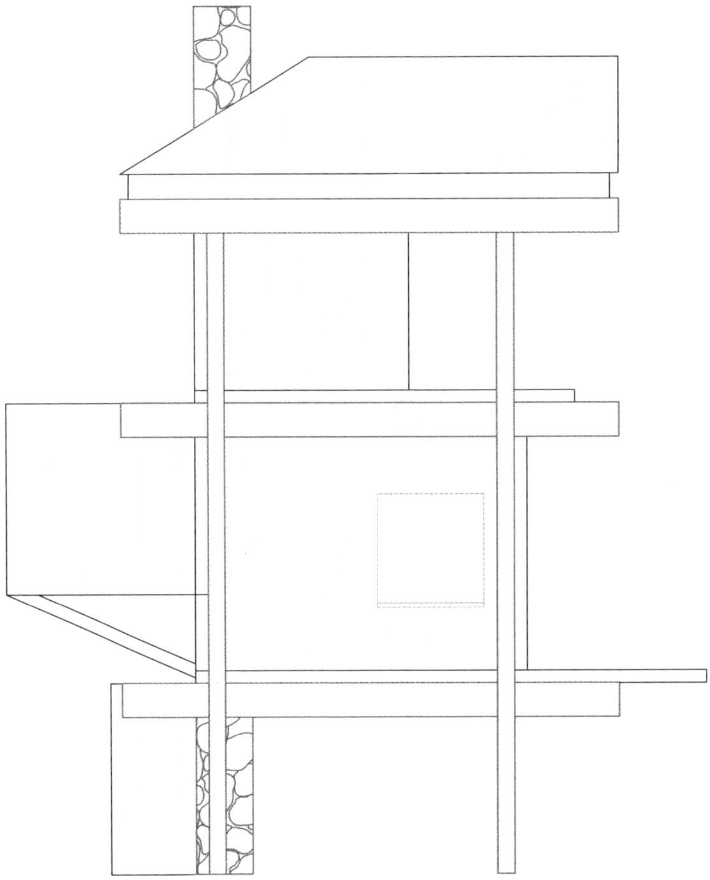

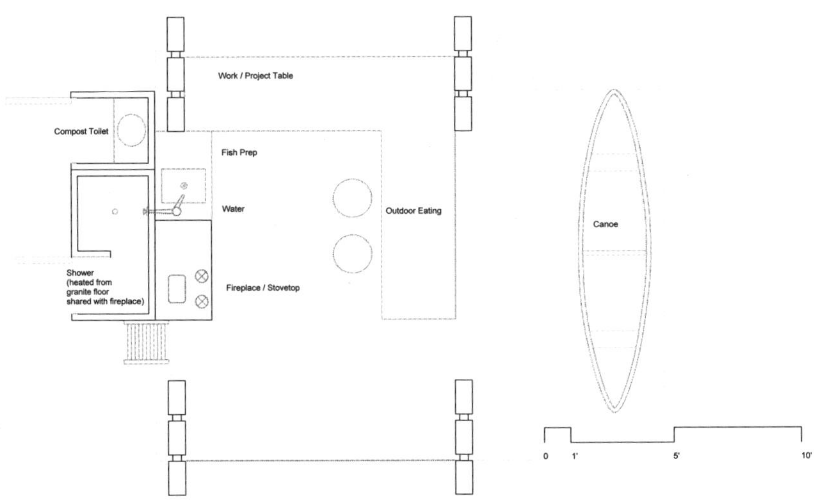
Work / Project Table
Compost Toilet
Fish Prep
Water
Outdoor Eating
Shower
(heated from
granite floor
shared with fireplace)
Fireplace / Stovetop
Canoe
0
1'
5'
10'

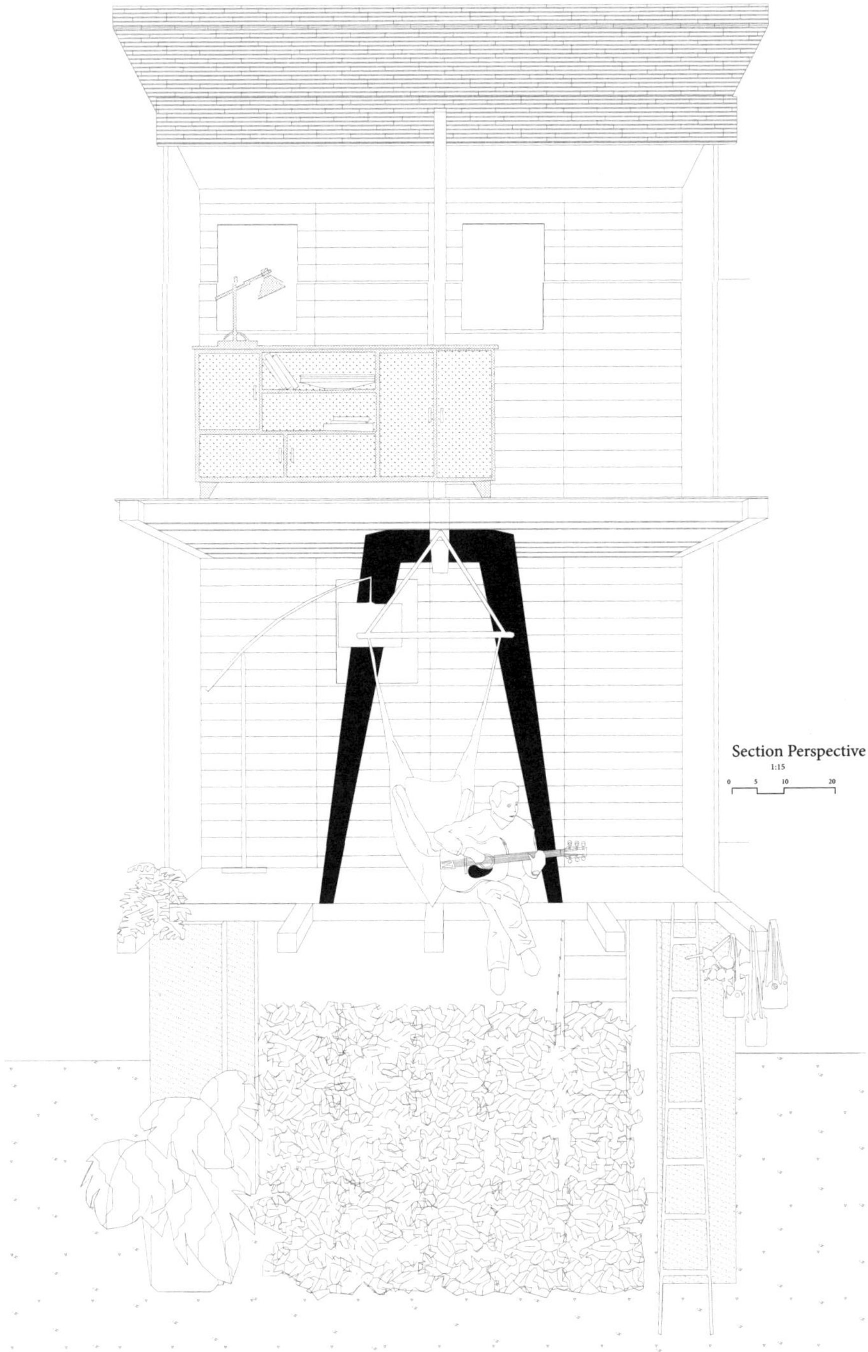
Section Perspective
1:15
0 5 10 20

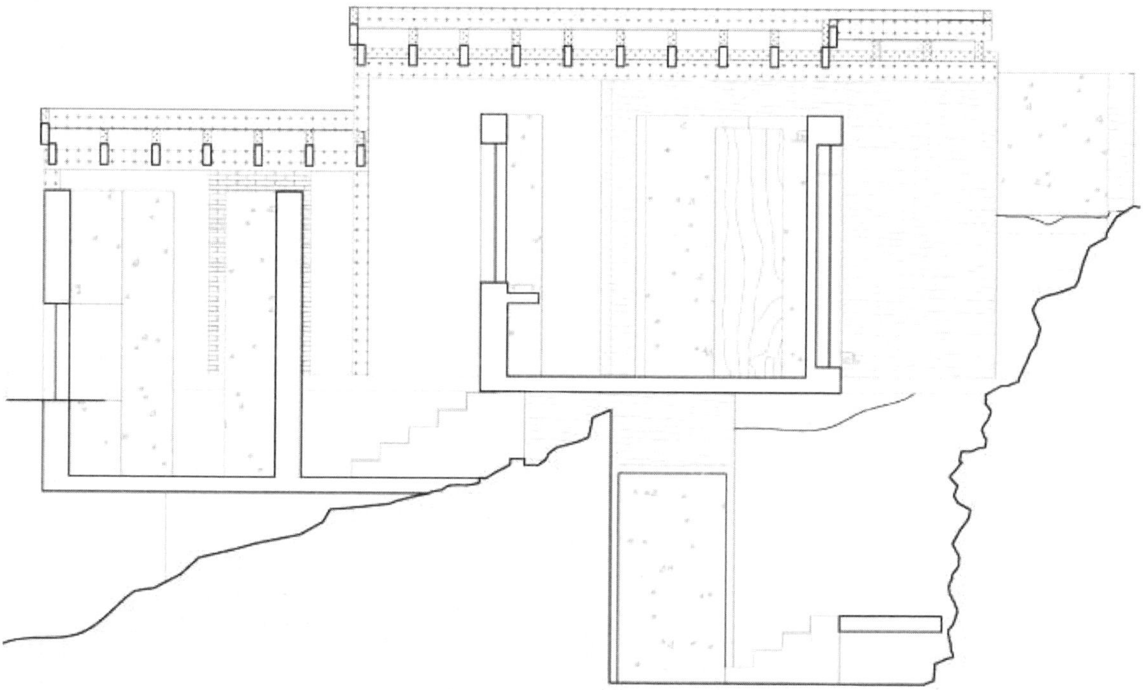

DESIGN STUDIO I

Hannah Berryhill [C]
Bruno Soria
Nick Perrin

In Design Studio I, students fine-tune skills of rigorous observation, analysis, and creative problem solving through the iterative design process. The studio begins with an analysis of organizational structures that occur in nature. Each student diagrams a plant specimen to understand its underlying geometry, symmetry, structure, and integral patterns. These diagrams are later translated into two-dimensional wallpapers and three-dimensional shadow models, exploring rules of symmetry, color theory, light, and shadow. These explorations in rigorous analysis, tessellations, iteration, well-crafted details, 2D and 3D representational drawing techniques, and light and shadow culminate with the students' main project of the semester: a hand-crafted light sculpture. Students work through multiple iterations of their light sculptures (either in the form of a pendant or table lamp) and embrace new technologies such as laser-cutting and three-dimensional digital modeling to create and represent their designs. Additionally, students learn about the wellness and spatial implications that light has upon the human body and the built environment and integrate human centered considerations into their designs. Although there is an emphasis on precision and craft in this studio, a dialogue about the idiosyncrasies found in hand-made objects and those found in nature are maintained throughout the semester. Environmental strategies are incorporated as a new area of study.

The work of the studio is developed in four projects:

1. **Diagrammed Specimens:**
After a visit to the New Orleans Botanical Garden, students choose one natural specimen to diagram at a micro, mezzo, and macro scale. The concept of a "diagram" is introduced through the students' analysis of elements such as structure, growth, symmetry, motif, depth, flow of nutrients, and geometry found in their chosen plant.

2. **Nature-Inspired Wallpaper:**
Students reconsider their drafted diagrams of natural specimens as motifs in a pattern. There is an emphasis on scale: how do these motifs reveal themselves at various scales throughout the wallpaper? The rigorous, digital study of their specimens is synthesized with handmade reactions: each design should incorporate both digital and analog media.

3. **Kirigami Shadow Models:**
Having thoroughly studied and represented the specimen's organizational structures in two dimensions, these concepts are then translated into formal and spatial relationships via paper models folded using a the kirigami technique.

4. **Light Sculptures:**
In the longest project of the semester, students design a "light sculpture," either pendant or table lamp. Using their diagrams and paper models as a point of departure, this analysis becomes a three-dimensional, functional object that must translate important elements of their chosen plant while also integrating considerations of utility, fabrication, and user experience.

s> Jillian Brennan

i> Hannah Berryhill

s> Will Fichman + Jacob Starr

i> Hannah Berryhill

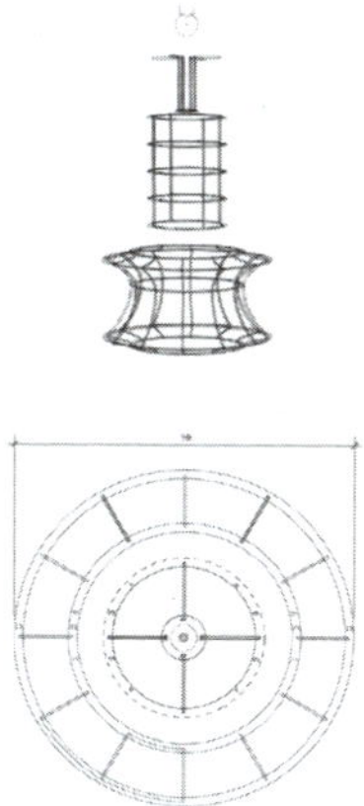

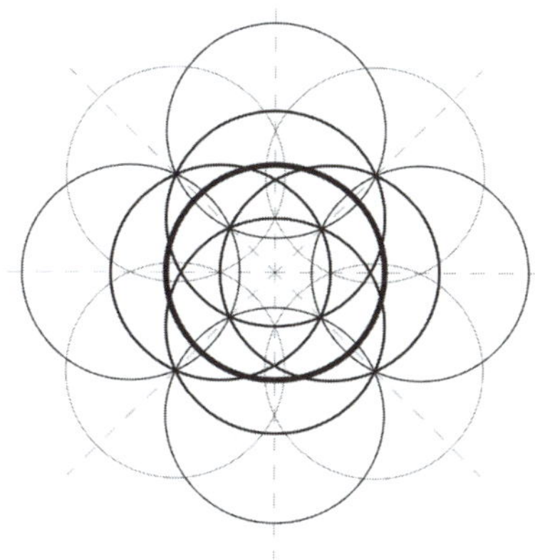

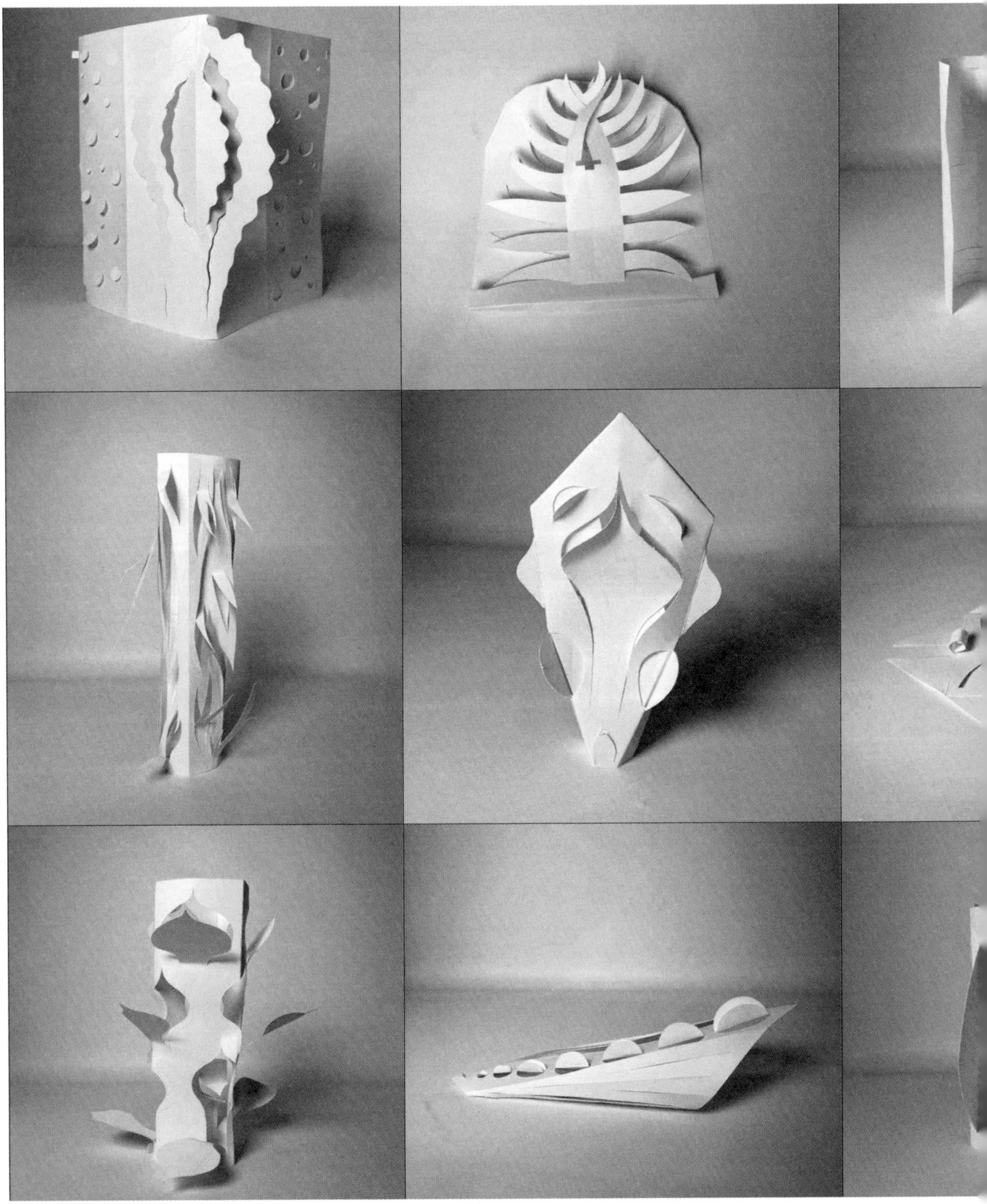

s> Sydney Lipoff

i> Hannah Berryhill

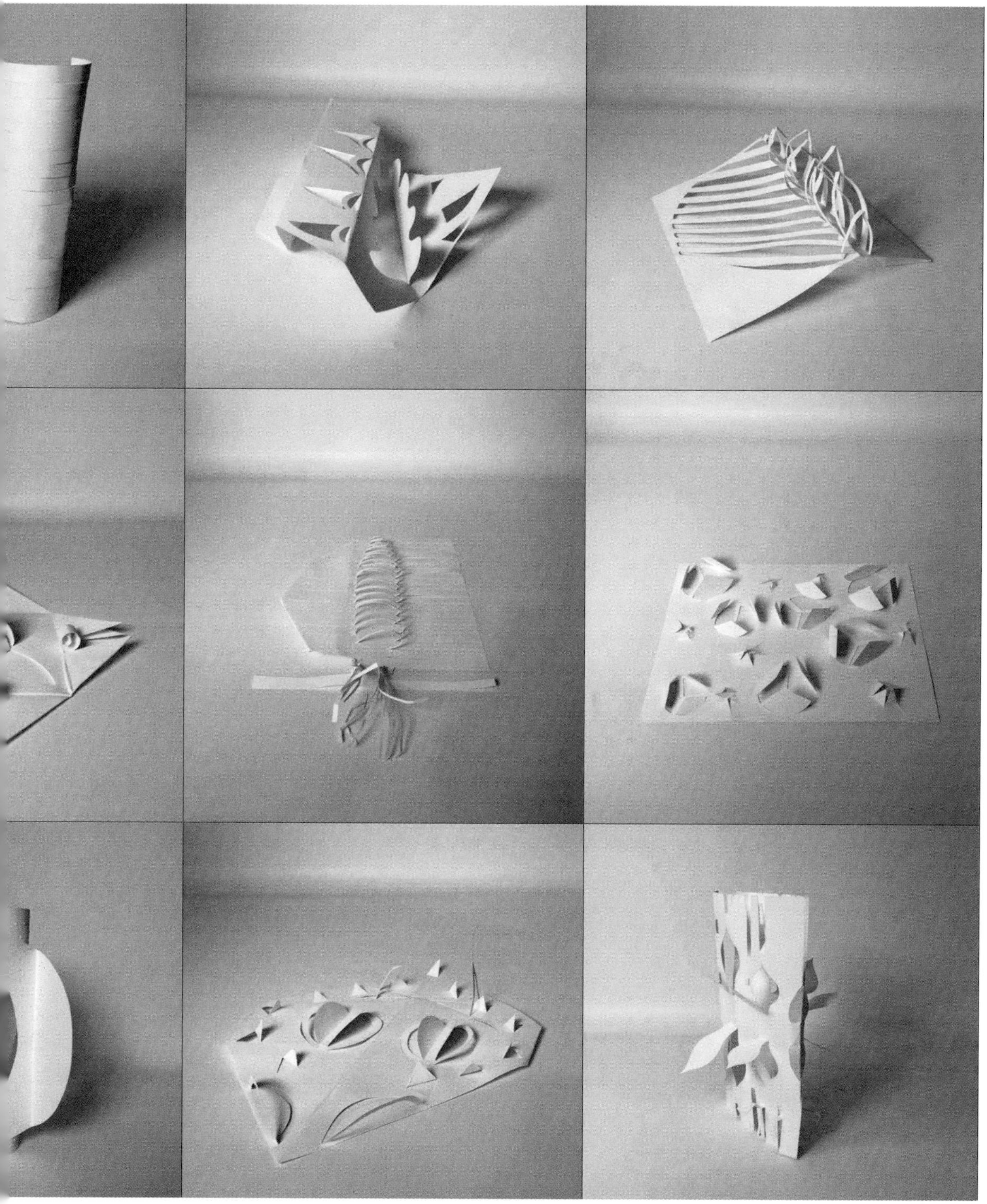

TEXTILE DESIGN

Bo Choi [C]

This course is an introduction to textile and fashion design. Students will develop their creativity and ideation skills as they explore the fashion design process, from concept development to final presentation. Students will be introduced to the history of textiles, fabric properties, color theory, and pattern design with industry-standard techniques. They will also learn how to generate and refine ideas for sustainability and technology in textile and fashion design. This course includes a lab component for production including garment construction, prototyping, textile engineer printing, and use of the Textile Lab's Juki industrial sewing machines.

PANTONE 19-1432 TCX Chocolate Fondant
PANTONE 11-0603 TCX Pastel Parchment
PANTONE 16-0550 TCX Lentil Sprout
PANTONE 16-1219 TCX Tuscany
PANTONE 17-1048 TCX Inca Gold
PANTONE 13-4804 TCX Pale Blue

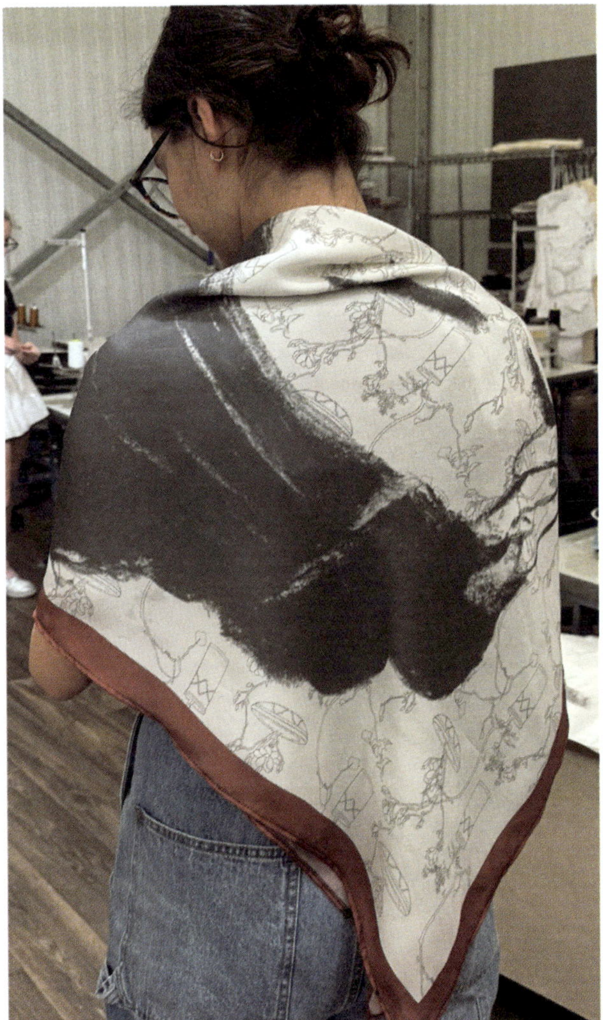

18-4028 TCX
Bering Sea
18-1553 TCX
Adrenaline Ru
19-3911 TCX
Black Beauty
11-0515 TCX
Lemon Icing

Cotton Patterned Material
Zipper Closure
Technical Flat: Back
Technical Flat: Front
Bunched Recycled Denim

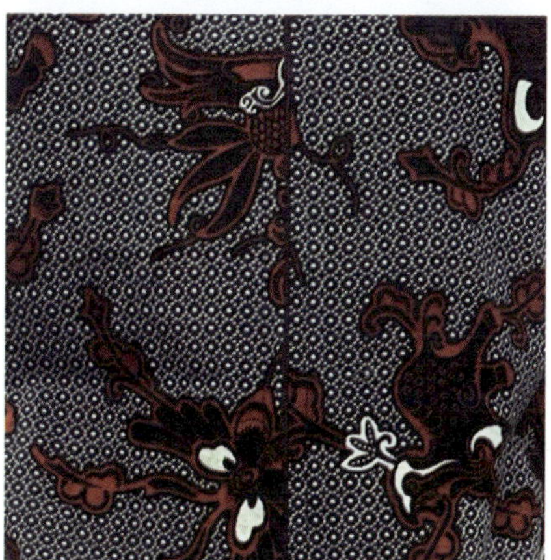

RESILIENCE REINFORCED

Kentaro Tsubaki
Charles Jones

New Orleans occupies the Mississippi Delta, the complex ecological system formed by the soil deposits interacting with the river and the gulf. In this soft, fluid ground, finding stability is a constant challenge. With a thoughtful application, modular precast concrete paving systems can reestablish a symbiotic relationship between the built environment and landscape as infrastructure and public amenities. By leveraging the digital design and fabrication techniques, the geometry of the ground surface can be performatively and aesthetically aligned; to detain, retain or permeate water, mitigate flooding, and aesthetically embrace the reflectivity of the water surface, contributing to the spatial quality of the context, encouraging the productive occupation of the ground plane.

The studio initiates students to the complex relationships between technology and society. By leveraging accessible open-source databases, digital modeling, and fabrication techniques, the studio systematically addresses the complex, social-cultural, ecological, and economic (infrastructural) relations through the performance of surface geometry of the ground plane relative to water in three distinct scales.

At the component and street scales, we aim to slow down the storm runoff by momentarily storing it where it falls. On an urban scale, we aim to detain overflow from the drainage system during peak storm intensity to maintain the demand at capacity.

The studio prioritizes the design process informed by visualized quantifiable parameters through simulations and prototyping to find "the equilibrium" beyond individual beliefs and aesthetic preferences. The design methodology and resulting solutions are internationally relevant, as the water-related issues are no longer unique to our region.

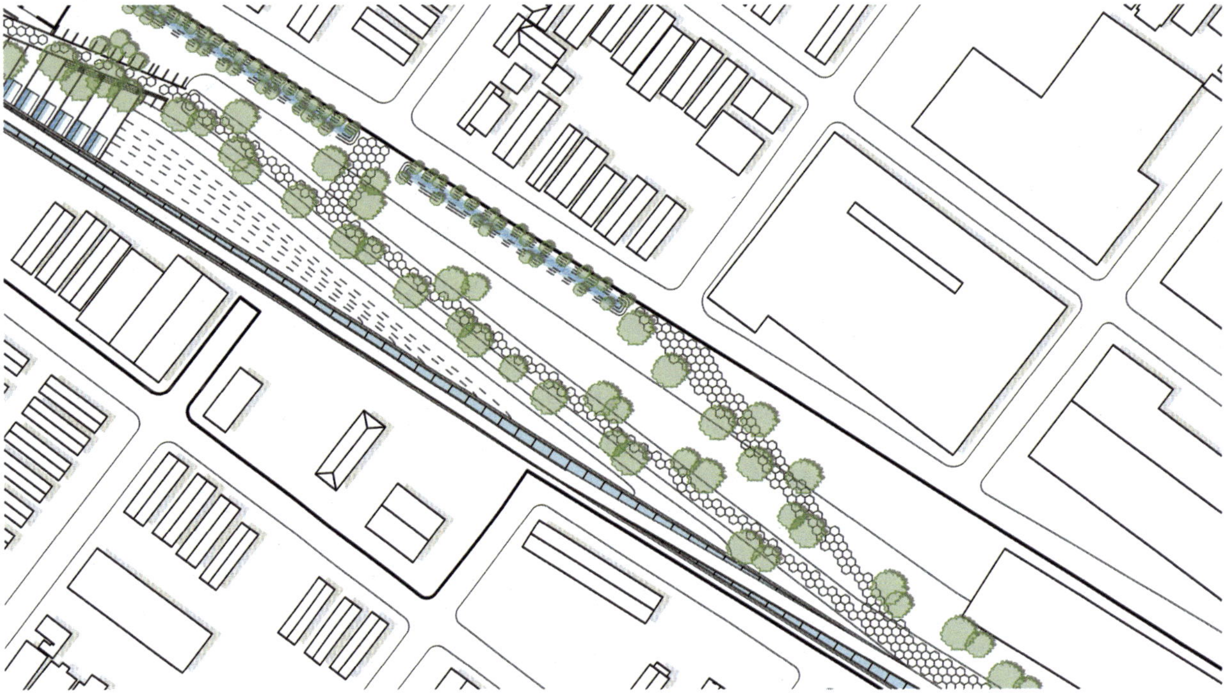

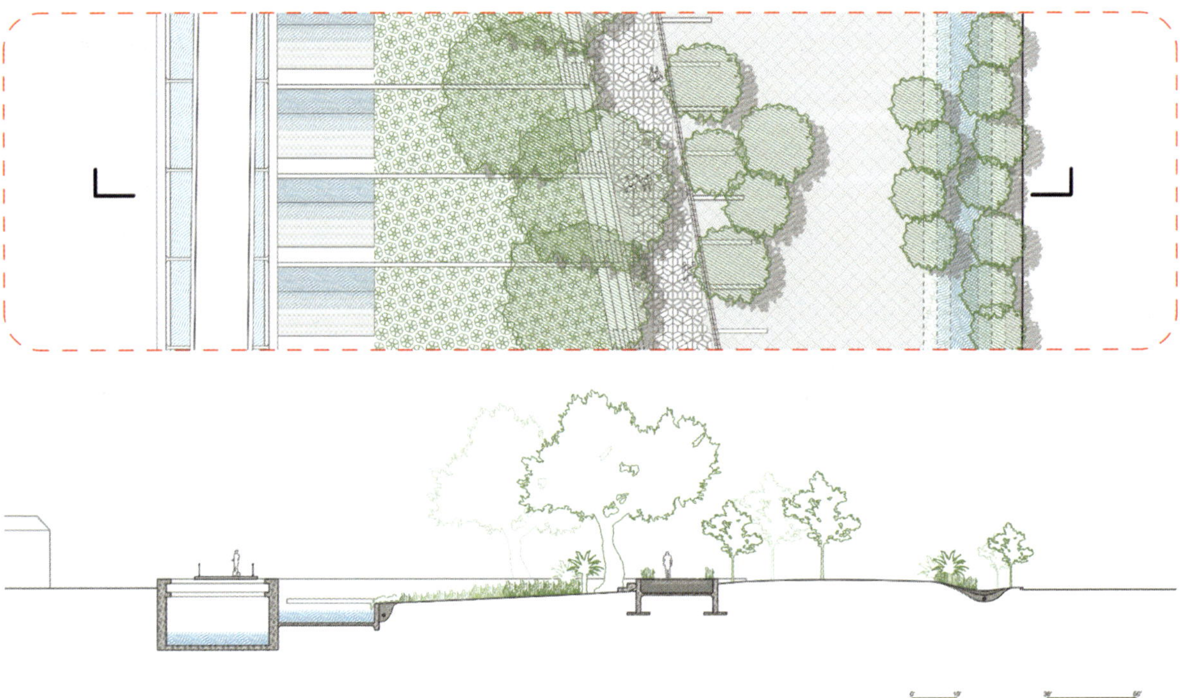

DETAIL PLAN AND SECTION // SCALE 1"=16'

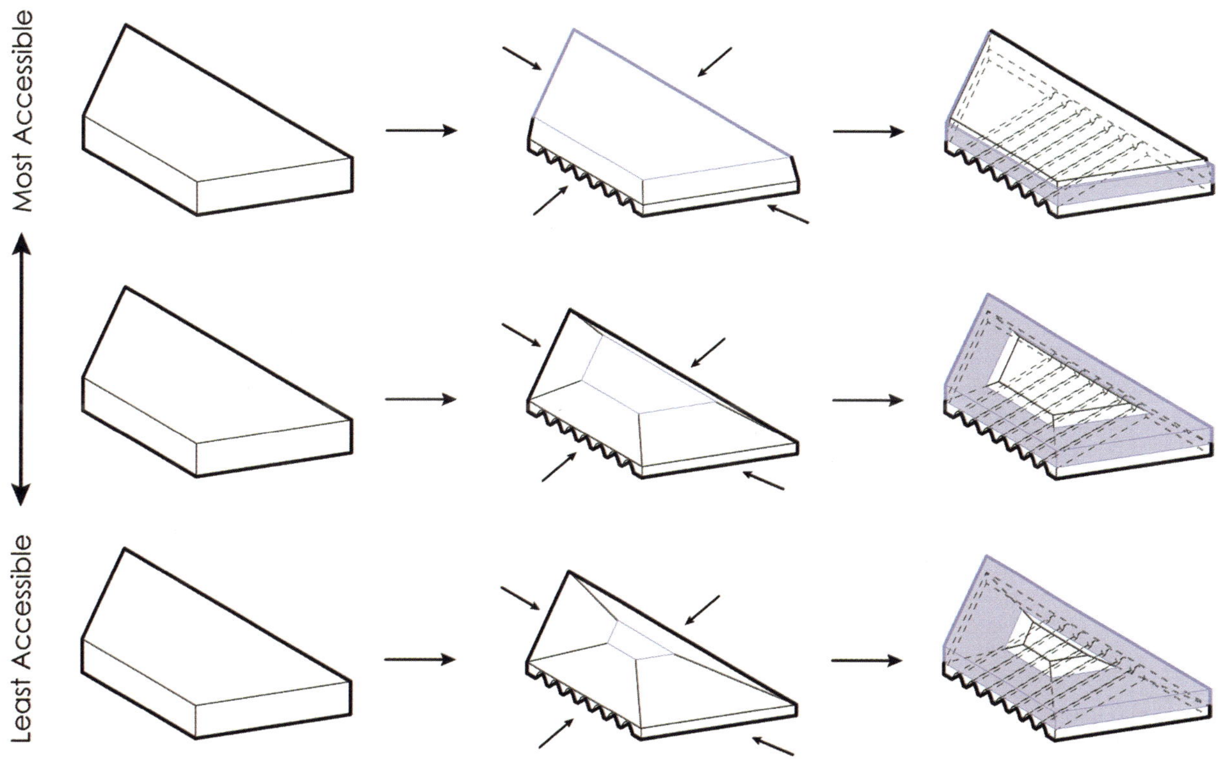
Most Accessible
Least Accessible

s > Alyssa Barber

i > Kentaro Tsubaki + Charles Jones

ALEX LANGLEY & SAM SPENCER

IMMATERIAL

New Sensations from Old Materials

ARCH 5990 Instructors
Cordula Roser Gray & Ammar Elouini

s> Alex Langley & Sam Spencer **i>** Cordula Roser Gray & Ammar Elouini

In today's building practices, the life-cycle of the building and its materials are treated as one and the same. We conflate the condition of the building as a whole with that of its components. Once demolished, what remains is not the building but the building materials. However, the damage has already been done, and what could have been reused is reduced to useless fragments.

Sifting through the wreckage of this material graveyard, the only materials with the hope of escaping are the ones that retained their factory edge or still possess enough size to be re-dimensioned. This is because form is equally as important as matter in informing how we perceive materials. Unfortunately, demolition does not discriminate between the form of the building and that of its materials, it destroys both. And yet we continue to demolish buildings and destroy their materials as well.

The standardization of building materials has caused them to become more ubiquitous and less precious than they once were. There is no formal difference between a new brick and one from the 1800s. Therefore, it is easy to demolish an existing brick and then purchase a different one. But this formal assessment does not consider immaterial qualities embedded in the materials. This project aims to broaden the criteria for the valuation of materials to include immaterial qualities such as perceptual and cultural significance. Myths emerge around objects because of the meanings people confer upon them. Could architects similarly orchestrate the talismanification of building materials?

Through a radical approach towards material reuse, architecture can reposition our perception of what is valuable, by bringing out latent immaterial qualities within used materials.

Fig. 1
The Pool

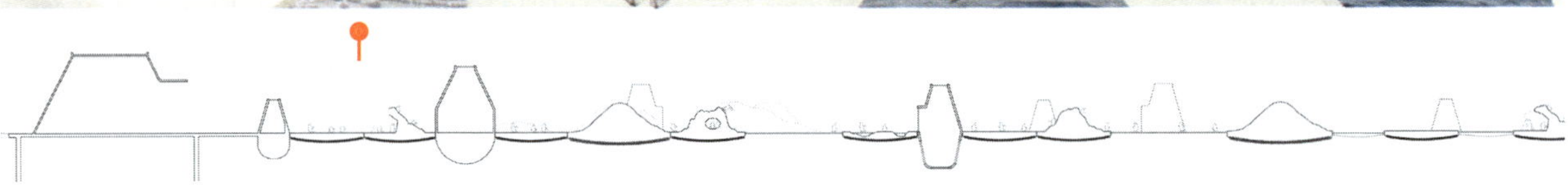

SETH LASKIN

POLAR SYNERGIES

A Replicable Model for Antarctic Preservation

ARCH 5990 Instructors
Cordula Roser Gray & Ammar Elouini

This research explores the effects of climate change on Antarctica, particularly the loss of sea ice and its impact on the local food network and wildlife, including penguins. The proposed thesis suggests the implementation of coastal interventions that will invigorate local food networks while integrating into the context of environmental research in the region. The thesis uses Adelaide Island, located off the western coast of the Antarctic Peninsula, as a test site. The island's rich ecosystem and research infrastructure create a synergy between scientific research, wildlife conservation, and environmental preservation.

By exploring advancements in 3D printing technologies, the thesis creates a self-expanding network that highlights a symbiotic relationship between human and animal, promoting healthy environmental preservation and wildlife conservation. This proposal can be translated to similar contexts throughout the continent, creating a sustainable solution to the ongoing environmental crisis in Antarctica.

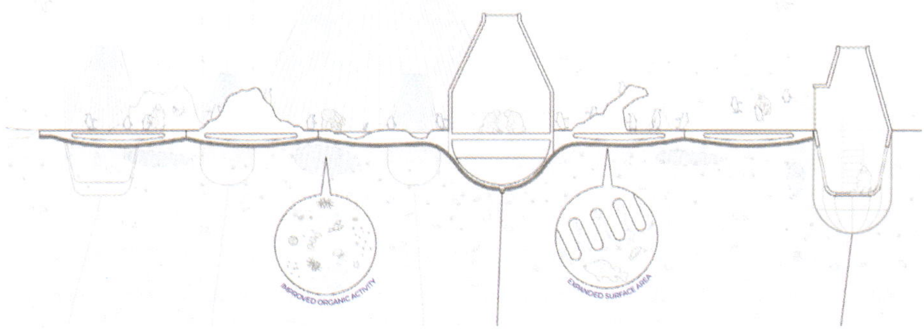

Fig. 1
Fabrication, Research, & Utility Section

Fig. 1
Platform Clusters

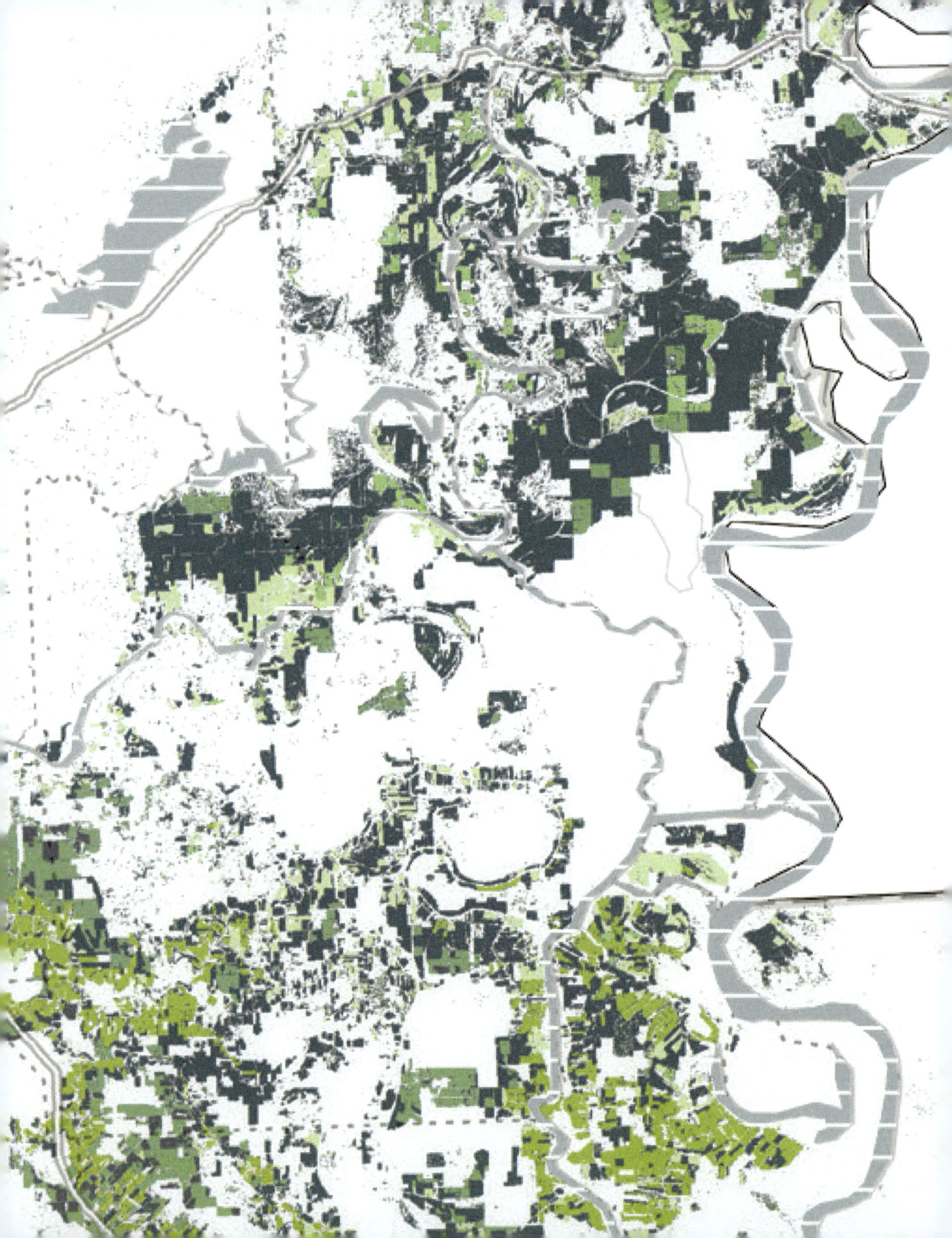

ENVIRONMENTALLY AFFORDABLE

Natural resources

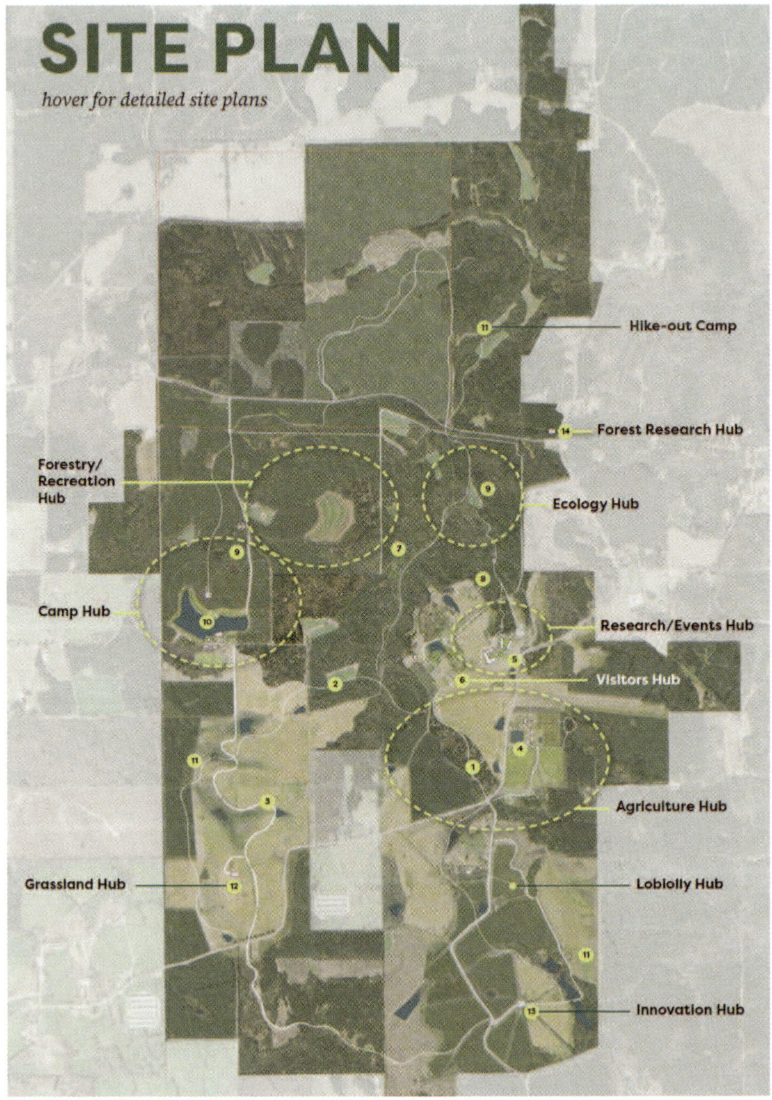

CIRCLE BAR RANCH

PROJECT PROPOSAL
AIA24 Conference
Washington D.C., 2024

Credits
Kenneth Schwartz, FAIA (Team Leader), Thomas Woltz, FASLA and Susan Nelson, ASLA. Nelson Byrd Woltz Landscape Architects. Phyllis Taylor Center for Social Innovation.

The owner of a 2,800-acre property in a poor area of southeastern Mississippi engaged the Phyllis Taylor Center for Social Innovation, a cross-disciplinary center for social innovation at Tulane University, to explore strategies for integrating a variety of remediation efforts to repair and restore the land and its ecosystem. The goal is to honor and celebrate the history of this place while charting a path for a more progressive future while engaging people who will work with and study this land in well into the future.

Over a three-year period, the team orchestrated visioning exercises and Design Thinking activities with members of the community including the workers at the ranch and other area landowners. From "Listening to the Land", the collaboration explored and identified multiple narratives of the land's history dating back to pre-colonial times. We also charted the owners' own legacy of both innovative management and ambitious leadership, identifying circles of interest and opportunities that overlap in suggesting new approaches and alternatives to the status quo of ubiquitous monocultural plantations of loblolly pine. Included in this complete master plan is a new "Camp Blazing Star", an outdoor education center whose curriculum and campus have been designed by the team as a place where New Orleans and rural Mississippi youth can engage a full range of issues tied to the land and the experience of outdoor education. The camp is named in recognition of the beautiful native flower in the region and for the liberating value of what this experience represents for these youth. A major agenda for this project includes economic opportunities through agricultural strategies and new job creation and wealth building strategies for the predominantly Black population in the area. These strategies serve as reparations and reconciliation for the devastating legacy of slavery and the closely related exploitive history of sharecropping in this area.

There are important Equity, Diversity, and Inclusion aspects that permeate this project. In addition, the interconnection of architecture, landscape architecture, planning, and research produces an approach that is more than the sum of its parts. As a process, we believe that there are lessons tied to the project that guide a more integrated, inclusive, and diversified approach to the land that can create a better future for the property.

In addition to the integrated design intentions, a fundamental agenda of social innovation holds the promise of creating a vibrant and inclusive future for Circle Bar Ranch. We believe that there are lessons to be learned from this case study that will open opportunities for other architects to consider. Collaboration and innovation can help to release the potential of a large property when it is seen as a site for ameliorating injustices of the past while promoting a more liberating, expressive, immersive approach to sustainability at this scale.

Fig.1
Existing longleaf as seed-trees
Photo. Circle Bar Ranch

Fig. 2 Forestry plan and camp trails

Fig.3 Site visit Photo

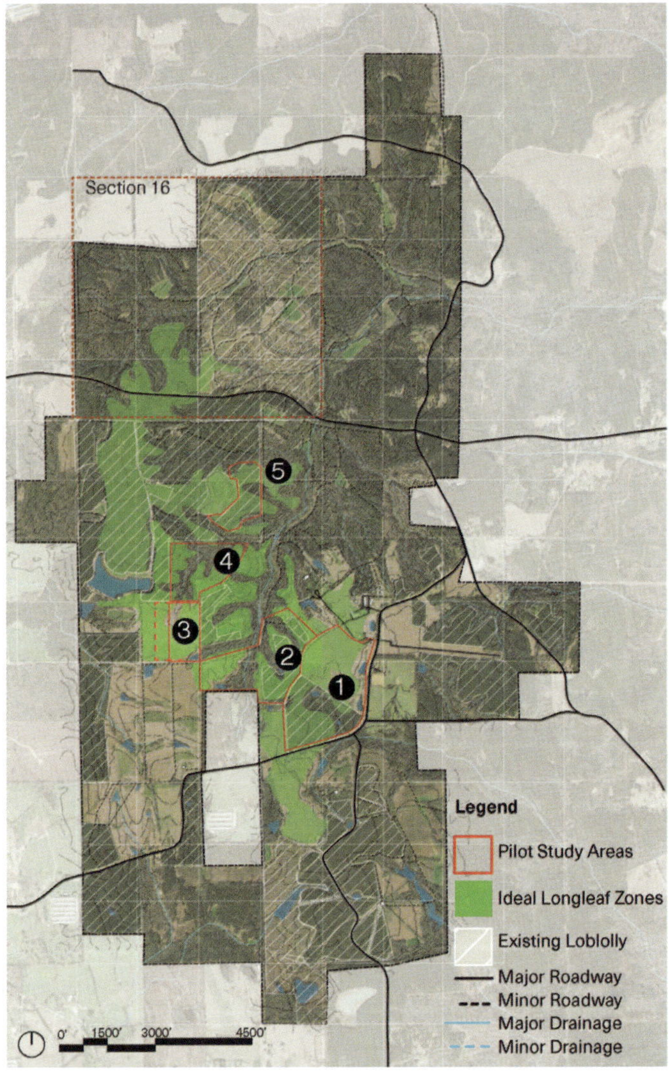

Fig. 4
Aerial rendering
Circle Bar Ranch

JUDITH KINNARD, IRENE KEIL
& NICK JENISCH

SHADE-WATER PAVILION

PAVILION
New Orleans, 2016

Credits
Judith Kinnard, Irene Keil & Nick Jenisch (Project Leads).
Dana Brown, Gaylan Williams (Landscape Architects).
Walter Zehner (Engineer).
William Slattery (Builder).
Jay Bonck (Fabric Fabrication).

The Shade - Water pavilion was constructed in 2016 on a vacant corner site in a struggling New Orleans neighborhood. The pavilion was part of an incremental urban design and engagement process that began in 2012 and continues today. The project was initiated by community leaders who sought to activate a neglected strip of public land cutting diagonally through the Hollygrove neighborhood. Located in a low-lying district between Carrollton Avenue and the Jefferson Parish line, this residential community was developed in the 1920s along the edges of the raised rail corridor connecting New Orleans north to Chicago. The line was decommissioned in 1955 and replaced with an underground drainage canal below the right-of-way leaving a 50-foot-wide swath of fenced vacant plots along the mile-long corridor above. The Carrollton/Hollygrove Community Development Corporation (CHCDC) & Hollygrove Market and Farm invited the Tulane City Center to help the neighborhood imagine how the area could be reclaimed as a place of gathering rather than rupture. Students and faculty engaged the project in a design studio and as summer fellows as the project took shape through several phases.

As the project has developed over several years, the fundamental intent has remained the same: to connect the neighborhood by providing an outdoor meeting space. The pavilion is designed to provide shelter while gathering and directing the rainwater into gardens using four elements distinguished by their form, their material, and their purpose: the porous rain gardens, a heavy concrete groundwork, a slender steel frame, and a series of translucent fabric tents. These Semperian elements each play a specific role in establishing shelter. The steel tube framework reiterates the rhythm of the trees and establishes the urban presence of the structure through its scale and proportions. The ground plane and the "roof" membrane work together to address the multiple demands of collecting, shedding, and directing water while promoting airflow. The project accepts and advances the dynamic forces associated with the physical environment, which include hydrology as well as sun and wind. The design team sought to move beyond the notion of a pavilion as a static object and to understand it as continually responsive to the highly dynamic forces of weather and weathering over time.

Fig.1
Dogs playing under pavilion
Photo.

Fig. 2 Waterworks Diagram.

Fig.3 Waterworks Photo.

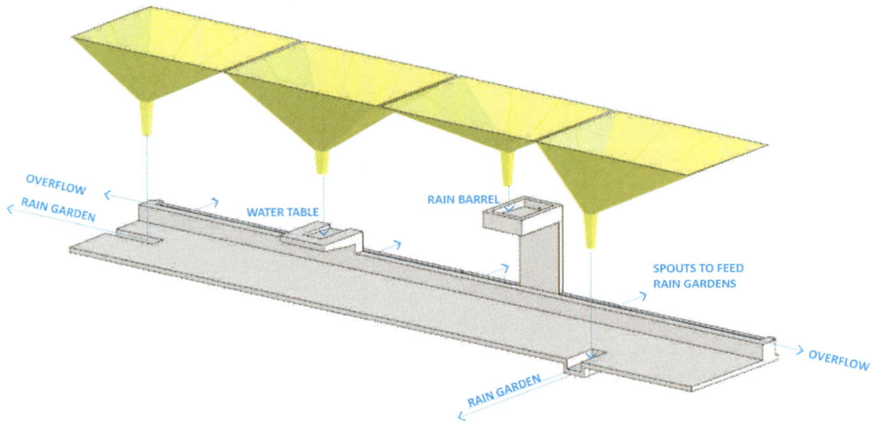

Fig. 4
Concept.
Sketch.

Fig. 5
Concept.
Section and plan.

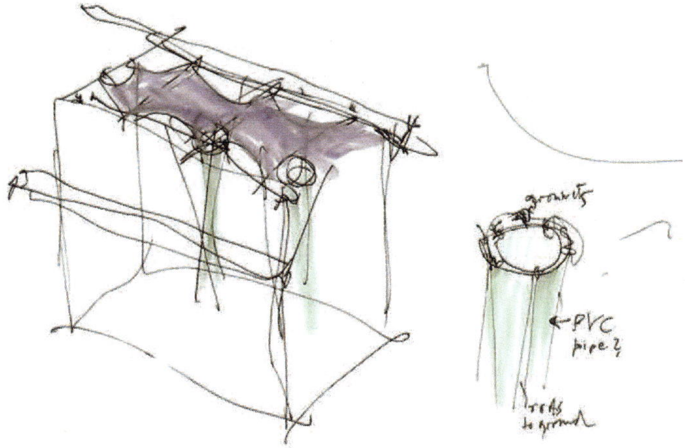

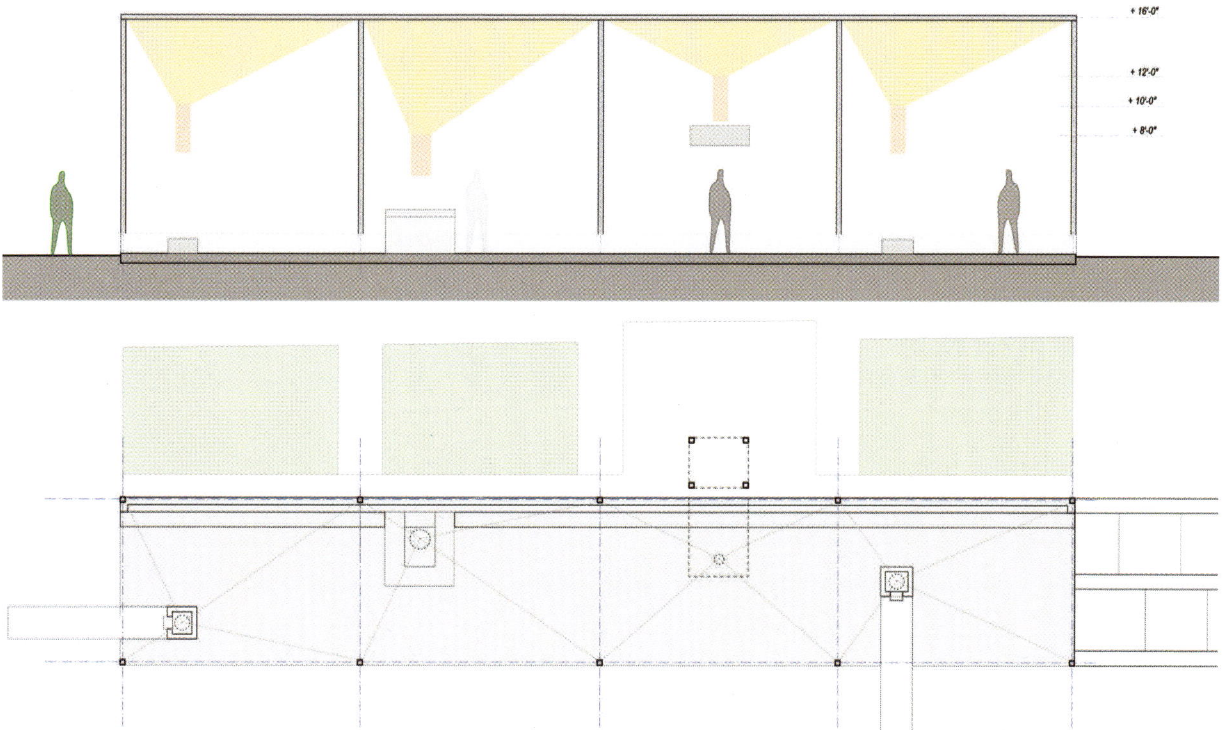

FA21

Carrie Norman [C]
Nimet Anwar
Nick Jenisch
Charles Jones
Andrew Liles
Wendy Redfield
Cordula Roser Gray [C]

FA22

Nimet Anwar
Cynthia Dubberley
Irene Keil
Emmanual Osorno
Wendy Redfield [C]
Cordula Roser Gray [C]

BUILDING & LANDSCAPE

Core studio 2021 focuses on the role of landscape in architectural design. Students develop basic skills in site analysis, design, and representation. They learn to understand site characteristics as both natural (a result of nature) and cultural (a result of people), and develop strategies for articulating the relationship between building and site. The course highlights the designer's ethical obligations to the larger network of social and ecological systems and conditions. Building design themes include spatial organization and hierarchy, circulation, structure, and enclosure. The studio is integrated with digital media classes to ensure that students gain fluency in computer-aided design processes, drawing, spatial modeling, and digital design techniques.

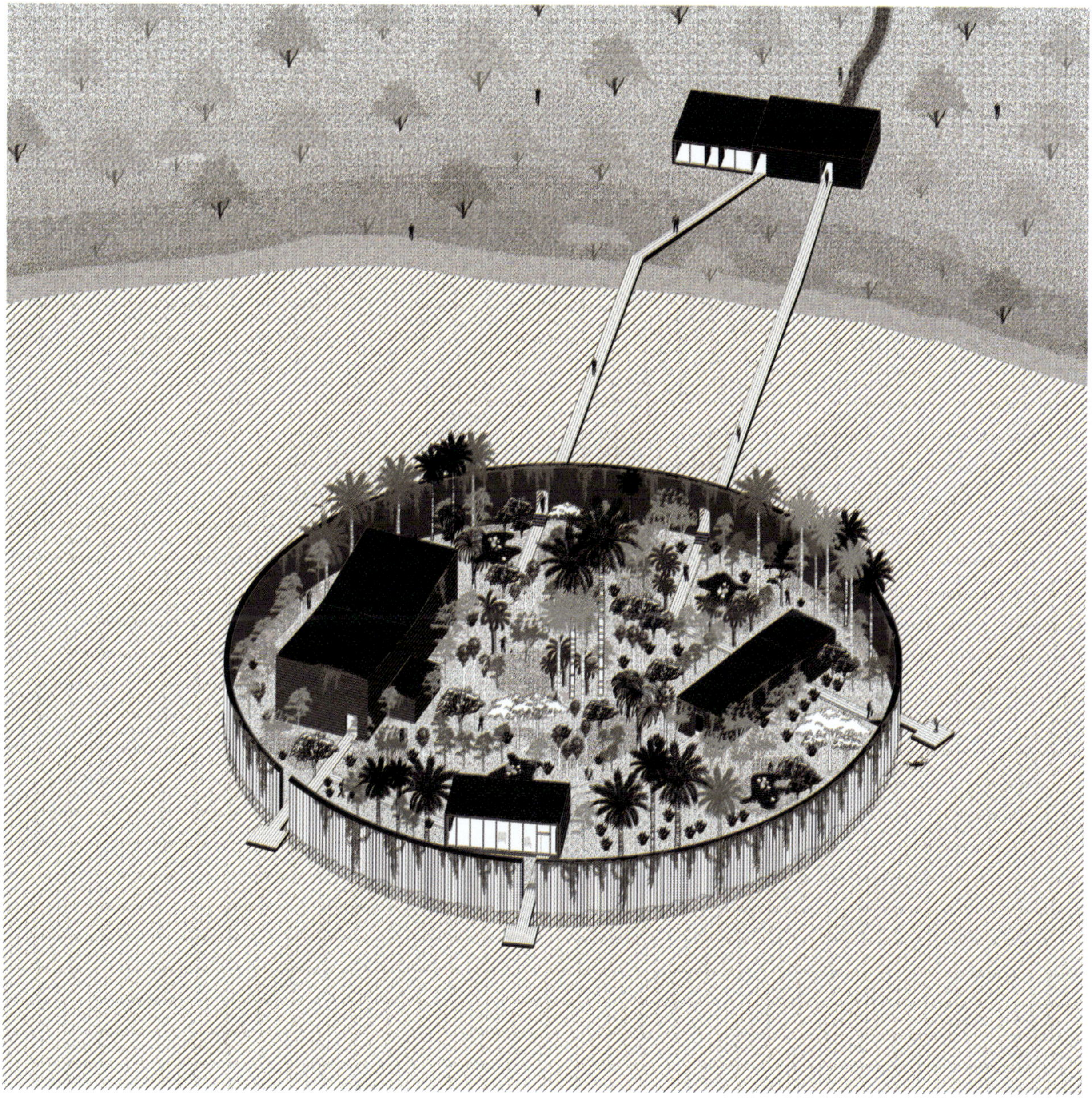

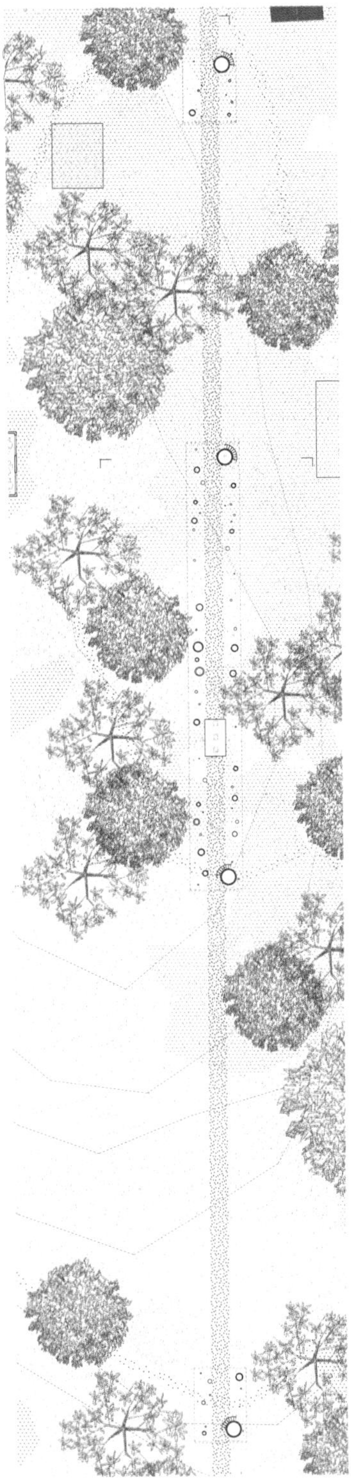

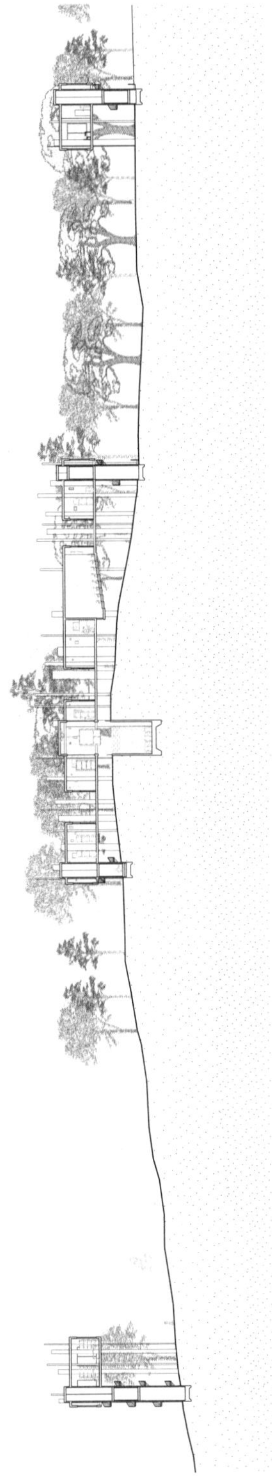

s> Cassie Kadish **i>** Carrie Norman

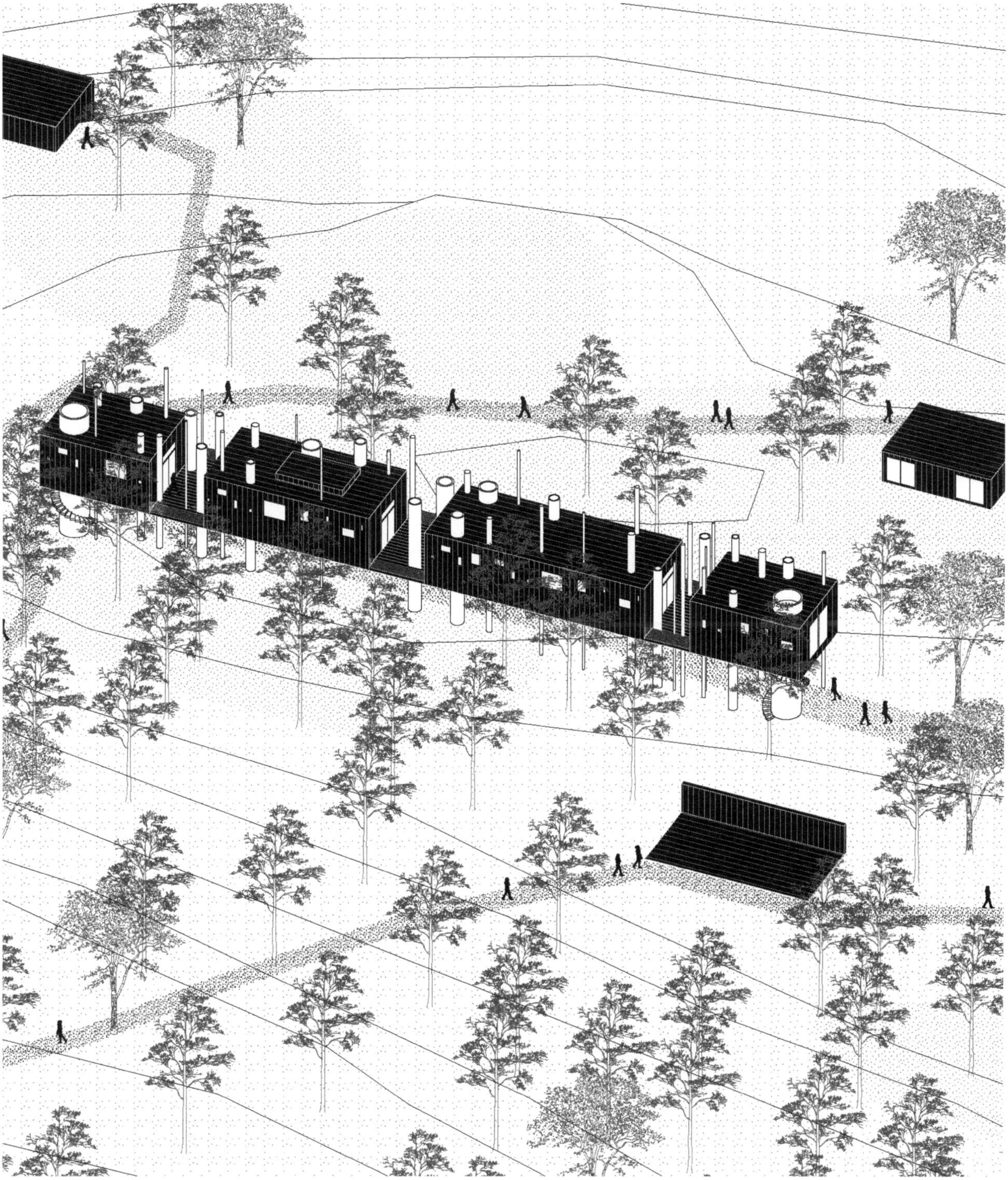

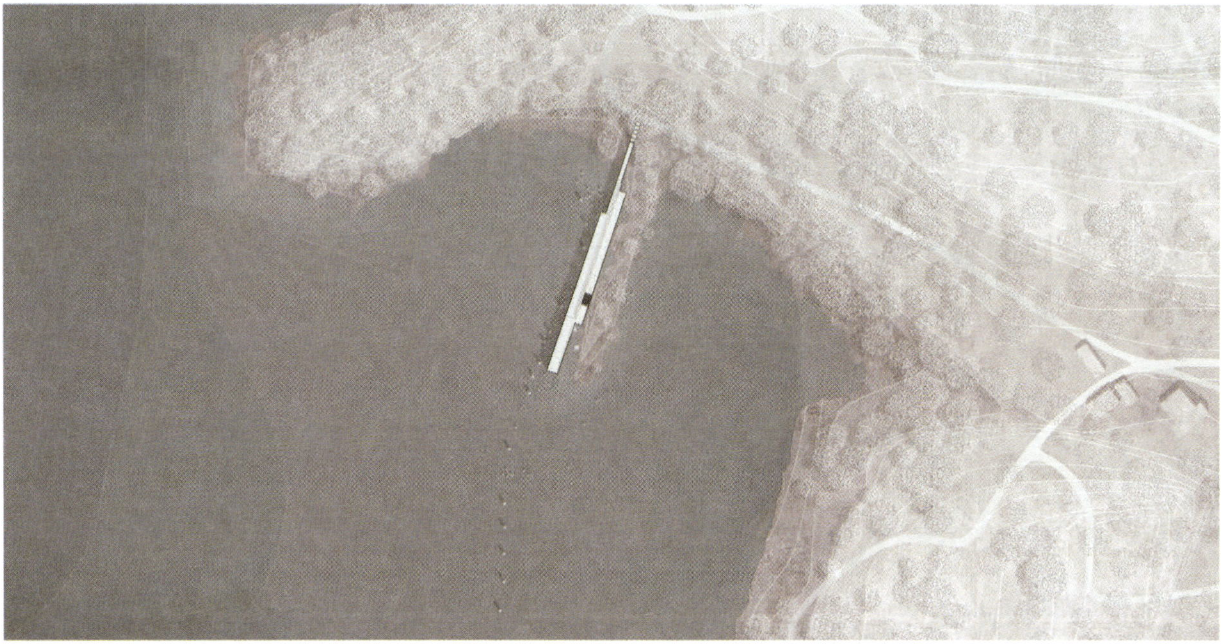

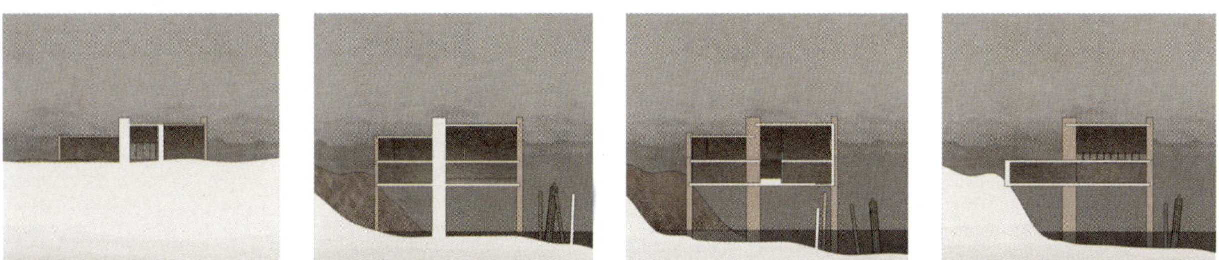

s> Zaynab Eltaib **i>** Andrew Liles

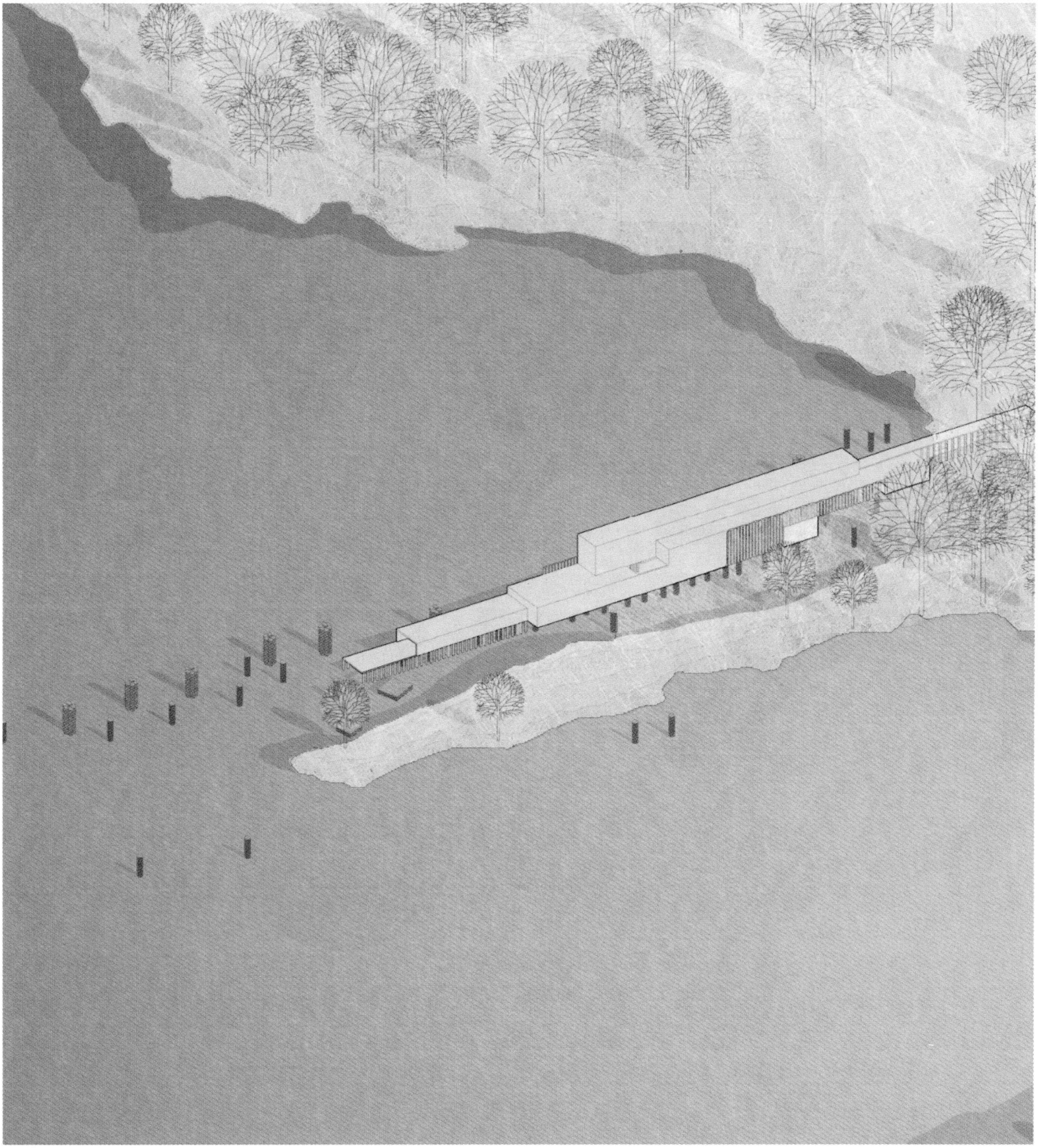

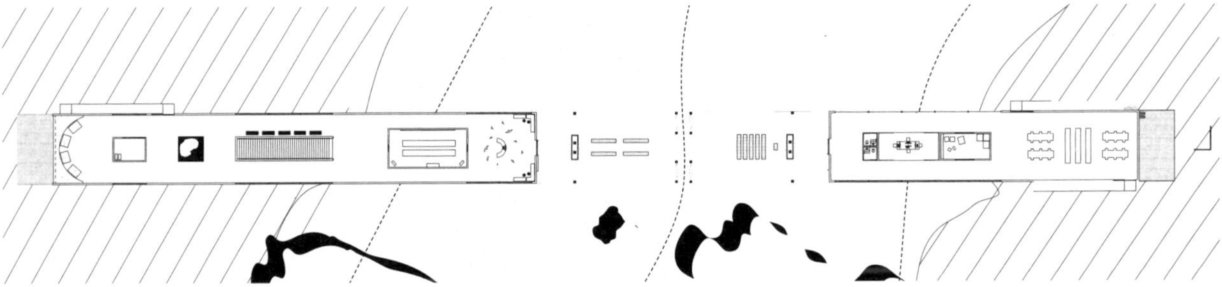

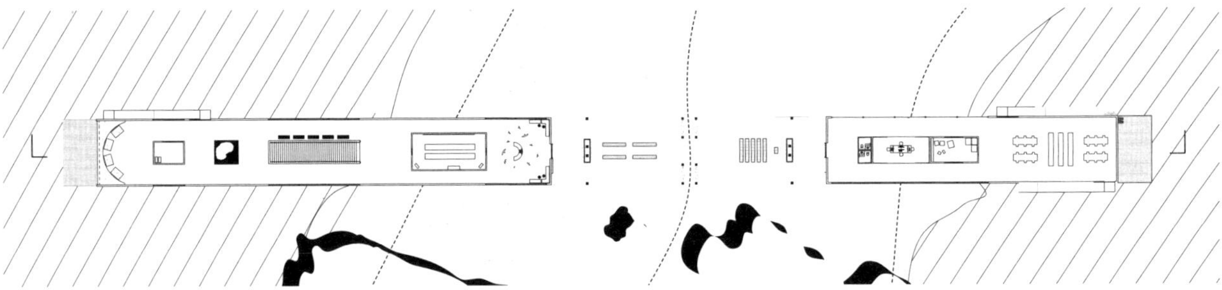

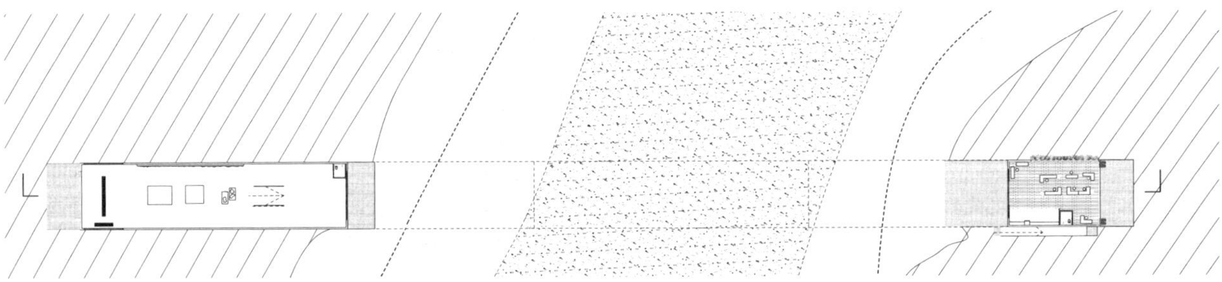

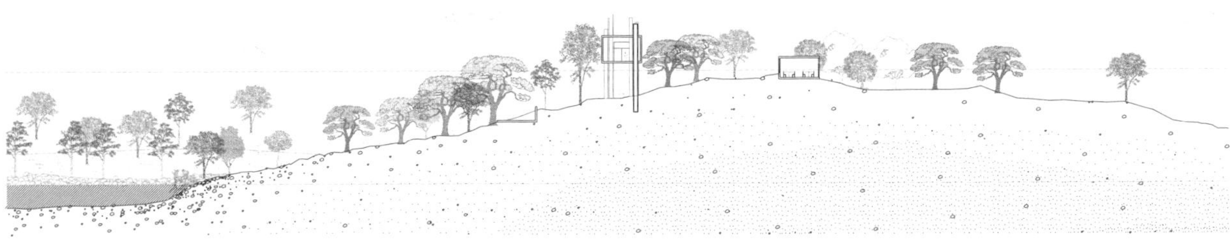

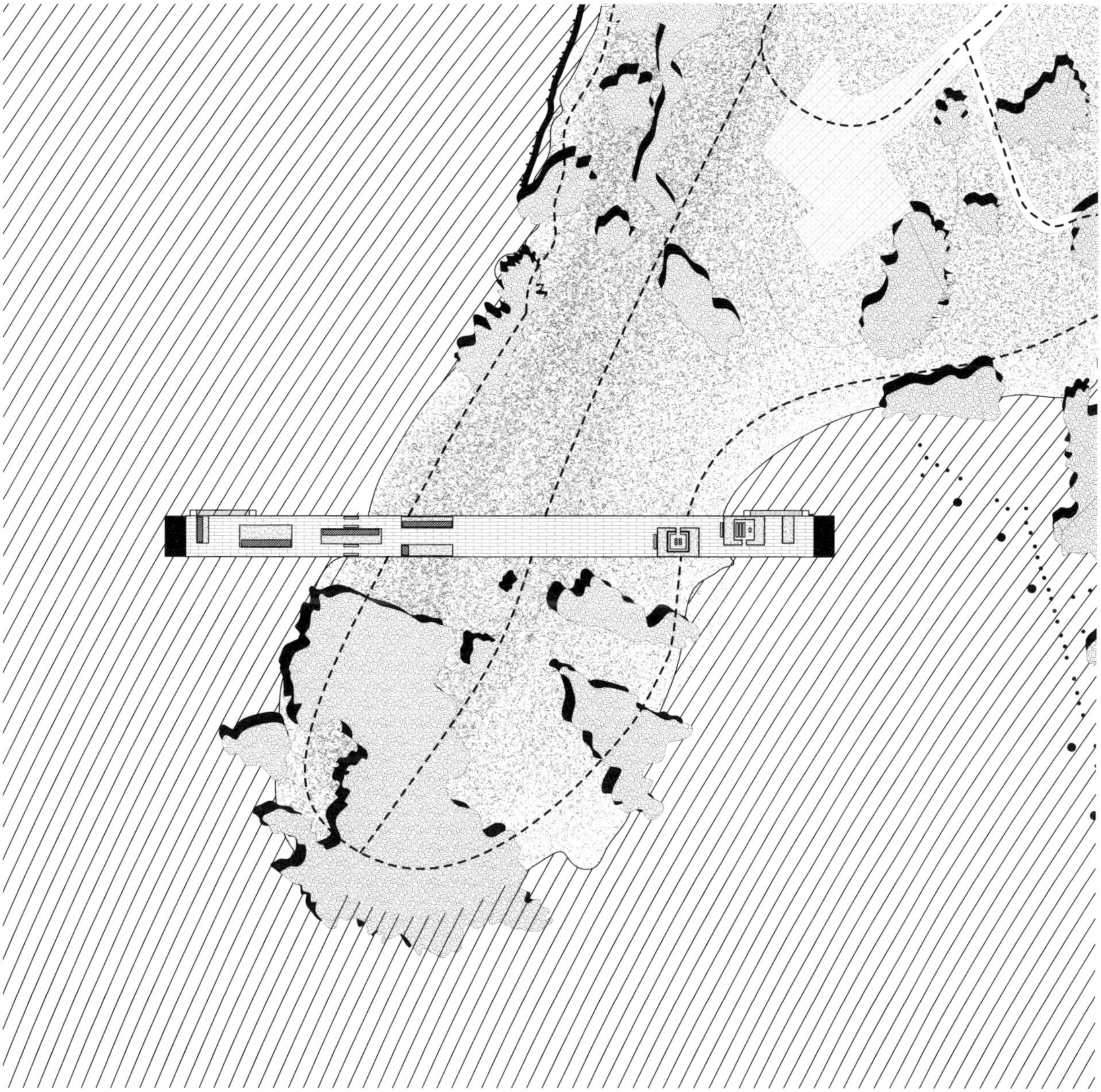

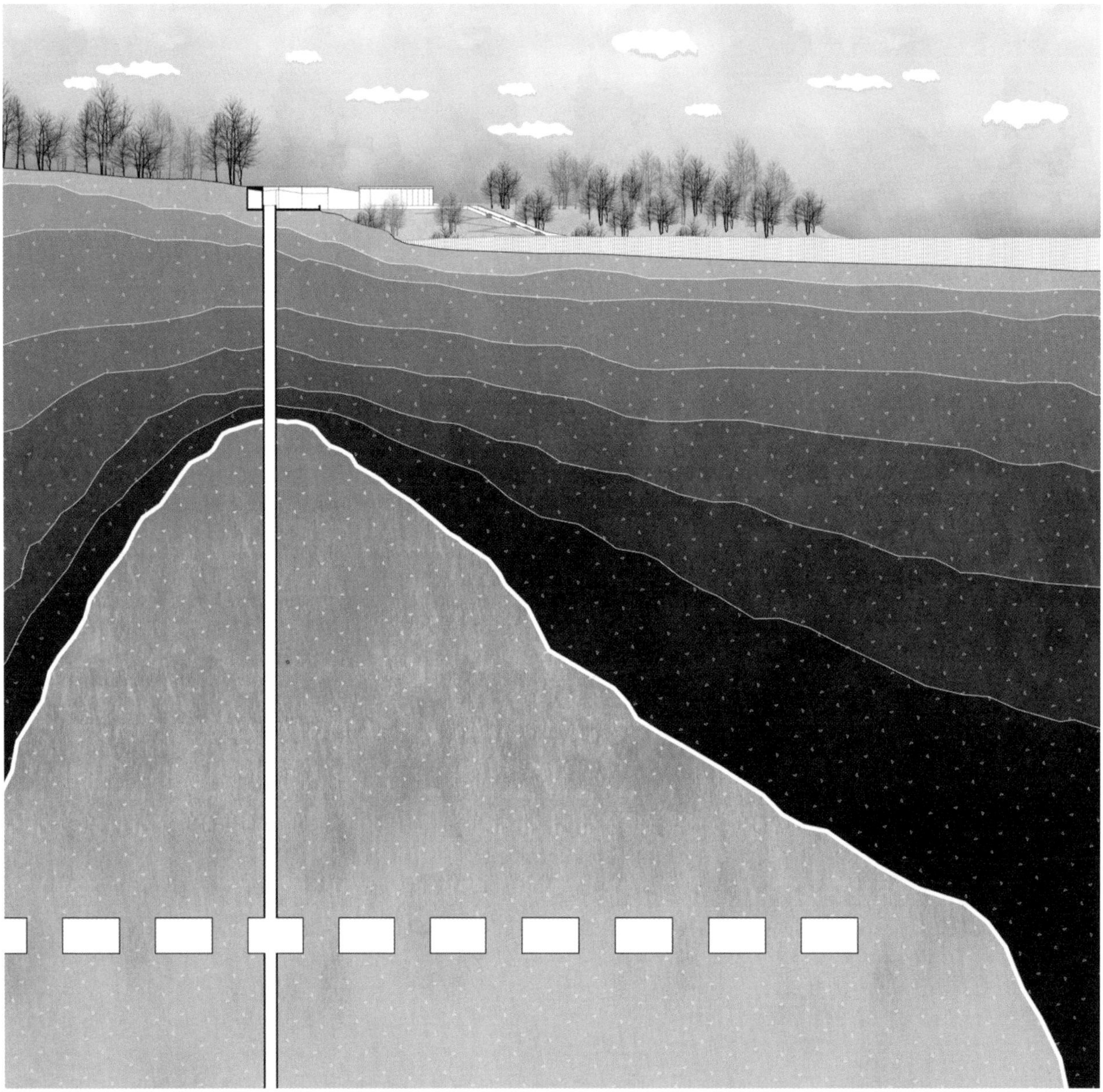

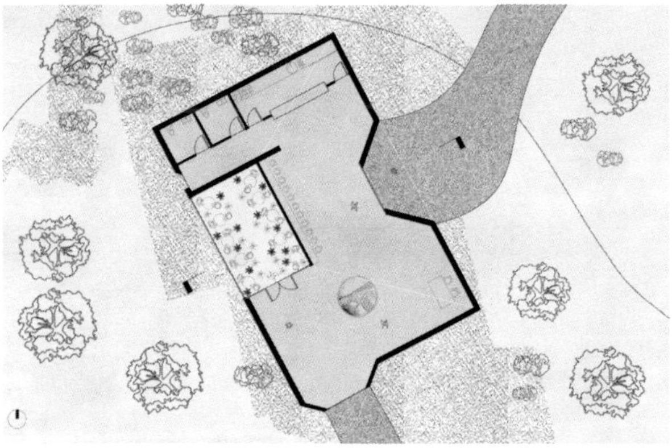

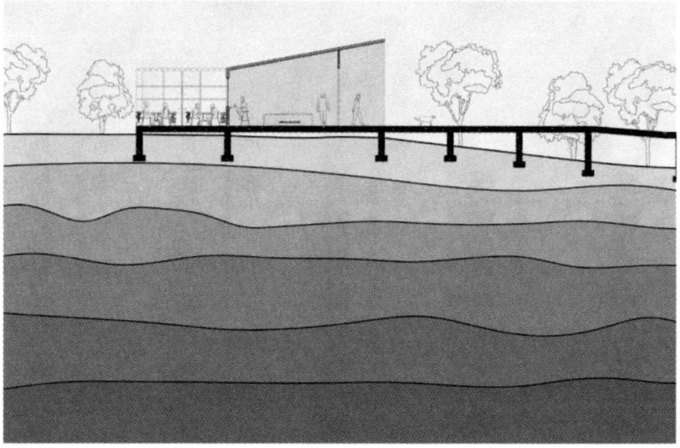

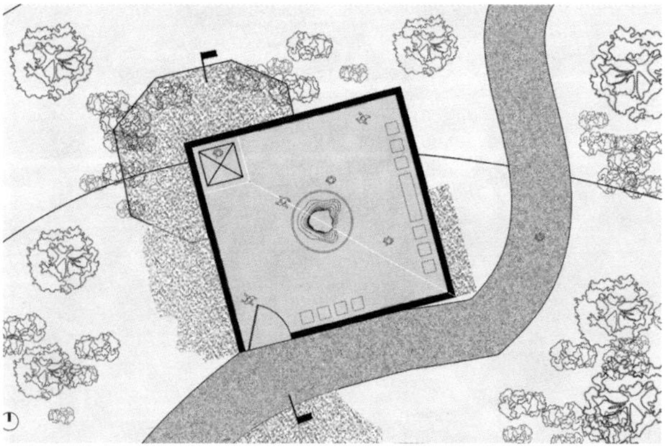

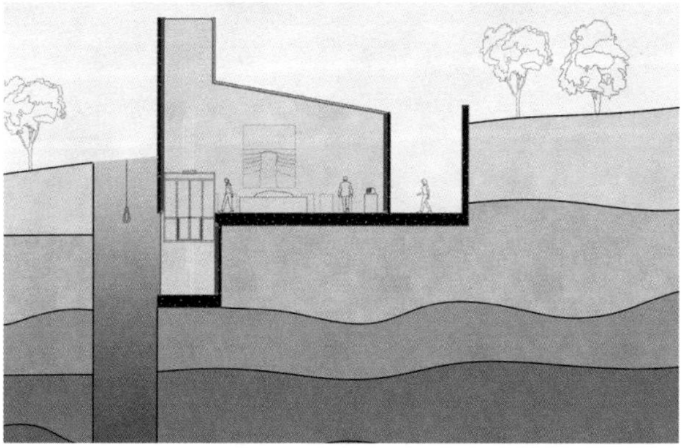

s> Emma Lindley

i> Cynthia Dubberley

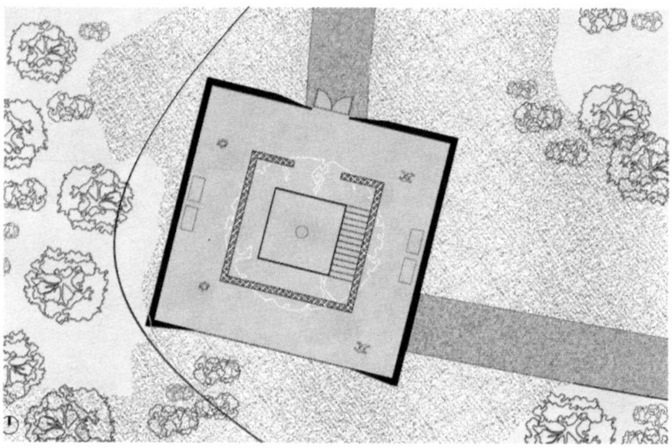

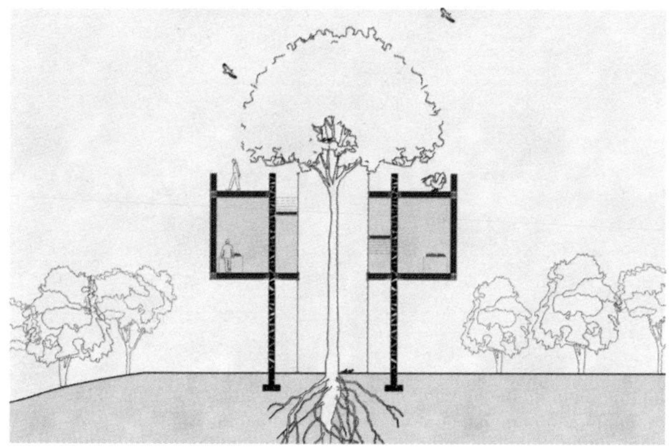

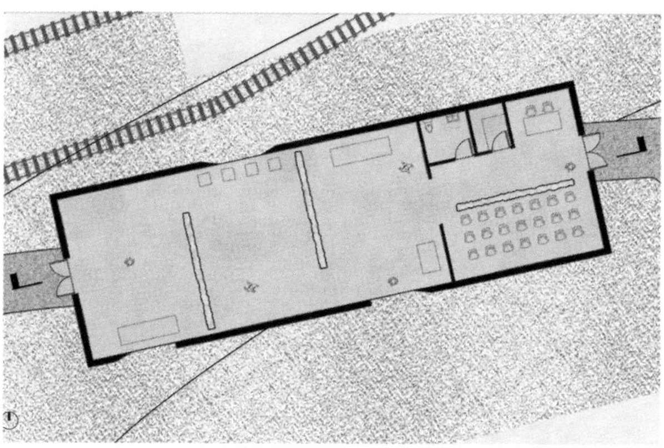

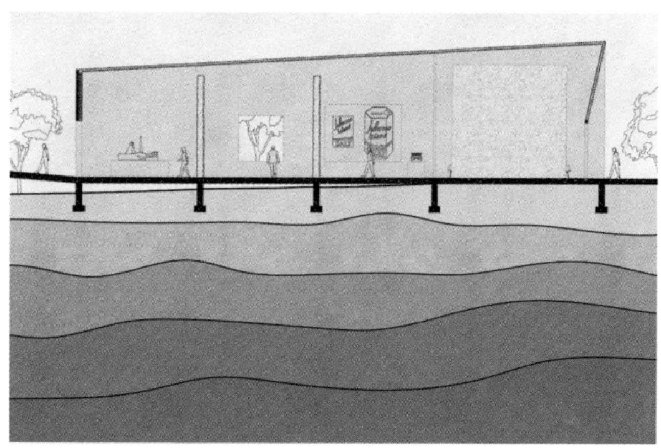

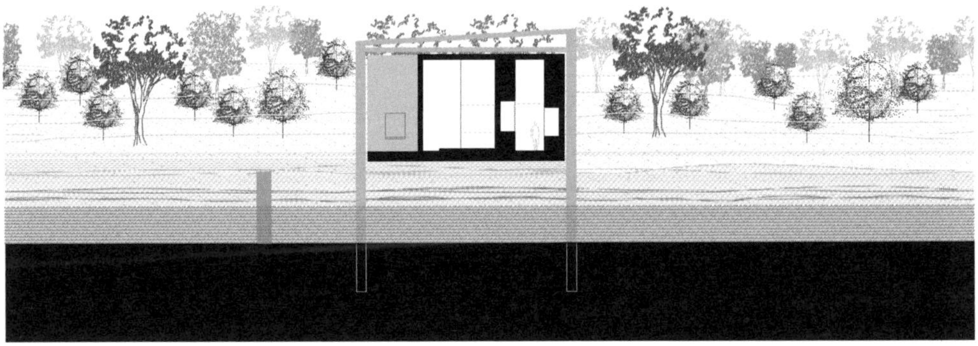

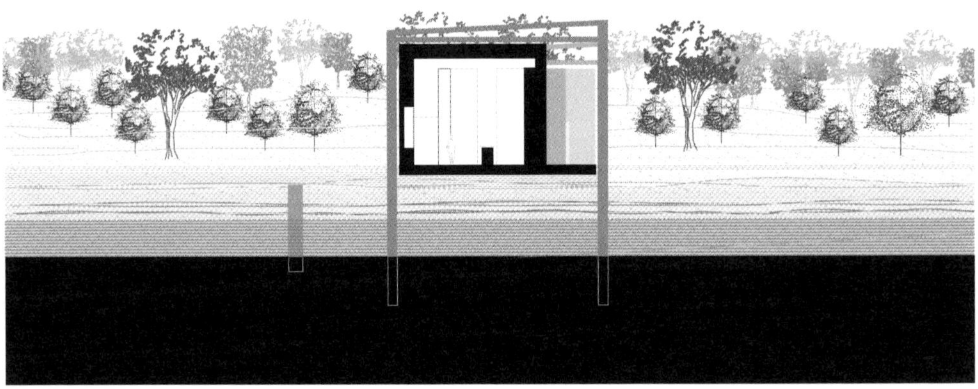

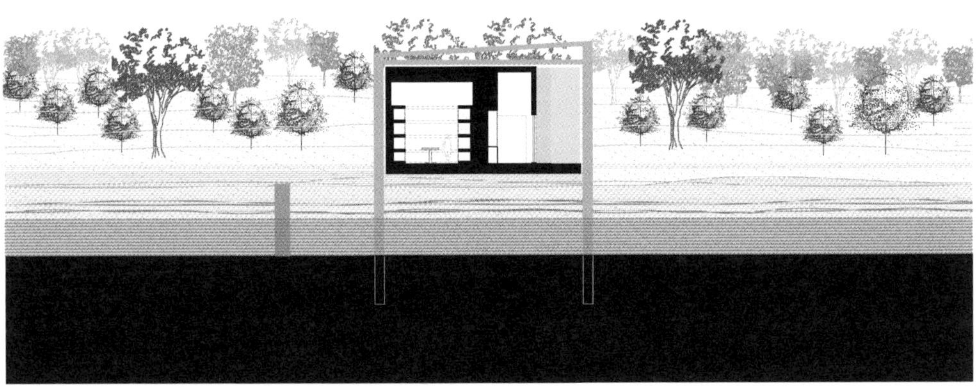

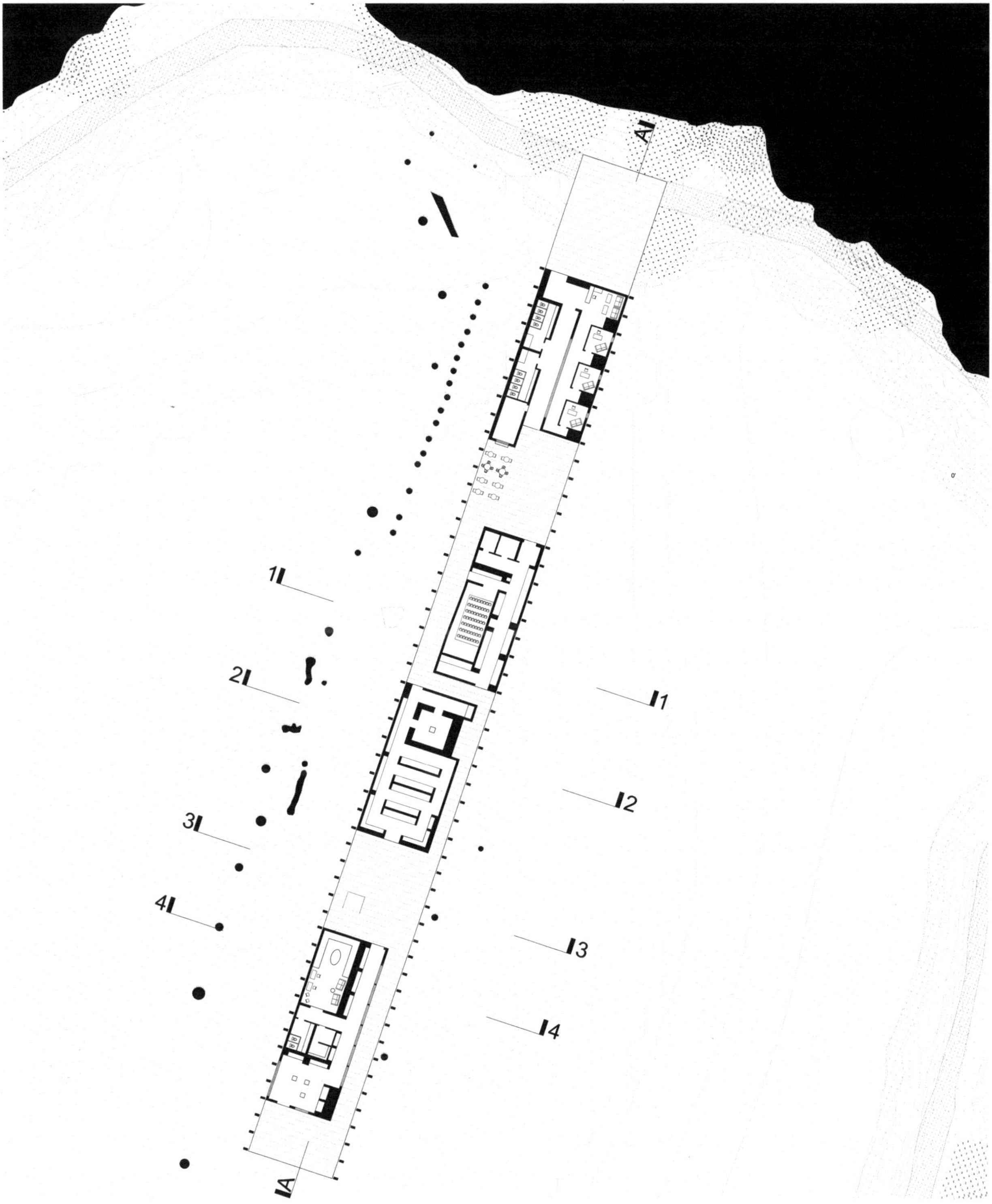
A
1
2
3
4
1
2
3
4
A

s> Jack Levine

i> Irene Keil

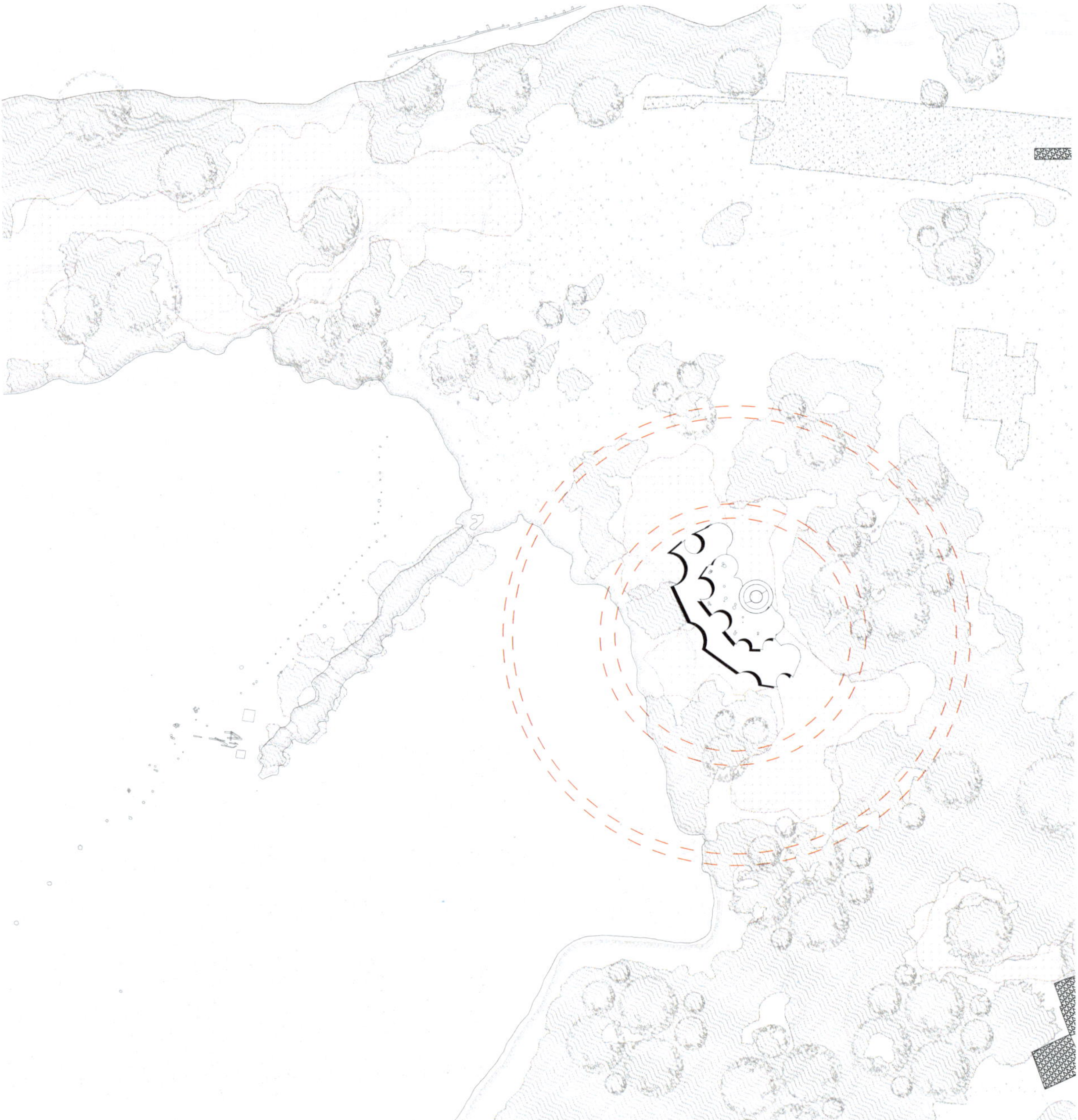

EXPLORING GULF COAST CLIMATE FUTURES

Margarita Jover [C]
Liz Camuti

The NAS Gulf Research Studio builds on Louisiana's Climate Action Plan, catalyzing partnerships and support for implementation. The timeline for this multi-year initiative is organized into three parts: conceptualization, engagement and knowledge development, and implementation preparation.

Fall Studio 2022 focuses on Phase 1: Conceptualization. Research focuses on the three directions marked by NAS: the historic and current conditions of Gulf Coast oil legacy infrastructures, the affected communities and their health, and the environmental challenges and opportunities for protection.

This research aligns with the broader goal of reducing fossil-fuel energy consumption and increasing opportunities for renewable energy and carbon sequestration in the built environment. Scales of research and design include the regional, the metropolitan, and the architectural scale down to 1:500.

Cities and regions in the western world are preparing for an already changed climate, implementing strategies for long-term climate recovery. Climate adaptation plans for 2030, 2050, and 2100 are growing for cities in developed countries, following the 2015 Paris Agreement. Today, Louisiana is not an exception, as the state has approved the first 'Climate Action Plan in the Gulf South. In August 2020, Governor John Bel Edwards signed an Executive Order that created the Climate Initiatives Task Force.
In this political context, the goal for universities is twofold. They aim to ground future professional architects with skills in urban design and landscape architecture, in existing realities and plans (including the Louisiana Climate Action Plan and the Louisiana Coastal Master Plan), while inspiring them to design innovative spatial strategies for 'urban reform' at several scales.

The Louisiana landscape presents an important case study, facing a changed climate, a landscape impacted by ongoing land loss, and centuries of inequity and poor communities susceptible to ongoing environmental justice issues.

This interdisciplinary research studio analyzes how ecological and hydrological restoration, climate mitigation, and opportunities presented by the energy transition can combine across the territory in different sites and scales. Through reimagining practices, the studio expands the concept of coastal Louisiana's working landscapes that considers public health, equity, and labor practices.

MONROE
JACKSON
MERIDIAN
ALEXANDRIA
MCCOMB
HATTIESBURG
BATON ROUGE
GULFPORT
BILOXI
LAKE CHARLES
I-10
LAFAYETTE
SLIDELL
NEW IBERIA
NEW ORLEANS
MORGAN CITY
HOUMA

Existing Problems
Monoculture

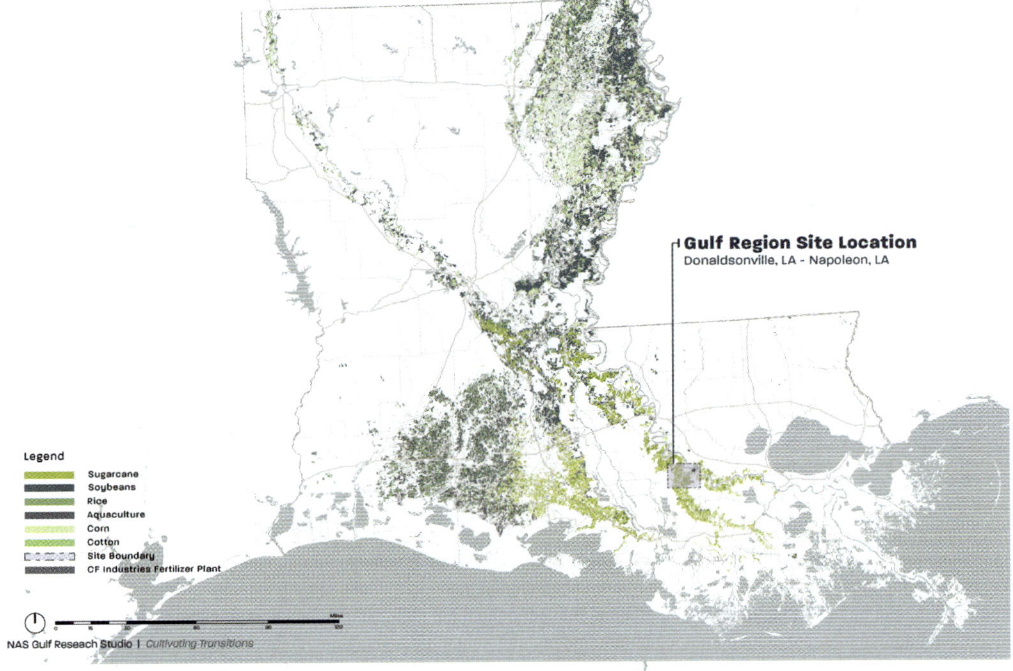

Louisiana Agriculture

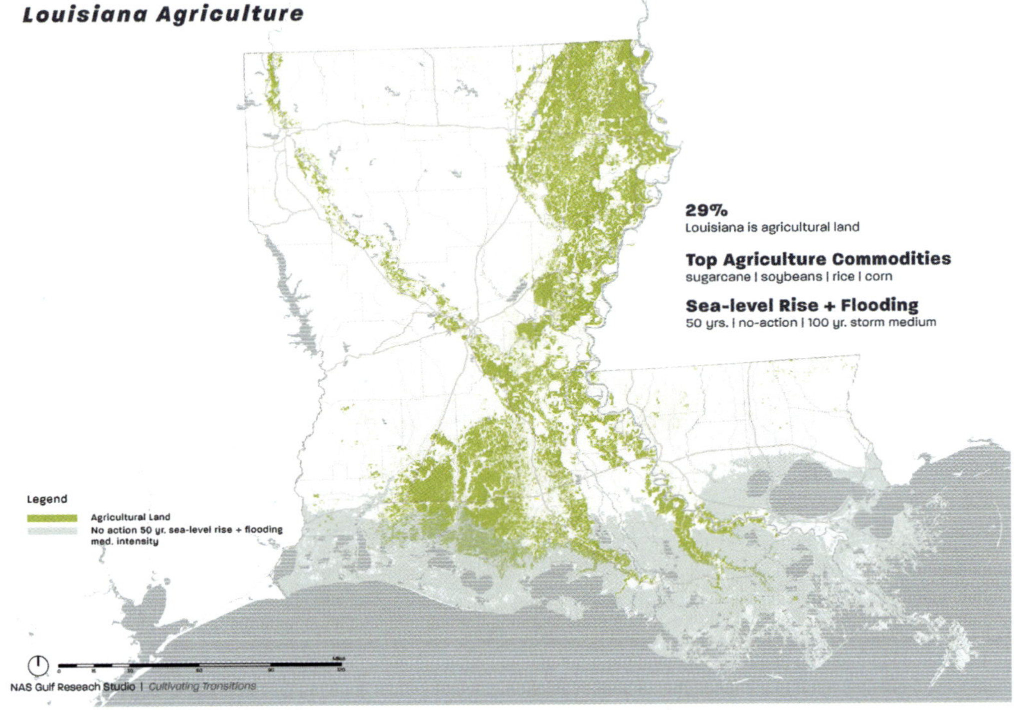

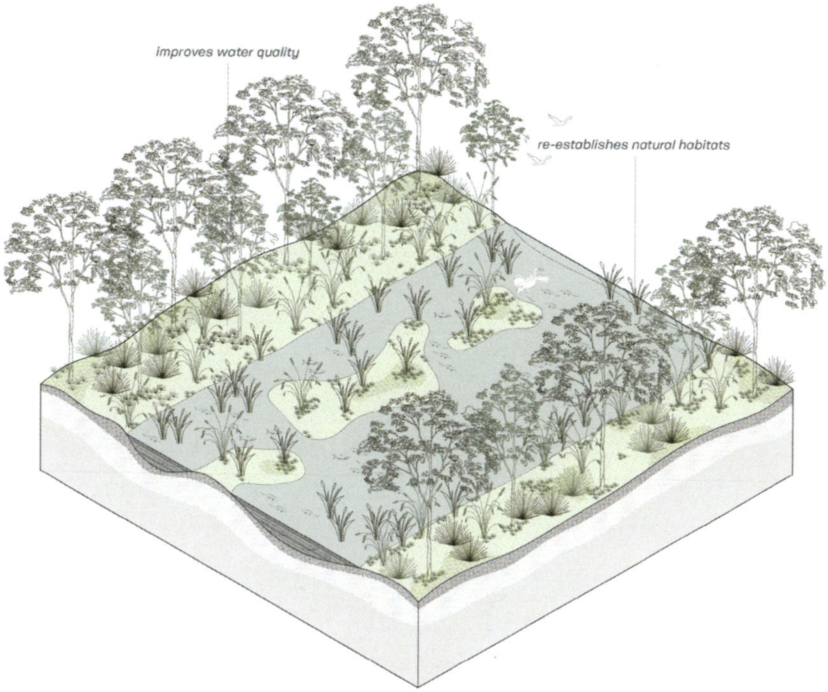

Agroforestry

year-round

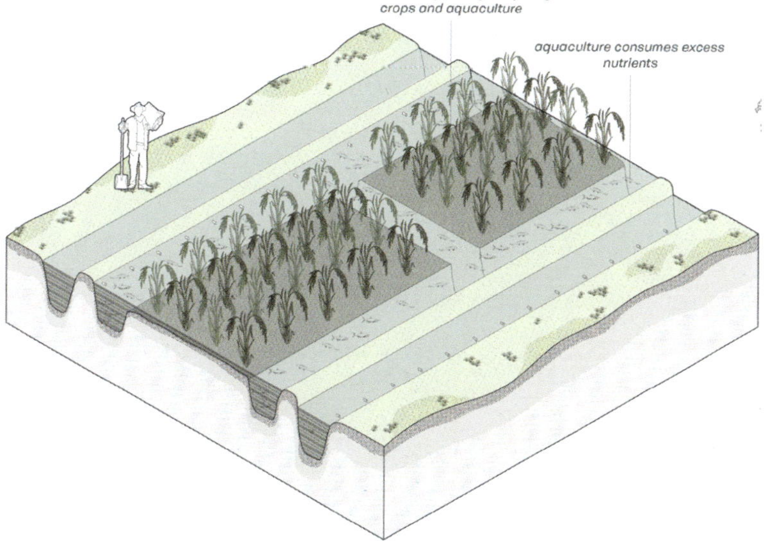

Rice + Aquaculture

May - Oct

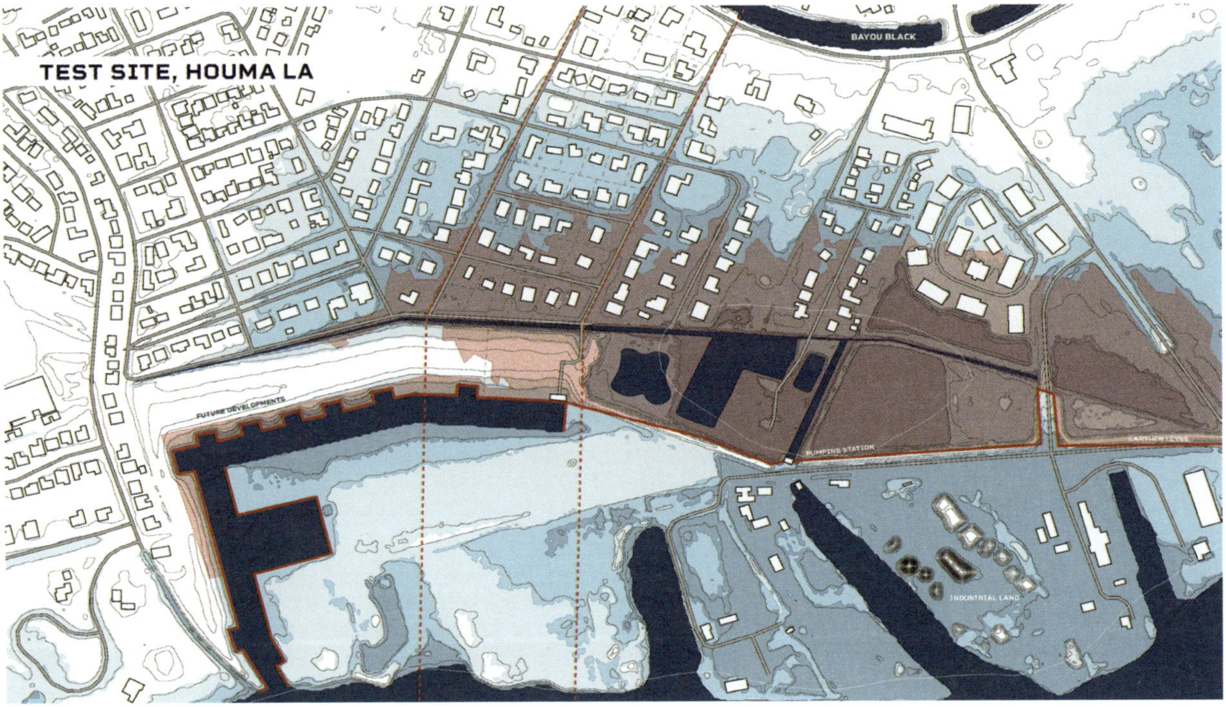
TEST SITE, HOUMA LA
BAYOU BLACK
FUTURE DEVELOPMENTS
PUMPING STATION
EARTHEN LEVEE
INDUSTRIAL LAND

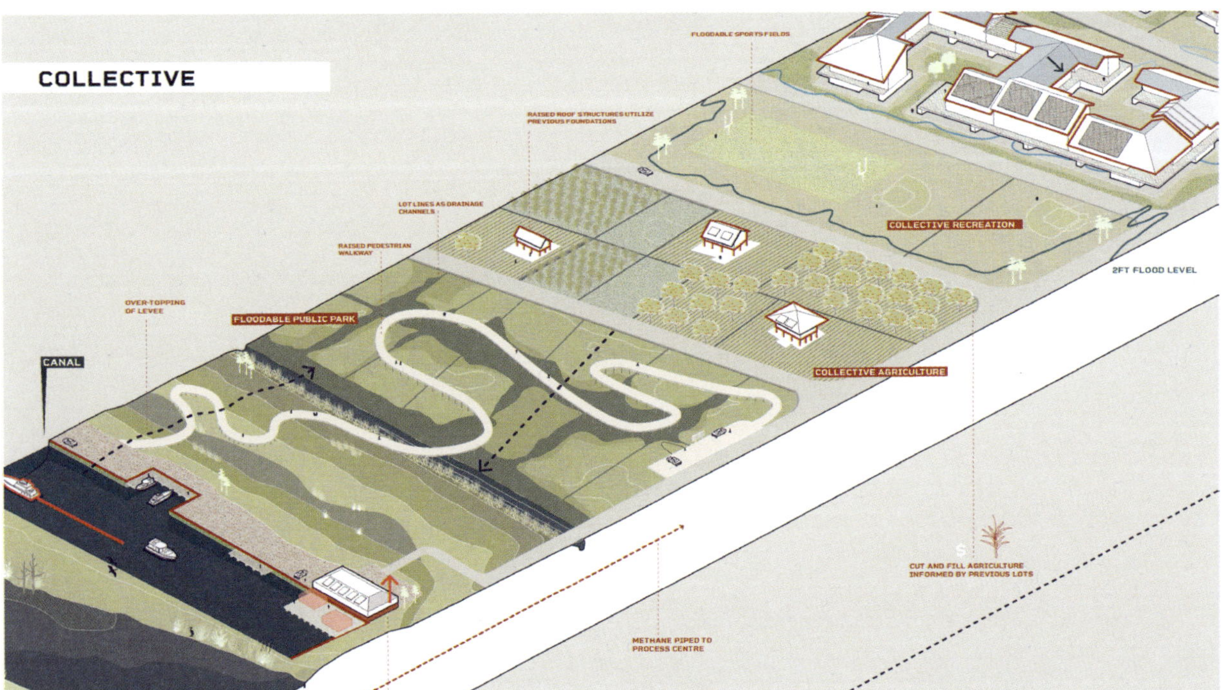
COLLECTIVE
FLOODABLE SPORTS FIELDS
RAISED ROOF STRUCTURES UTILIZE PREVIOUS FOUNDATIONS
LOT LINES AS DRAINAGE CHANNELS
RAISED PEDESTRIAN WALKWAY
OVER-TOPPING OF LEVEE
FLOODABLE PUBLIC PARK
CANAL
COLLECTIVE RECREATION
2FT FLOOD LEVEL
COLLECTIVE AGRICULTURE
CUT AND FILL AGRICULTURE INFORMED BY PREVIOUS LOTS
METHANE PIPED TO PROCESS CENTRE

THE AMPHIBIOUS EDGE

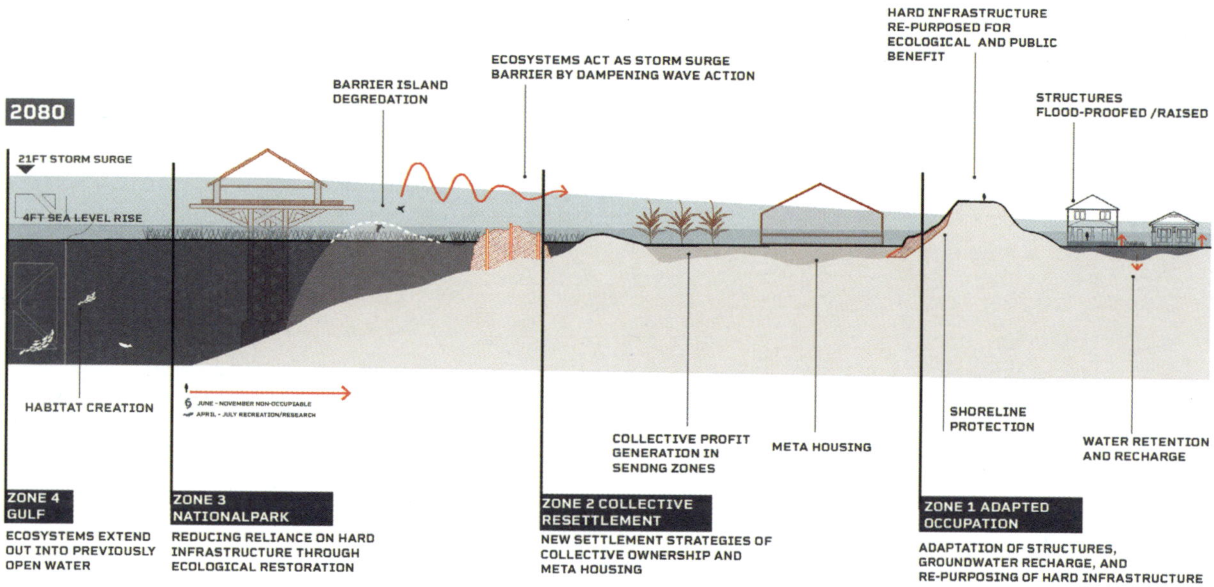

META HOME

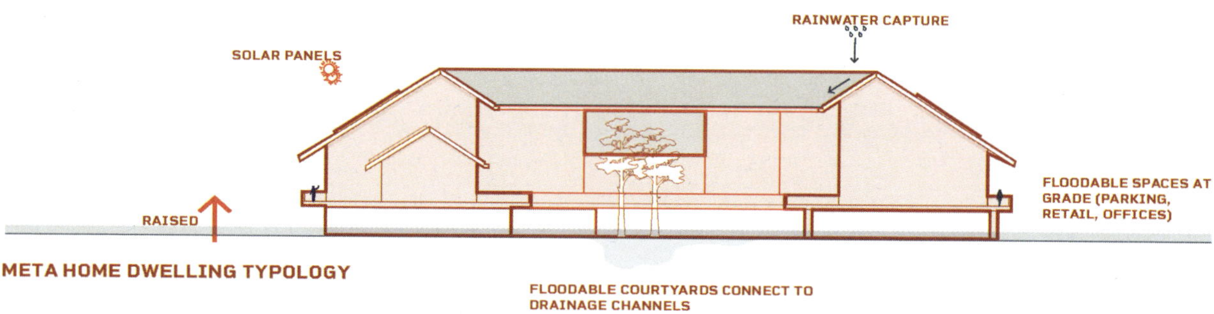

ARCH 4041-5051/ 6041 **Research Studio** UG + GR **FA22**

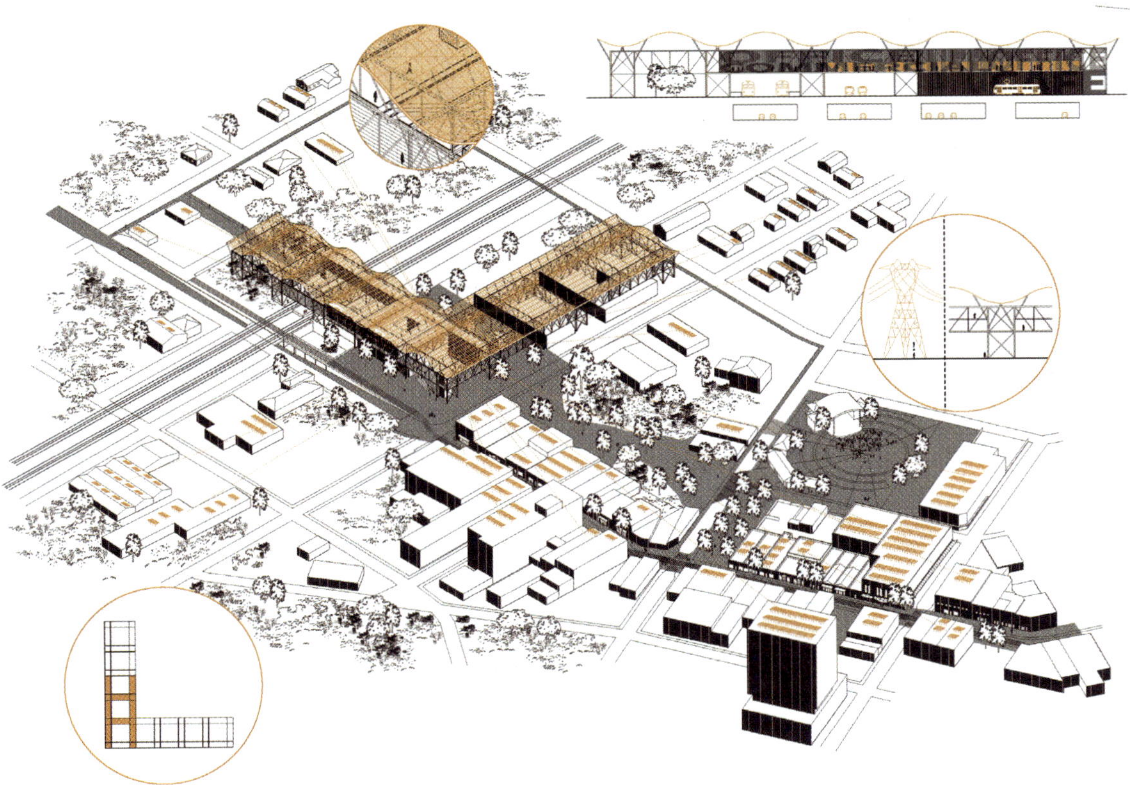

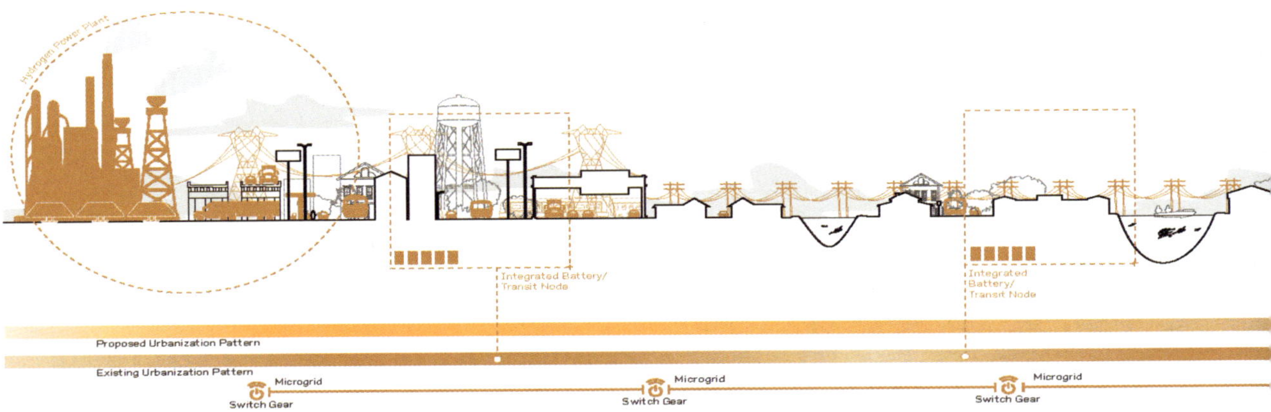

s > Merrie Afseth + Connor Little **i >** Margarita Jover + Liz Camuti

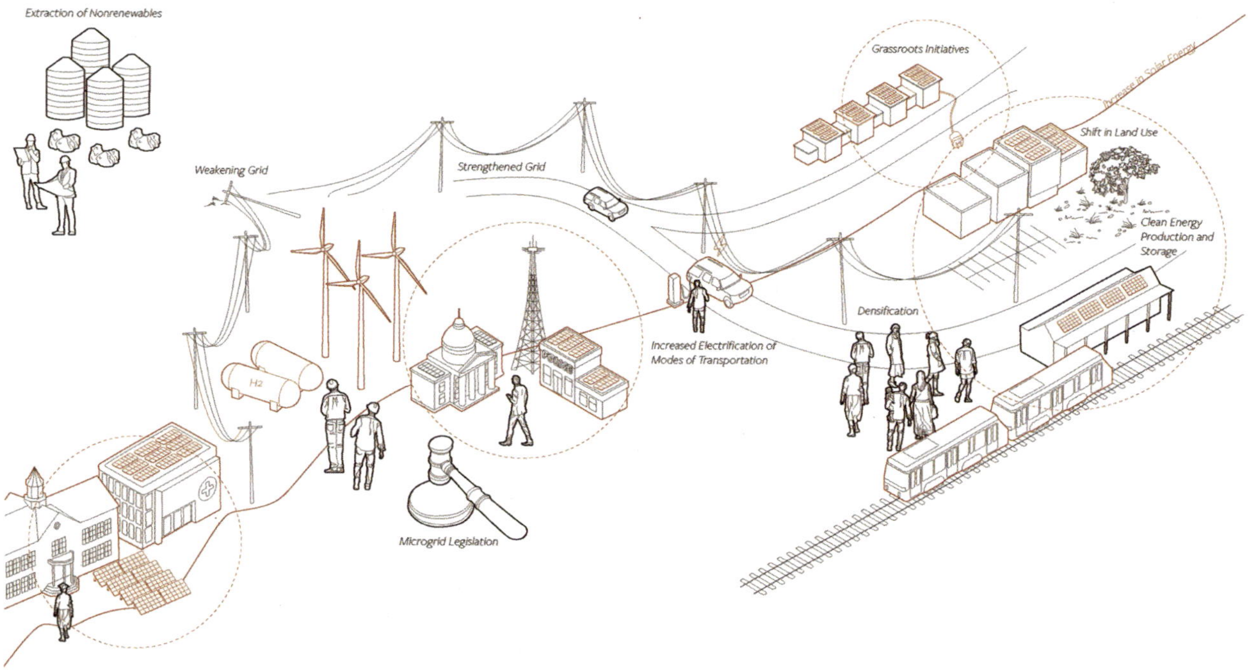

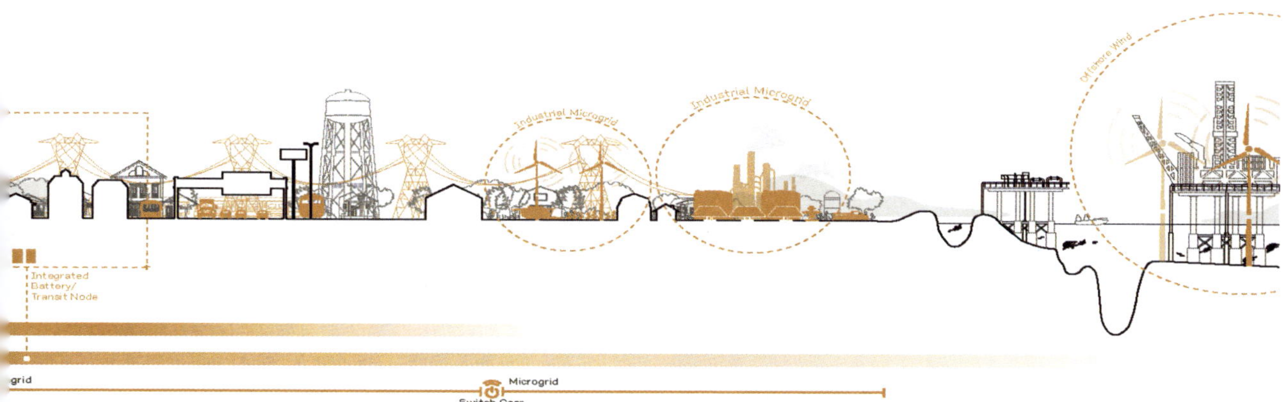

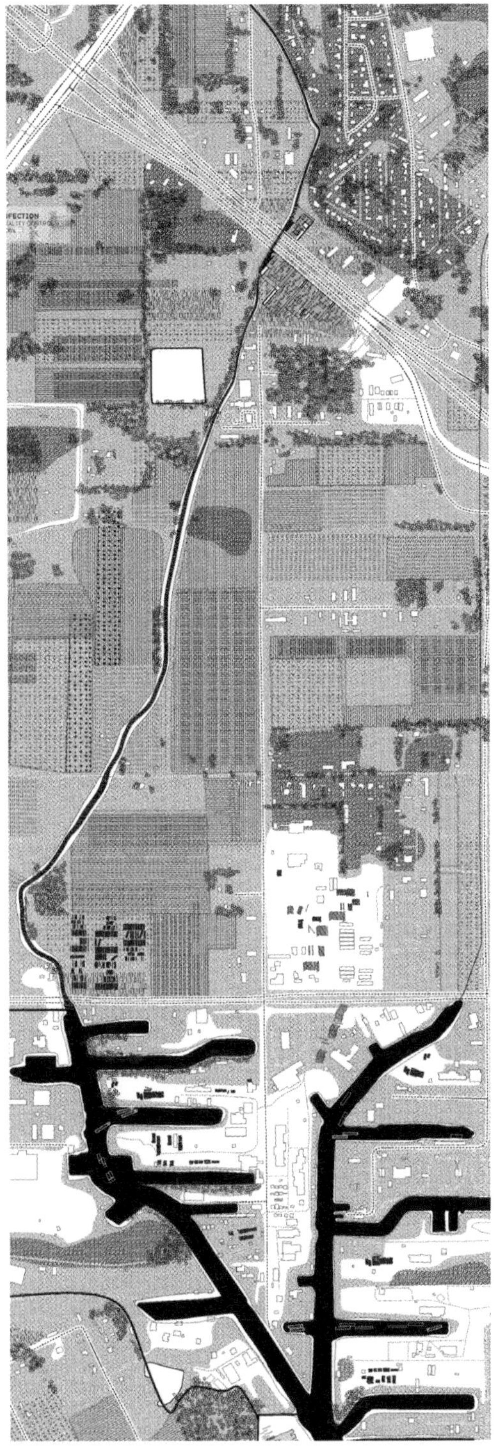

s > Leah Bohatch

i > Margarita Jover + Liz Camuti

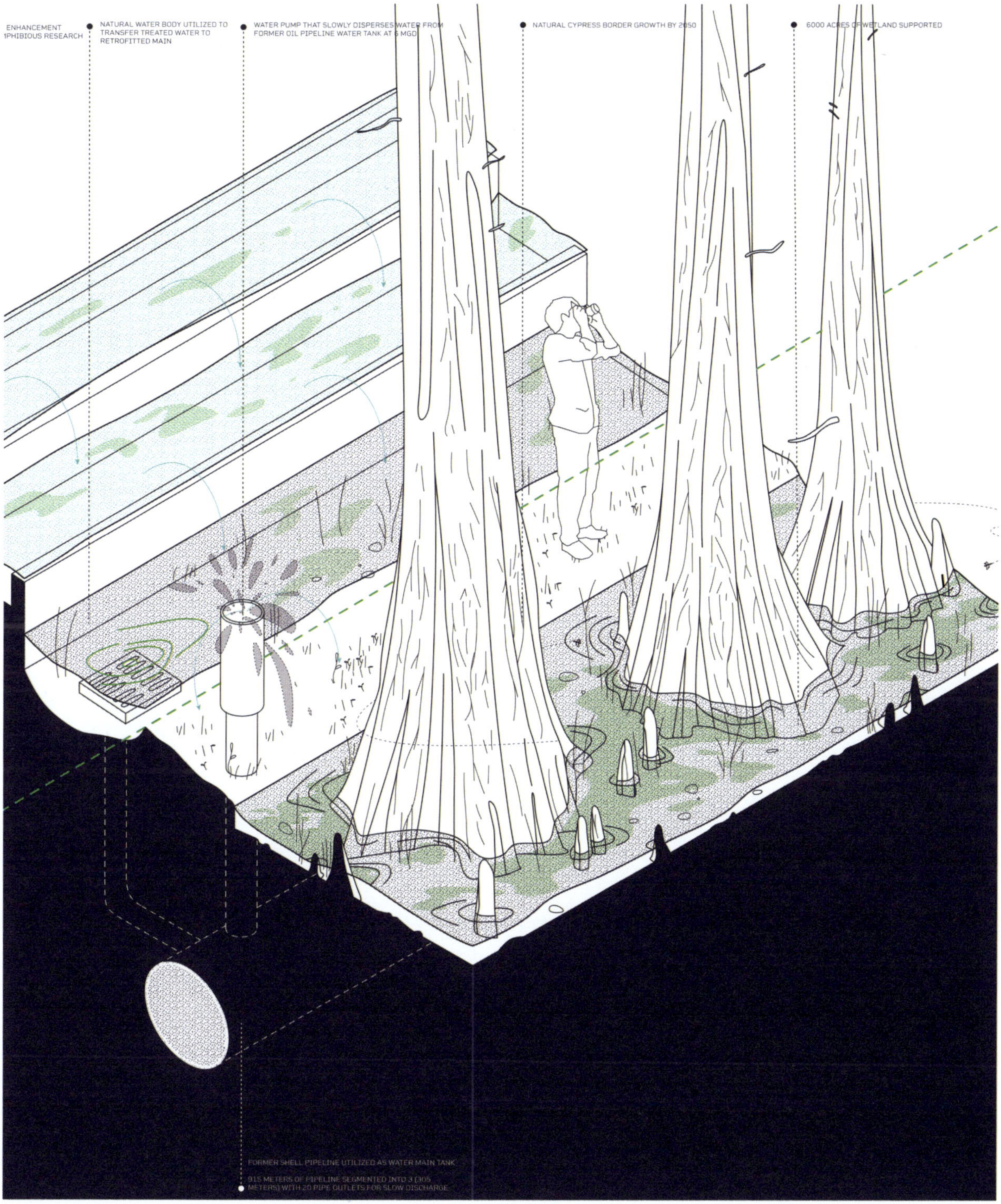

RAJASTHAN CITIES PROJECT. JAIPUR

Inaki Alday
Pankaj Vir Gupta
Andrea Bardon de Tena

Jaipur - the first planned Indian city, founded in 1727, retains an extraordinary built heritage and serves as the capital of Rajasthan - the largest state in India. It also epitomizes the most urgent crisis facing the country: the fast-depleting supply of drinking water for what will be soon the world's most populated country. In a dynamic and growing city of currently 3.65 million inhabitants, expected to reach 5 million in 2030 (UN World Urbanization Prospects), architecture, landscape architecture, preservation, and urbanism are essential to redefine a built environment. We propose to deploy them to support a thriving urban life, recover the relationship with water sources, facilitate the use and reuse of water, and engender socio-economic and ecological equality among the residents in the city.

The studio is organized as a lab, with three research directors (faculty) and seven team members working in different group configurations and individually, exerting individual agency while pushing a collective endeavor. Individual creativity, interests, and abilities are focused on identifying the critical issues and proposing the best comprehensive answers. Each student is challenged to face the entire complexity and produce unique responses, responsible and adventurous at the same time. The current team builds on the knowledge produced by the previous generations of the Yamuna River Project Studio.

The methodology is organized into 3 main steps: Search, Research (multi-scale Design Research), and Evaluation. It is critical to note that this series of steps are not cleanly delimited and there are overlaps and a process of going back and forth. The design process is not linear when searching for information, speculating, proposing, testing, and evaluating. From each of these moments, the design process moves to the next and, almost always, to the previous at the same time. Similarly, the scales of design are multiple and the process includes the movement from each one to the more detailed and the more general alternatively, if not simultaneously.

The outcome is a collective design research that produces a coherent proposition of multiple coordinated pieces, a common strategic plan, and individual/small team-specific propositions for specific sites/situations. Each of the last ones is a test of the critical interventions necessary to implement the strategic vision and a detailed design response.

s > Megan Spoor + Quinn Marie Kelsey

i > Iñaki Alday + Pankaj Vir Gupta
+ Andrea Bardon de Tena

EXISTING SITE CONDITION

1:5000 ISOMETRIC

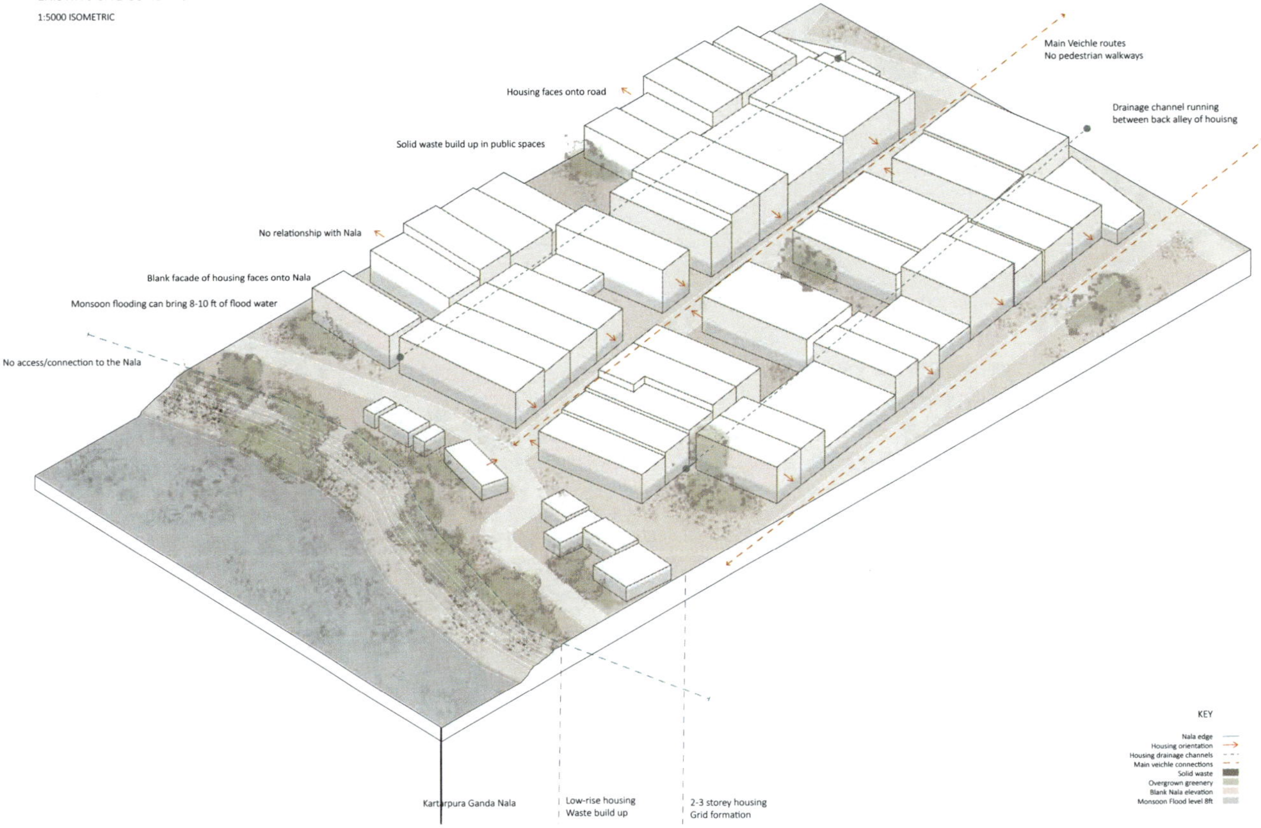

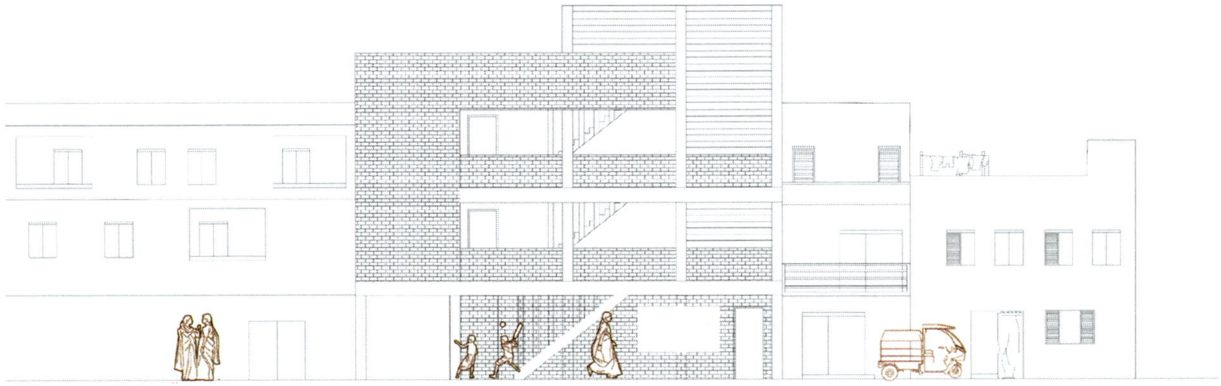

s > Megan Spoor + Quinn Marie Kelsey

i > Iñaki Alday + Pankaj Vir Gupta
+ Andrea Bardon de Tena

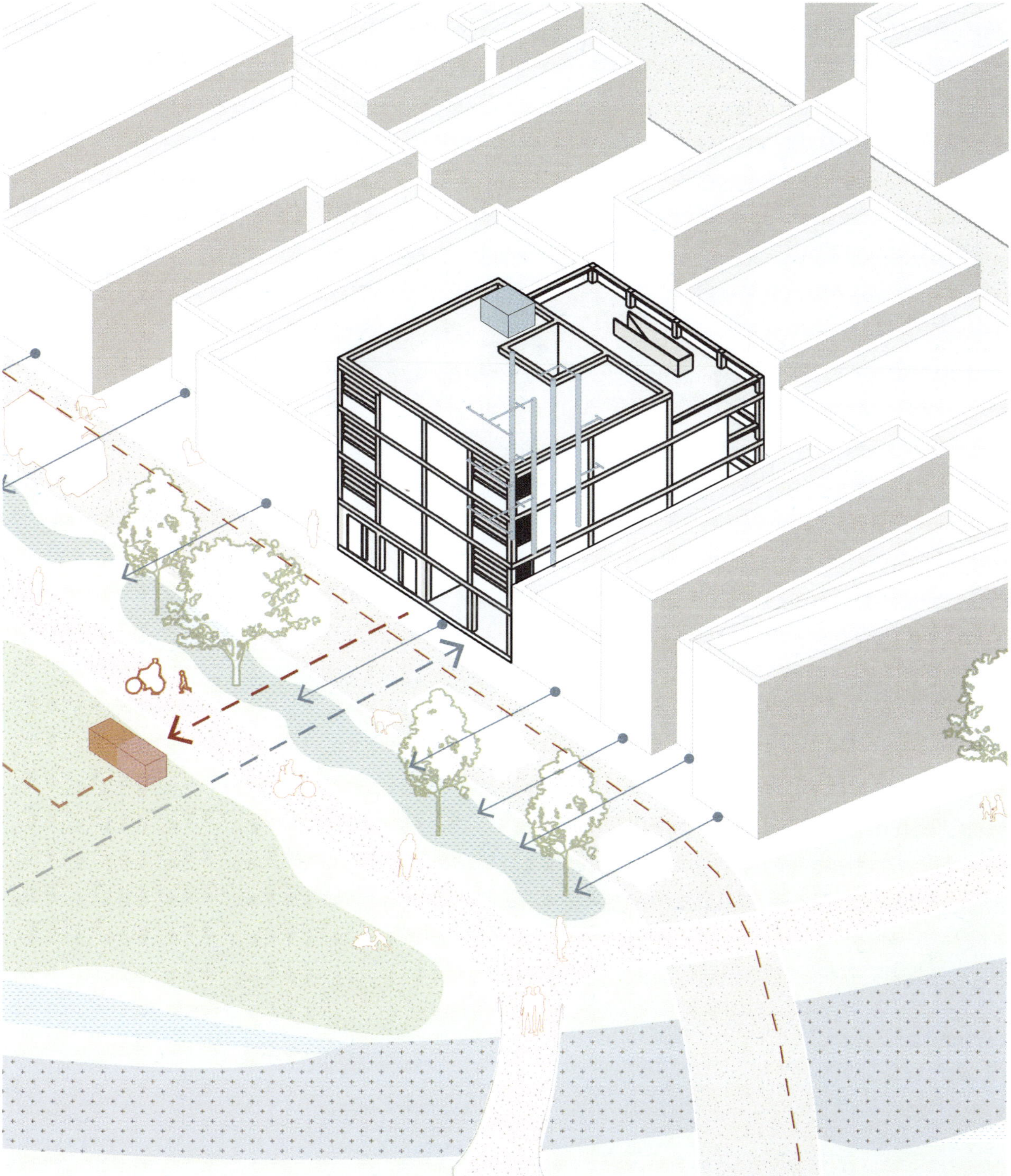

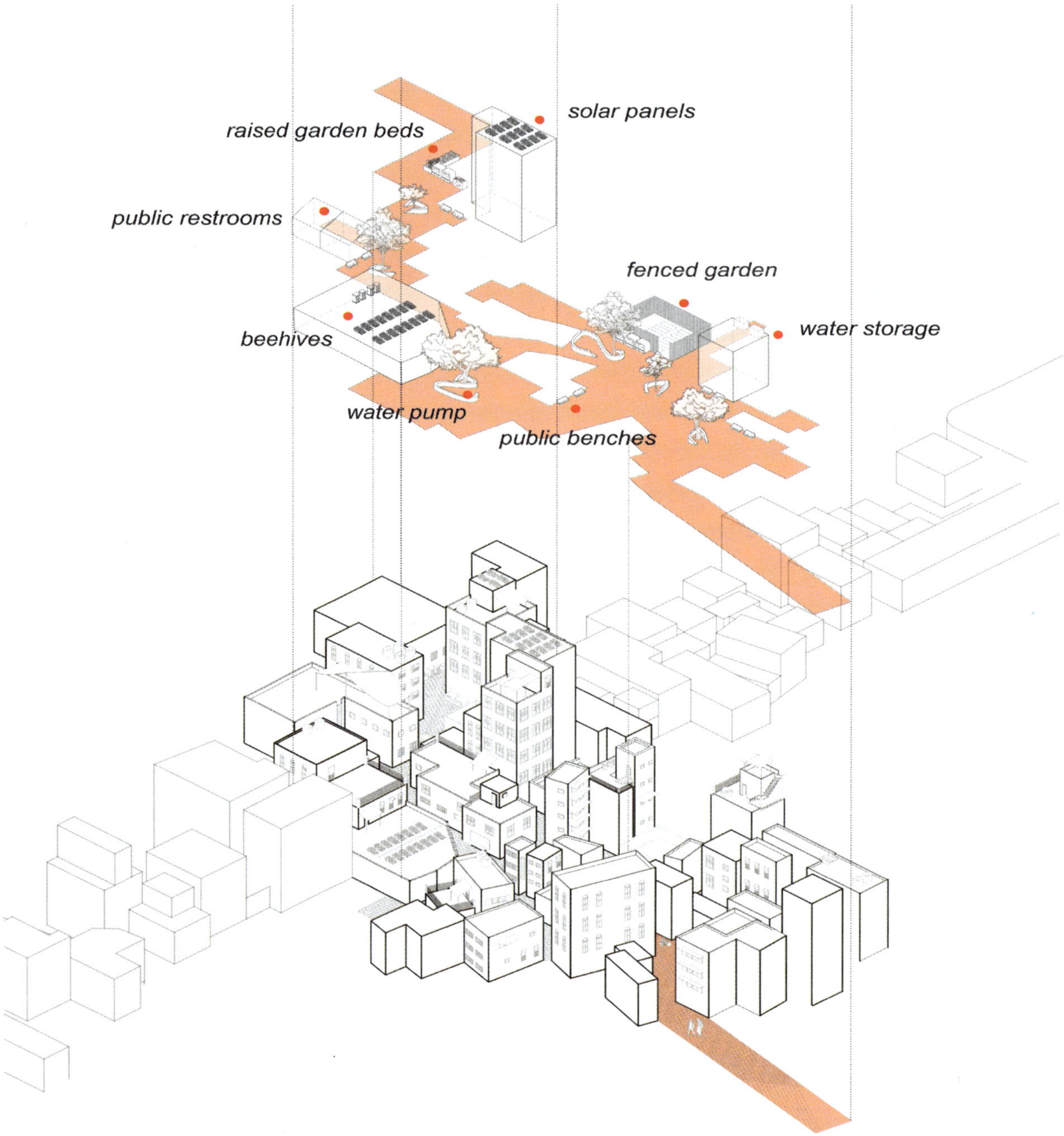
raised garden beds
solar panels
public restrooms
fenced garden
beehives
water storage
water pump
public benches

s > Hayley Burroughs

i > Iñaki Alday + Pankaj Vir Gupta
+ Andrea Bardon de Tena

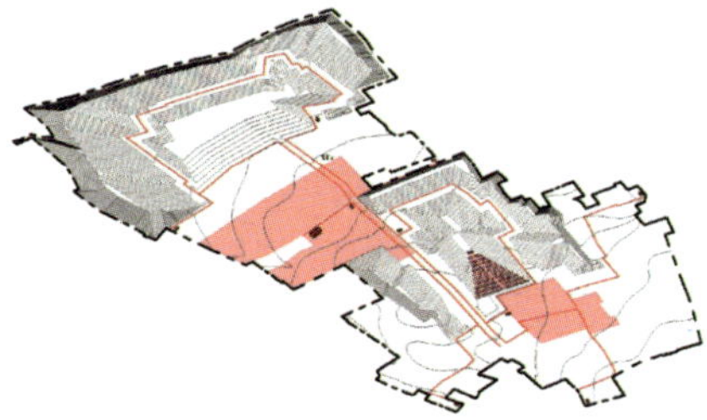

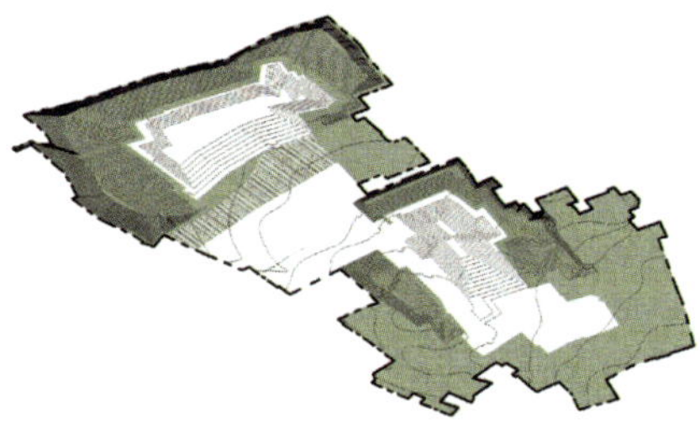

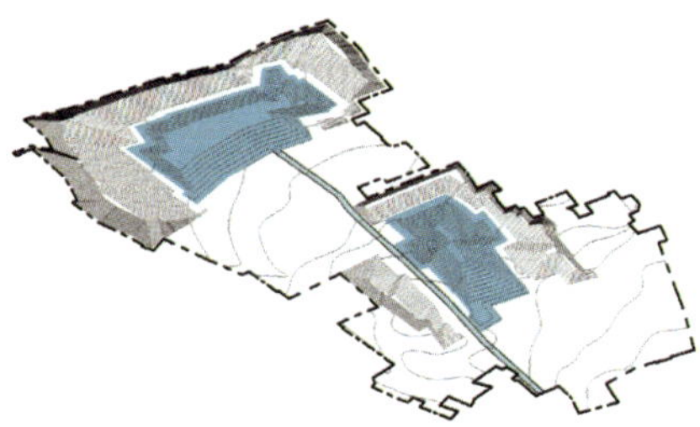

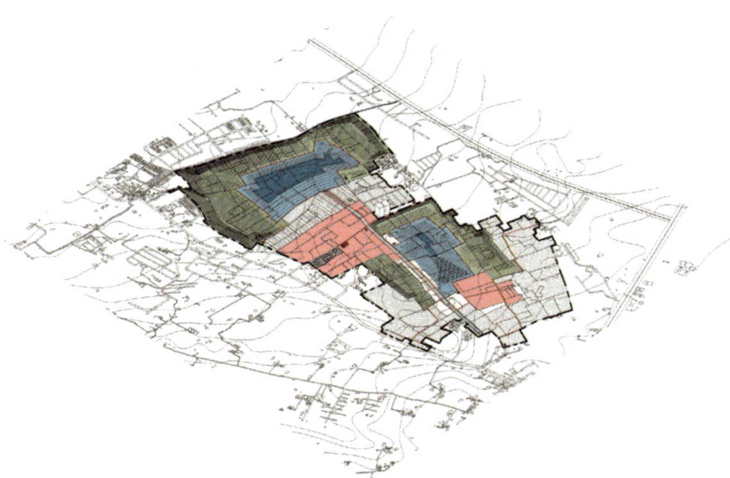

s > Sean Fowler

i > Iñaki Alday + Pankaj Vir Gupta + Andrea Bardon de Tena

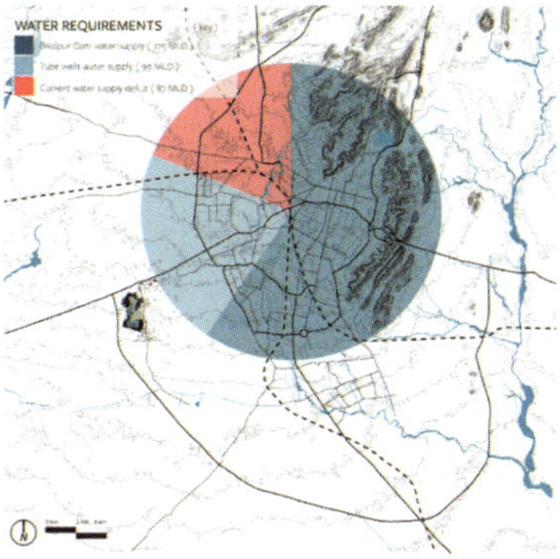

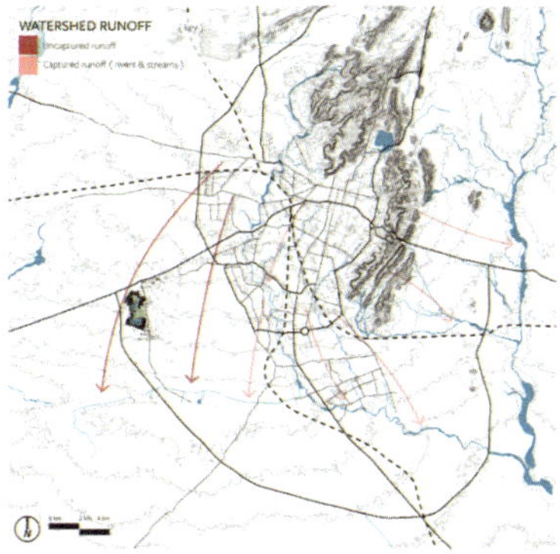

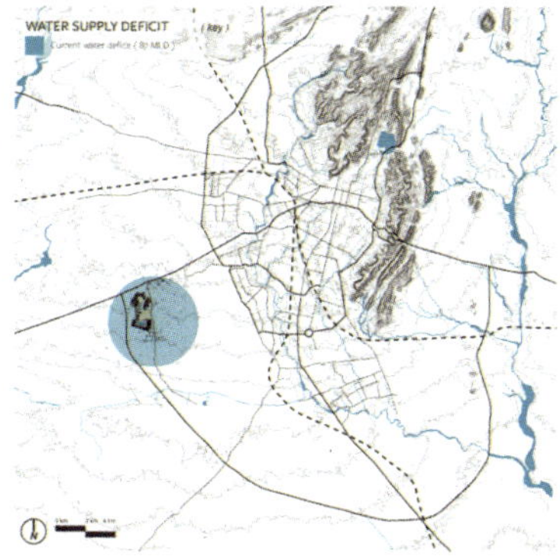

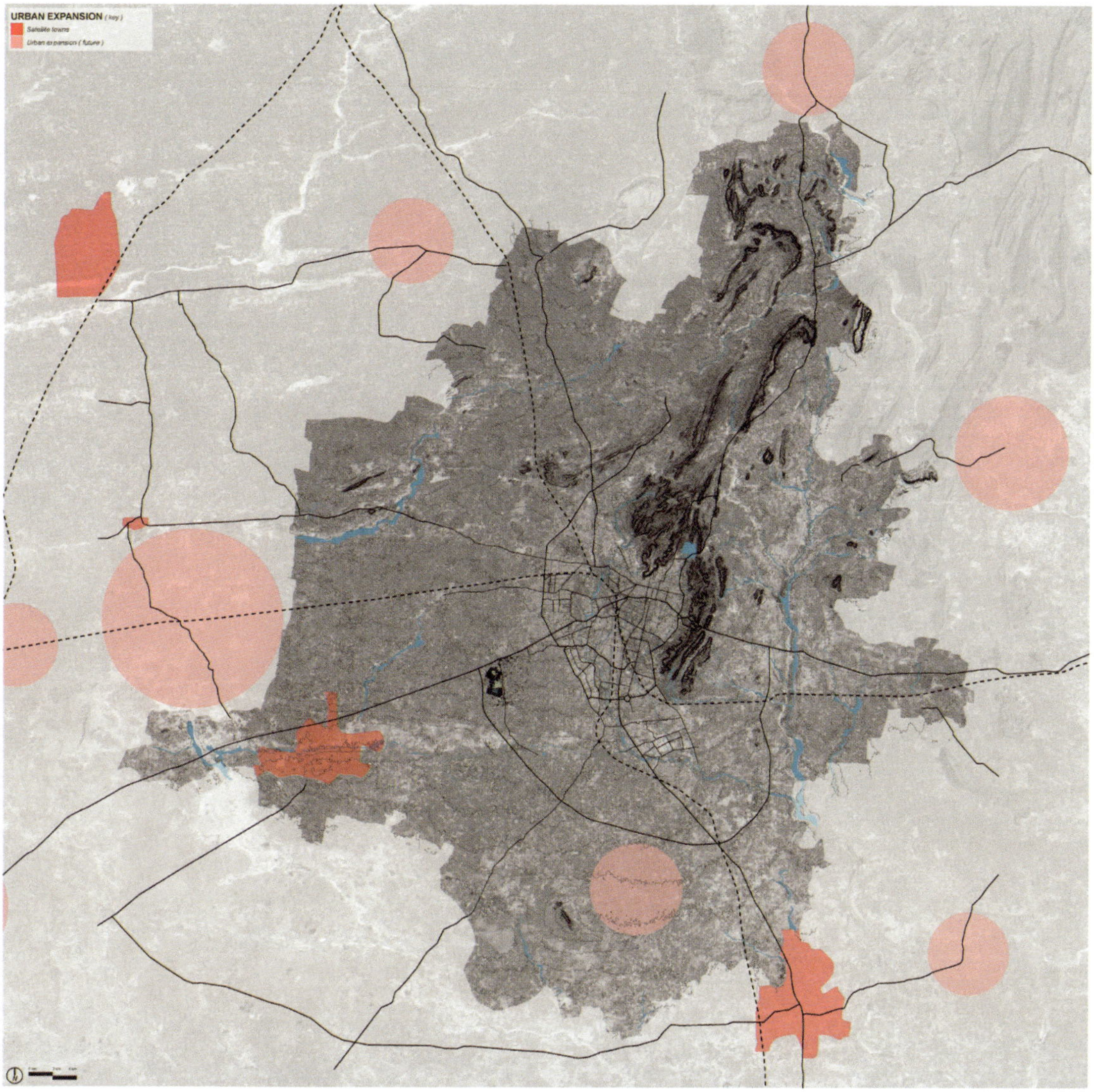

ADDIS ABABA THE UPPER KEBENA RIVER PROJECT

Ruben Garcia-Rubio [C]

Addis Ababa has rapidly evolved from a small military outpost to a sprawling metropolis, doubling its population to over 4 million in recent years. This unprecedented demographic boom has brought uncontrolled urbanization to the city, testing the administration's ability to provide basic services to the population: water pollution, water scarcity, flooding, lack of basic infrastructures/facilities, etc. If these urban issues are not addressed in the present, climate change will add to the continuous growth of the population, and both will exacerbate the already present problems.

This academic research aims to design a holistic urban resilience and regenerative strategy for Addis Ababa, whose main elements are the more than 50km of river tributaries that weave the city. For that, the project proposes a new "ecological infrastructure" that uses the rivers as the main element to tackle the city's most urgent challenges and, given the city's morphology, to reach most parts of the population.

The methodology of the project is organized into three different nested scales. The first analyzes and diagnoses the city's development from different perspectives. The second scale proposes solutions to the urban issues found in the analysis through a comprehensive strategic plan, in this case, for a specific area: the Upper Kebena River watershed. Finally, the third, and most minor scale, includes specific architectural interventions at the most urgent and significant sites.

The published work shows the outcomes of the "Addis Ababa River City" research project, an academic program created to address these challenges through an urban resilience strategy and architectural design methodology. Together, this body of work demonstrates how to elucidate a holistic understanding of the complex urban realities that Addis Ababa, and many other growing worldwide metropolises, face today and into the future.

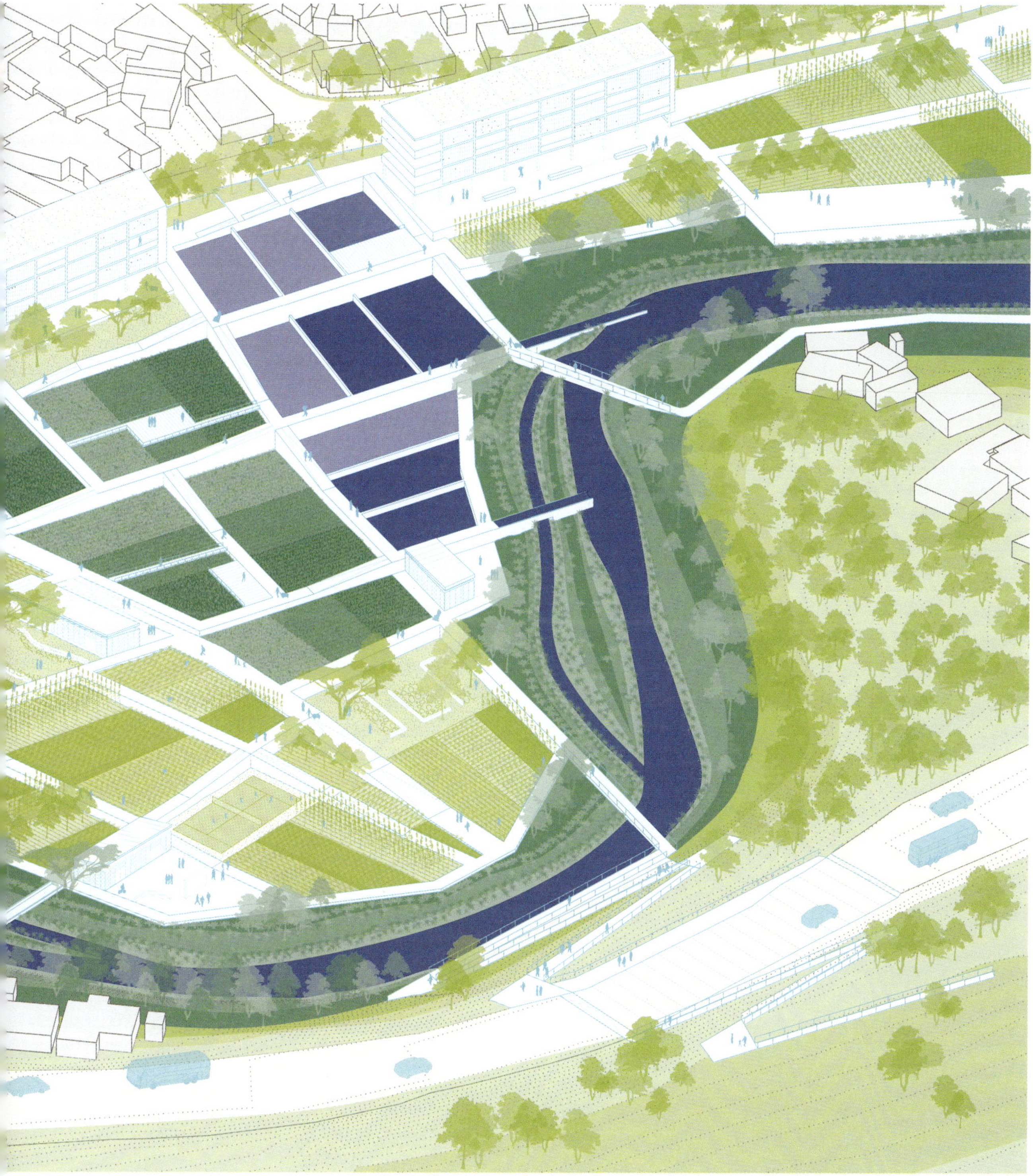

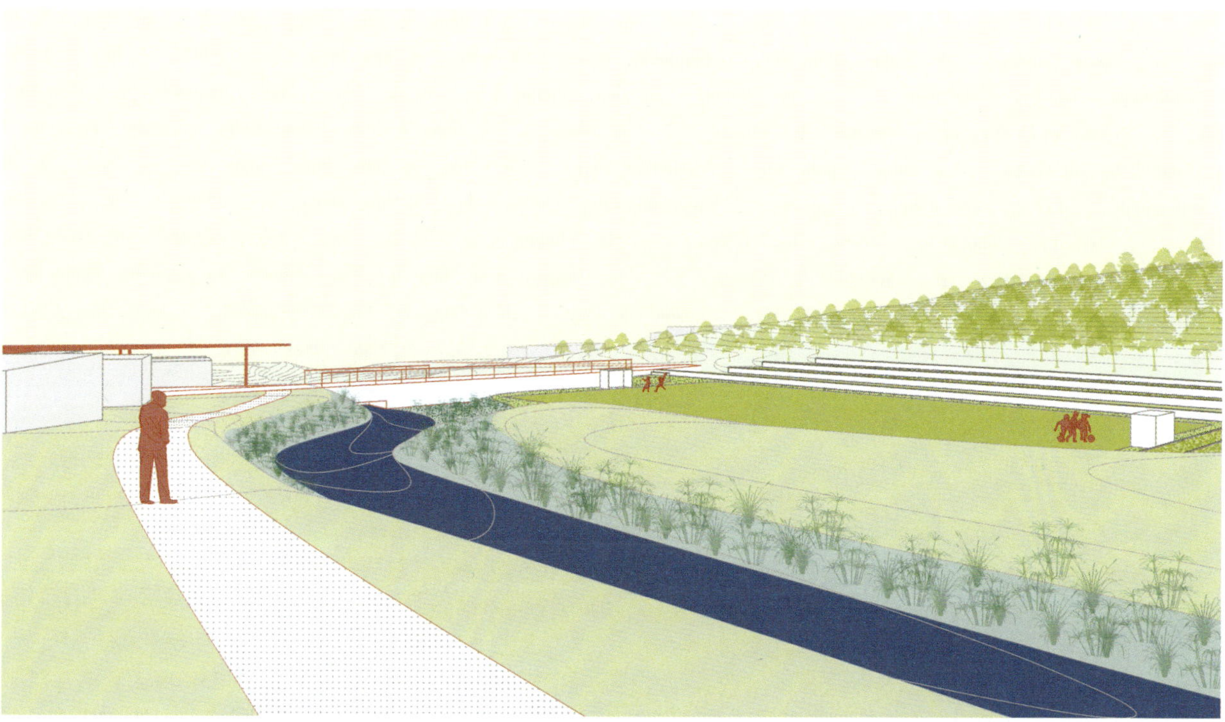

SEAN FOWLER

NOT ONE DROP

Urban Water Management in New Orleans

ARCH 6990 Advisor
Iñaki Alday

s> Sean Fowler

i> Iñaki Alday

New Orleans faces storms, flooding, subsidence, and threats to fresh drinking water, which are consequences of the City's relationship to water. These threats have existed since early in the City's history and are worsening with climate change. Structures restrict, manage, and move this water at the cost of effort and money. Considering water a resource, rather than a threat, can help New Orleans adapt to these changes and address these long-standing issues.

The hypothesis of this research is that stormwater is a potential source of fresh water, as rainwater is a sustainable and resilient alternative to outside water sources and related infrastructure. This could help make New Orleans more independent from infrastructure solutions to systemic threats. Water should not be concentrated in the centralizing drainage system and ejected. Instead, a distributed system should collect and infiltrate rainfall into the water table.

This proposal reconsiders the current drainage and domestic water systems to address threats to these systems, including local flooding and subsidence. Redirecting this water would limit the need for drainage canals and pumping stations to only the largest storms. If drainage infrastructure is needed only during severe weather events, then these spaces can be reclaimed by communities.

The proposal suggests reshaping high-topography neighborhoods to slow and stop runoff, infiltrating captured stormwater into groundwater, and lowering topography neighborhoods to drain stormwater and infiltrate this captured stormwater into groundwater. This infiltrated groundwater can help slow or reverse subsidence and provide a resilient source of fresh water for the city.

Fig. 1
Residential Block

Fig. 2
Street Section Comparisons

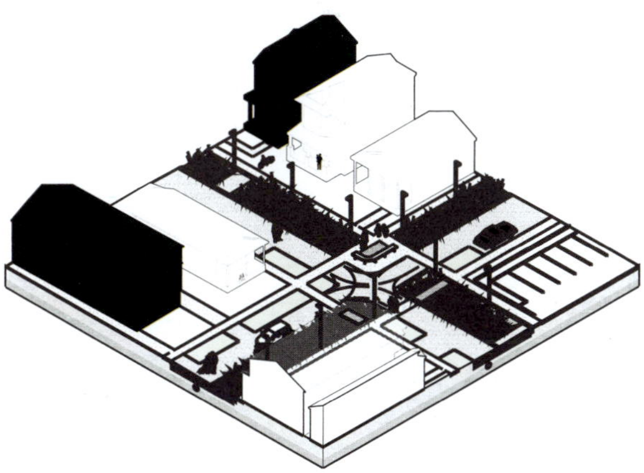

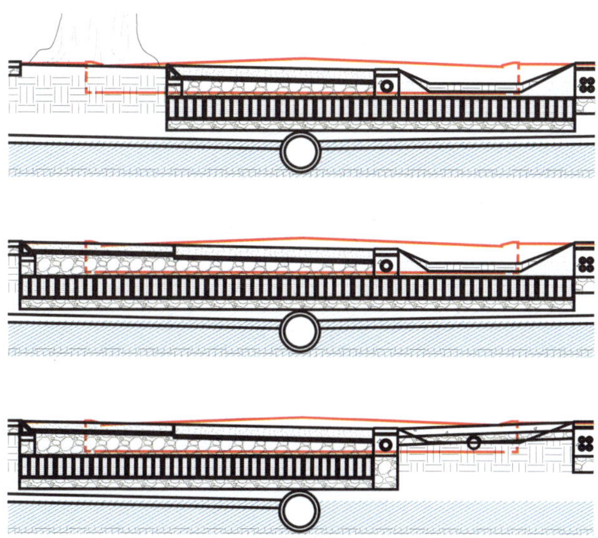

MEGAN SPOOR

BEYOND RETREAT

Realigning the Welsh Coastline for Resilient Habitation

ARCH 6990 Advisors
Iñaki Alday & Liz Camuti

s> Megan Spoor **i>** Iñaki Alday & Liz Camuti

Fairbourne, a small village on the Welsh coastline, is zoned for coastal realignment as part of the UK's Shoreline Management Plan. Driven by compounding flood risks and outdated structural defenses, the existing government plan proposed the removal of all residents and infrastructures, and the return of the land to salt marsh.

Wales is a country whose identity is tied to its coastline, With most of the population living and working in the coastal zone, the realignment of at-risk communities will contribute to a large-scale loss of culture, identity, and livelihood. Fairbourne will be the first of many communities placed at increasing risk due to sea level rise, and although scheduled to commence in 2025, relocation efforts are yet to materialize, as residents fight to remain in place. This creates an opportunity to establish new coastal occupation strategies that have the capacity to operate in future conditions of uncertainty and support the continued habitation of Wales's coastline.

"Beyond Retreat" presents an alternative settlement strategy for Fairbourne, that would enable the community to prepare for, and adapt to, the impacts of sea level rise, whilst minimizing community displacement, restoring coastal ecosystems, and regenerating local tourism. Operating in a scenario in which the Shoreline Management Plan for realignment is approved, the project presents a phased village level master plan. The strategy is driven by hydrological and ecological processes and grounded the unique culture of UK seaside holiday towns.

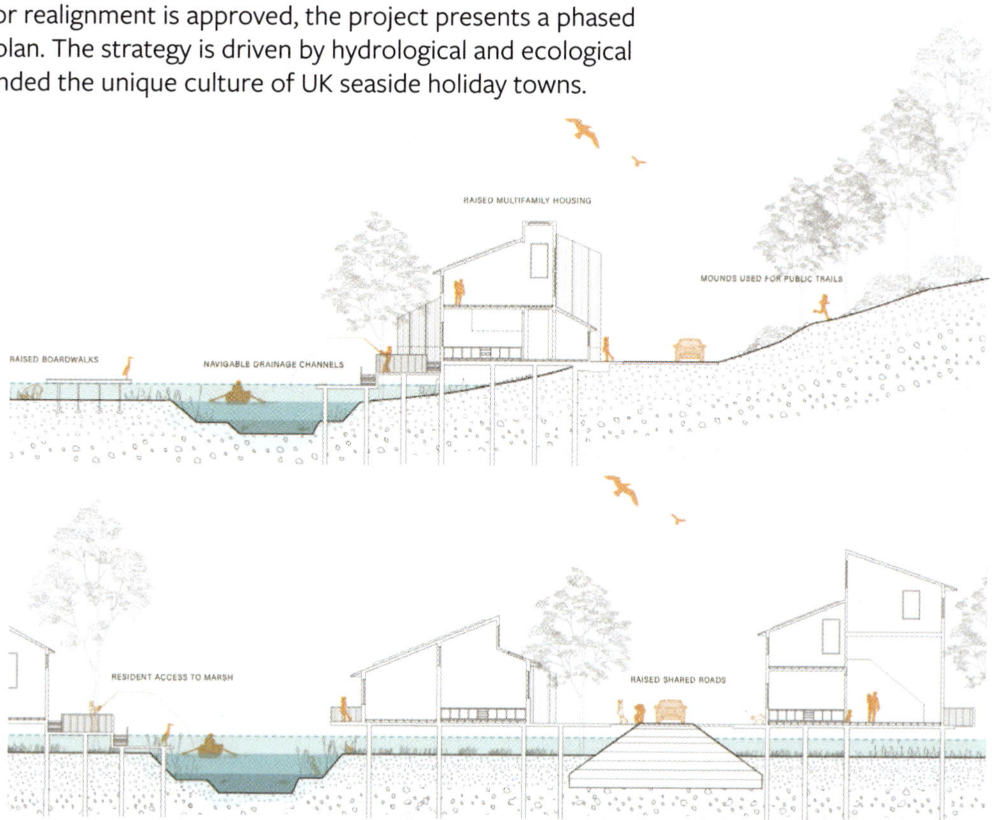

Fig.1
Regional Potential

Fig.2
Existing Conditions

Fig.3
Proposed Conditions

ESTUARY
REGENERATED MARSH
IRISH SEA
NEW MARSH
RECONNECTING HYDROLOGY
1 The embankment is lowered to allow water and sediment to shift inland and reestablish degraded marsh.
2 An opening in the sea wall feeds water and boats into a tourist marshland.
3 The sea wall is lowered to enable high-tide overtopping, to support regeneration of marshy field into marsh habitats, allowing shingle to wash inland as engineered defenses degrade.
4 Removed housing foundation slabs used for fishing piers and pavilions.
5 Inland of the railway, existing drainage channels are dug out to connect mountain run off with the new marsh conditions, creating an open tidal system that increases water capacity on site, and provides navigable channels for resident that lead from the mountains edge out to the Irish sea.
6 Existing water passages under the railway are widened for pedestrian and boat access, forming a permeable edge to the new settlement, as marsh creeps inland.
RAISING THE DATUM
7 Within the new tidal landscape, raised roads extend like fingers from the existing road that borders the mountains edge. Utilizing the higher elevation, small peninsula-like cul-de-sacs are formed, enabling raised housing to plug in.
8 A network of pedestrian board walks connect the new settlement with the marsh site.
9 Small tidal beaches and public parks sit within marsh.
10 Higher ground used for temporary holiday trailers, caravans and camping.

LEAH BOHATCH & CAMILLE KREISEL

INDUSTRIAL INTERFACE

A Transparent Relationship Between Wastewater and the Human

ARCH 5990 Instructors
Cordula Roser Gray & Ammar Elouini

s> Leah Bohatch & Camille Kreisel **i>** Cordula Roser Gray & Ammar Eloueini

The wastewater industry is currently an isolated system despite its great importance in serving the city and its inhabitants, creating a linear relationship that wastes a limited resource while harming the health of its source: the body. Looking towards the future of water infrastructure, a wastewater treatment plant in Miami utilizes a cyclical system that treats sewage water and uses the reclaimed water within the building to support the heat-stricken city and pedestrian, through the programming of a breathing, cooling aquatic center. Within the site, water becomes an interface between industry and the body, in which programs such as bath houses, pools, and mechanical rooms use reclaimed water to support the inhabitant. Public programming and circulation is introduced through a curated ribbon that weaves throughout the wastewater treatment systems, creating an interwoven and dependent relationship.

As a result, there occurs both a physical and social boundary between these systems and the person. Due to this separation, wastewater treatment systems are seen as dirty and unclean, leading to a lack of willingness amongst many to consume treated and reclaimed wastewater. As an increase in the demand for water grows in major cities, the reuse of treated wastewater can help to alleviate this need, but reclaiming wastewater does not occur due to misperceptions that the process is unsafe. By inserting wastewater treatment systems within everyday life on a smaller scale, this creates a network of water treatment that will address local water problems and increase optimized management of the water types. This inversion brings the city infrastructure and domestic space together to form a new industrial commons, generating new relationships and programs for a hybridized future.

Fig. 1
The Pool

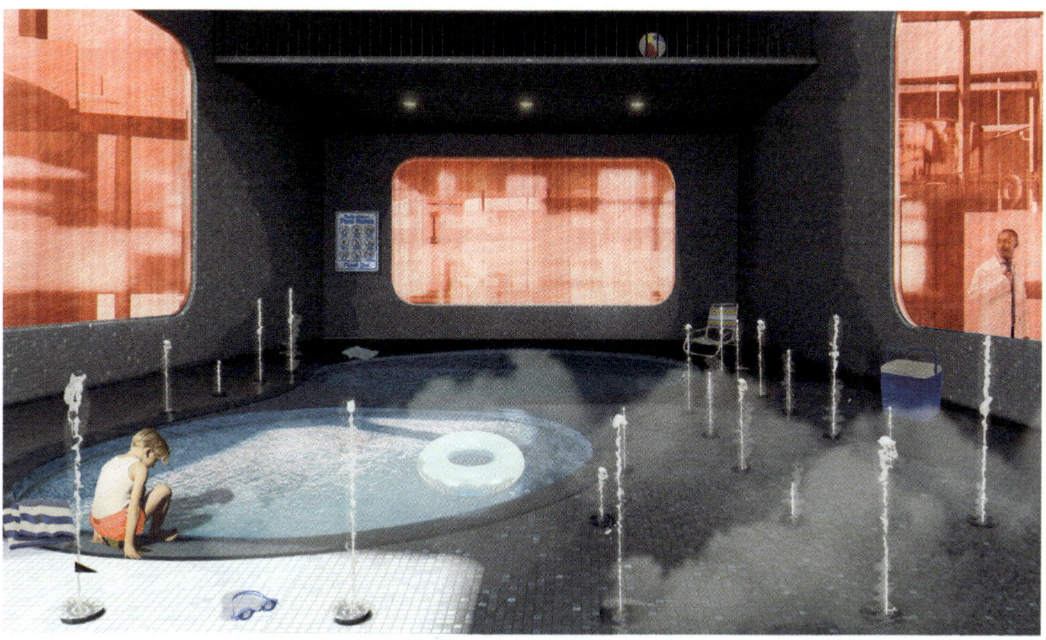

Fig. 2
Wastewater Treatment & Public Program

Fig.3
Amphibious Plaza

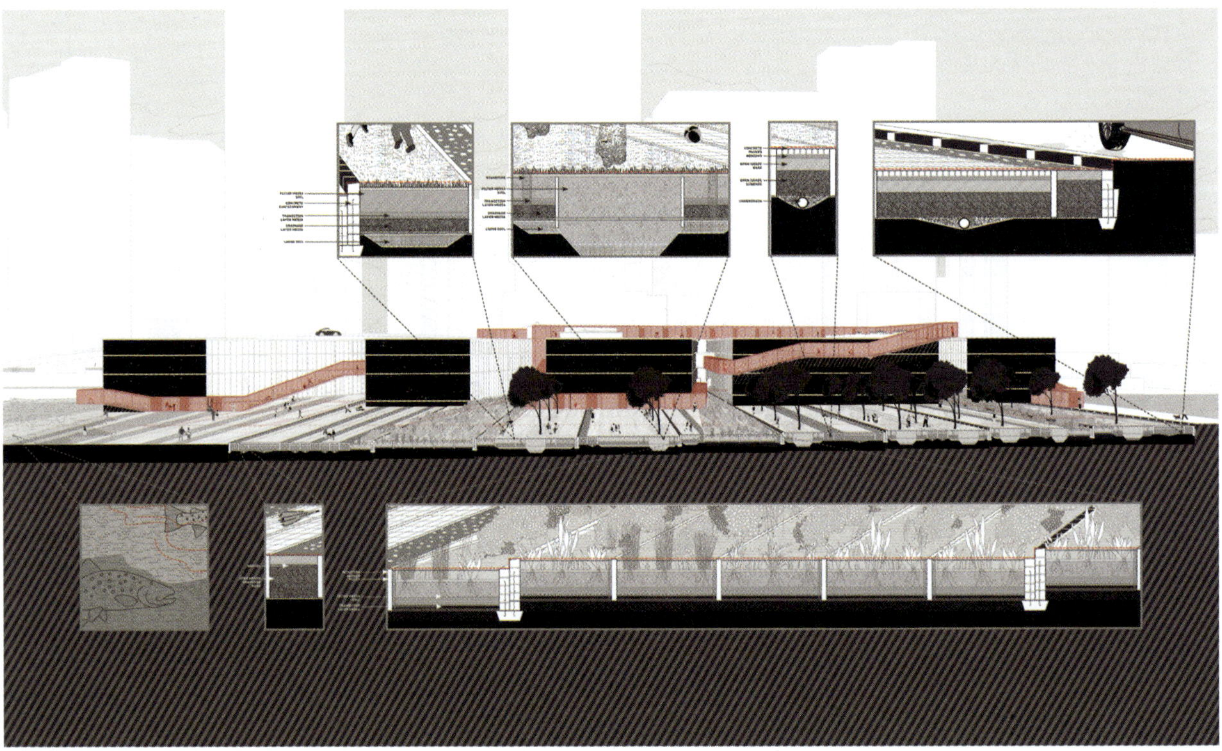

Fig. 4
The Plaza, Volume, & Ribbon

3

CULTURALLY AFFORDABLE

CAROL MCMICHAEL REESE, Ph.D.
Favrot IV Professor of Architecture
Ailsa Mellon Bruce Visiting Senior Fellow, summer 2022

RACE, HOUSING, AND COMMUNITY DESIGN IN THE COMPANY TOWNS OF THE US CANAL ZONE, 1920–1970

For almost a century (1904–1999), the United States governed the Panama Canal Zone as a colonial dominion inside the Republic of Panama, holding sovereign rights over the roughly 50-mile-long, 10-mile-wide swath of territory established by the 1903 Hay–Bunau-VarillaTreaty. Much of the zone was unbuilt and existed as tropical forest, providing a watershed that insured the canal's operation and replenishing the flow of water through its locks. But within this lush landscape, a coordinated and highly productive system of civilian communities and military bases was established to house and support those who first built the canal and later administered, serviced, and defended it. As a geopolitical enterprise, the Panama Canal, which opened to traffic in 1914, had global resonance. Its building, operation, and maintenance produced millions of documents—both textual and visual—that were carefully organized and archived from the beginning of the 20th century until the United States released the canal's operation to Panama on January 31, 1999. My work with these documents, which are housed in the National Archives, has

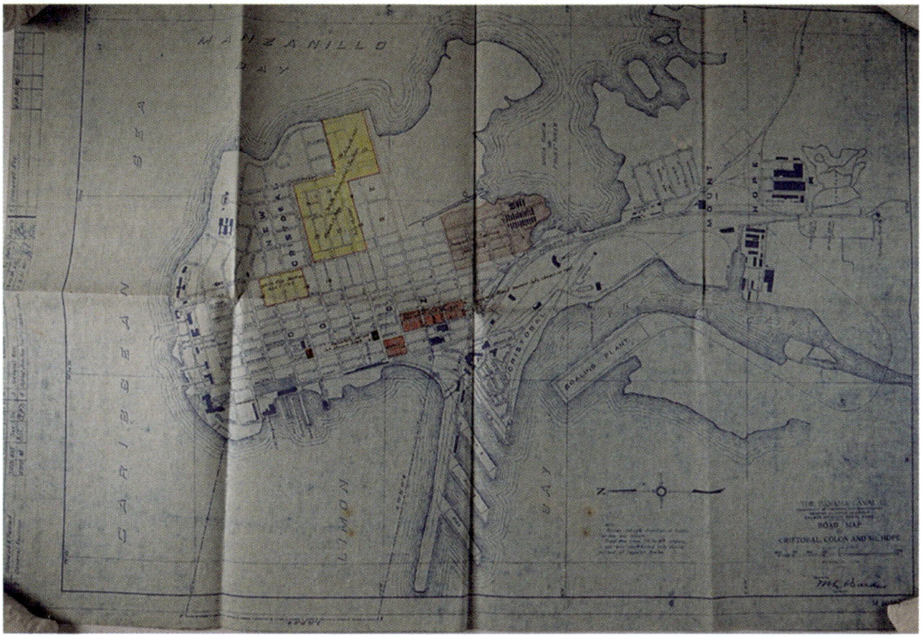

Fig.1
M. L. Walker,
Road Map of Cristobal, Colon and Mt. Hope
(showing "white residential section" in yellow),March 15, 1926, hand-colored blueprint, National Archives and Records Administration, Archives II, RecordGroup 185, Textual Records Division

been ongoing since 2002; it has focused on the socio-spatial design of civilian and military communities in the Canal Zone and its relationship to issues of 20th-century urbanization in the United States, Panama, and Latin America more broadly.

The United States exercised absolute control in the zone from its "tropical" capital of Balboa, which served as the administrative center for the canal's belt of specialized labor camps, industrial facilities, supply divisions, town sites, and military reservations. These installations evolved to meet the changing industrial, economic, and political conditions of the canal's operation; bilateral agreements between the United States and Panama; and multilateral international affairs. Initially planned according to prevailing Jim Crow racist ideologies, the zone's communities have received scant study, yet their architectural and social history yields important new knowledge about the US government's sanction of racial segregation, its reification in terms of community design, and the ensuing urban effects in neighboring Panama.

The Canal Zone was a territory in which international politics, labor relations, racial constructs, and ecological interdependencies created a fraught field of urban development. Confrontations between Panama and the United States, where the imbalance of power was an omnipresent issue, had important urbanistic consequences in both domains. For the United States, the zone provided a space to test sanitation measures, building materials, supply chain rationales, pre-built and portable structures, tropical landscapes, architectural adaptations for hot and humid climates, and, not least, planned communities with concomitant social and racial hierarchies. In Panama—a nation created in 1903 by treaty inconjunction with the US purchase of assets from the failed French canal project—the expression of modern nationhood through civic architecture and urban planning became paramount. These projects were undertaken as xenophobia in Panama mounted, stimulated not only by the tensions of US occupation, but also by the largely unwelcome presence of Black, predominantly Caribbean, canal laborers. Their status was constantly in flux, as theUnited States took responsibility for their welfare only so long as they were in canal employ. Shifting labor demands in the zone led to marked demographic and urban change in the Panamanian cities at either end of the canal, Panamá and Colón, where unemployed workers relocated to find jobs and housing, and pressured the Panamanian government for the rights of citizenship. Thus, relationships between the United States and Panama were both symbiotic and antagonistic, exacerbated by nationalist sentiments on both sides of the zone's mostly unfenced borders.

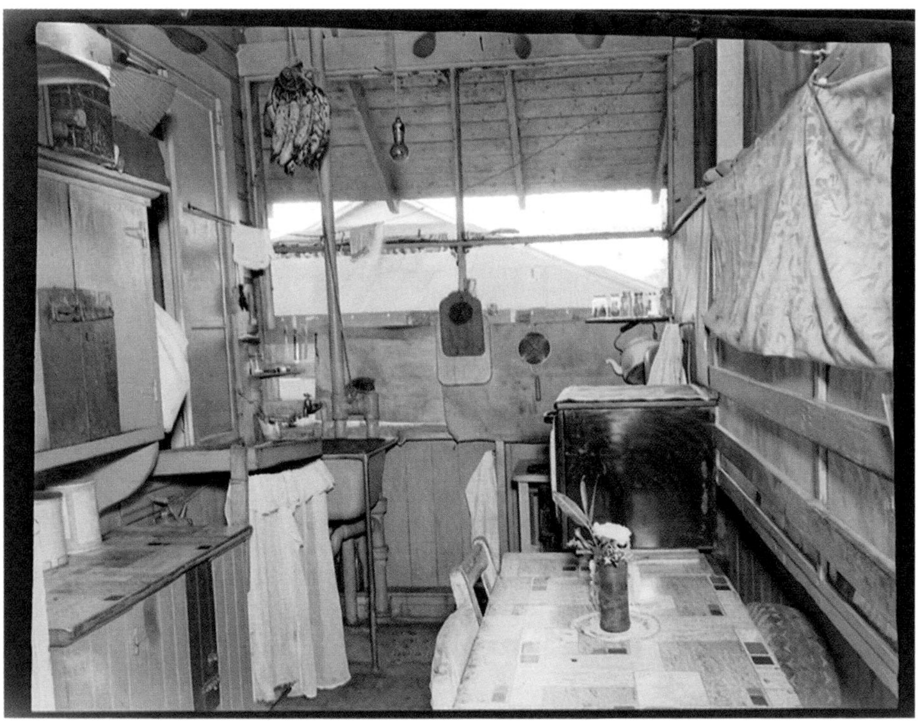

Fig.2
Unknown photographer, House 515, Apartment A, Old Type 18, 12 Single-family Apartments, Red Tank (kitchen in a designated Black community of the Panama Canal Zone), January 11, 1944, black-and-whitephotograph, National Archives and Records Administration, Archives II, Record Group 185, Still PicturesDivision

My research has parsed the Canal Zone's history in three eras. In 2013, with my coauthorThomas Ford Reese, I published El Canal de Panamá y su legado arquitectónico (1905–1920) / The Panama Canal and Its Architectural Legacy (1905–1920), a bilingual volume regarding the first period of the zone's development. My recent fellowship at the Center enabled me to return to the National Archives to seek carefully targeted documents related to key hypotheses concerning two subsequent eras of the urban development of the Canal Zone, which are the subject of our new book. In this well-advanced manuscript, we study the interwar period from the early 1920s to the early 1940s—when the economic effects of the Depression reduced appropriations for the canal and political conflict escalated during Arnulfo Arias's presidency—and the ensuing period of growth and prosperity from the early 1940s through the 1960s, when the facilities of the zone were mobilized to respond to the United States' entry into World War II and its operating entities were reorganized.

While in Washington, I found seminal reports on improvements in Canal Zone housingauthored by US advisors working in Latin America in the 1940s. These reports and related interventions include critical documents prepared by Jacob Leslie Crane (1892–1988) and Wallace Gleed Teare (1897–1989), advisors to the Federal Public Housing Authority, such astheir "Housing Program for the Panama Canal" (1945), which provided design solutions for upgrading the housing of the canal's Black laborers. These men, as well as other international consultants, worked with federal development agencies in the United States as well as Panamanian and other Latin American colleagues to support housing initiatives and sanitary urban improvements in the zone and elsewhere in the Americas. Their reports evidence international concerns about minimal standards for urban housing, health, and safety, moving concepts of urbanization in the direction of equity. In Washington, I found additional documents concerning the attitudes toward racial segregation in the US armed services thatwere impactful in the Canal Zone. From the earliest years of canal construction, the planning and building of military and civilian communities in the zone were closely interconnected. These newly consulted documents now allow me to connect the post–World War IIintegration of US military forces, the proliferation of military family housing in the zone after 1945, and the gradual dismantling of the racist Jim Crow protocols on which the Canal Zone communities were founded.

RUBEN GARCIA RUBIO
& SONSOLES VELA

THEATER & CULTURAL CENTER IN LUANCO

COMPETITION
2nd Prize
Spain, 2022

Credits
studioVRA | Ruben Garcia Rubio and Sonsoles Vela Navarro

How can socio-cultural space be created in a congested neighborhood? How can an old building be integrated with a new one? How can traditional sustainable strategies be updated in contemporary architecture? These three questions drove a social and cultural center proposal in Luanco (Spain).

On an urban scale, several strategies are proposed. The creation of a public square open to the main street is proposed, which gives entity to the area and acts as a prelude. The design does not modify the current volume but extends it, freeing up the ground floor, so the square and ground floor are conceived as continuous entities.

The proposal uses the registered building as a starting point on an architectural scale. On the outside, the new construction part takes certain lines and proportions and its appearance. Hence, the entire proposal is resolved with a family of aquamarine tones while relating them to the rest of the city and the environment. The interior of the proposal is divided into two parts. The existing building is conceived as a multipurpose part of the center. The ground floor is freed to act as a lobby/exhibition room. Meanwhile, the upper floors are reorganized into multiple and varied spaces. At the same time, the part of the new building is focused exclusively on the auditorium.

Another strategy is the interpretation of the glazed gallery at a construction scale. The proposal values this element of the region and translates it into accents on the façade to dialogue with the environment.

Fig.1
Inside view
Rendering.

Fig. 2 Main level plan
Access, Auditorium

Fig.3 Section
Street connection

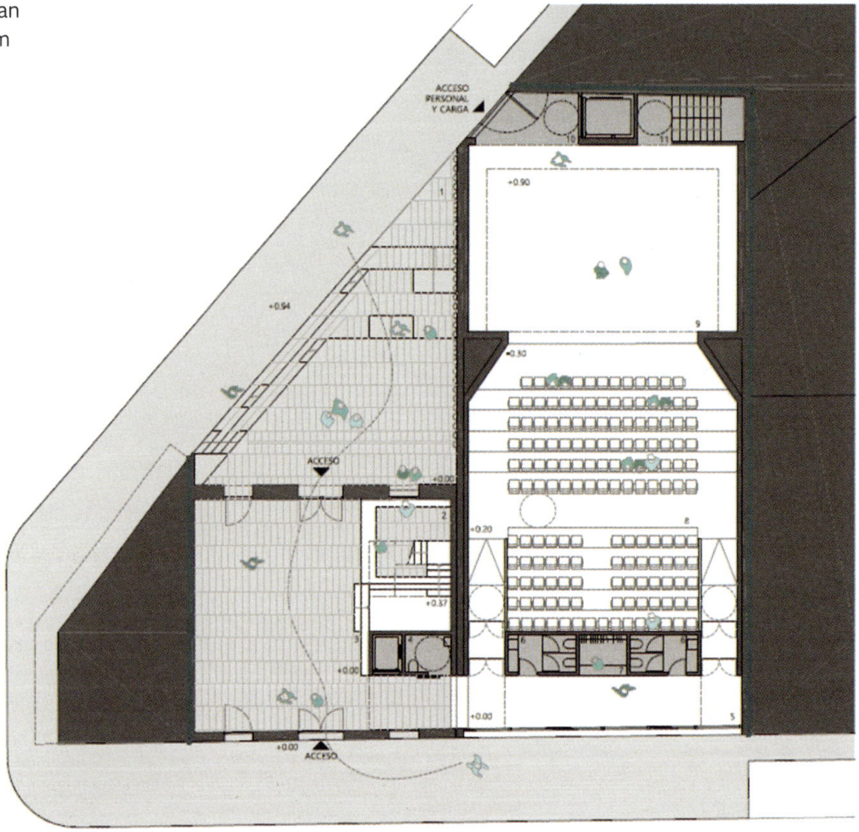

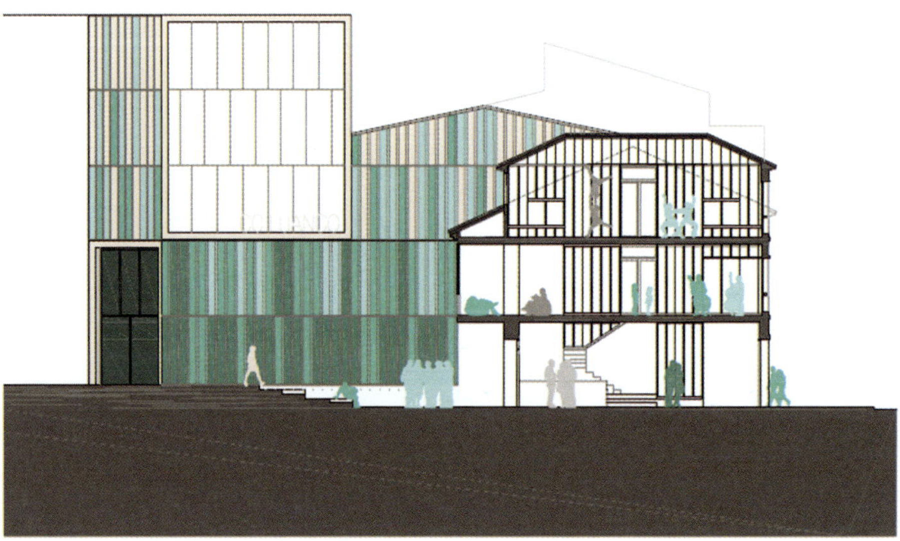

Fig. 4
Street view
Render (STUDIOKMH).

DISCOVER

Judith Kinnard [C]
Elizabeth Chen
Andrew Liles
Carrie Hunsicker
Kenneth Schwartz

The first core studio addresses foundational concepts and methods explored through drawing and making. To understand the shaping of place in an architectural sense, students carefully study the world around them and document what they find. The studio embraces inherent diversity that the students bring to class, with each student coming from a different lived experience, as a resource in understanding different cultures, backgrounds, and perspectives. Each student develops their own project that relates the human body to natural and constructed environments. The course introduces and employs a specific vocabulary to describe the nuances of the physical world, and students use this lexicon to link qualities of space to design operations.

Drawing comprises an architect's primary mode of seeing, thinking, and creating. It forms the cornerstone for a studio-based process of critical thinking. Each technique serves as a springboard for students to develop new approaches to finding solutions. Studio work implements an analysis of composition through observing the interrelated parts that make up a whole. Building an awareness of the relationships between parts propels the design process by generating ideas and defining physical properties to work with.

The studio's structure includes lectures and workshops on free-hand drawing, drafting, constructed perspectives, and manual 3D modeling. These lessons provide tools to approach design both empirically and experientially, allowing students to participate in the creative formation of change as they progress from recorded information to generative inspiration.

s> Skye Schofield-Saba **i>** Elizabeth Chen

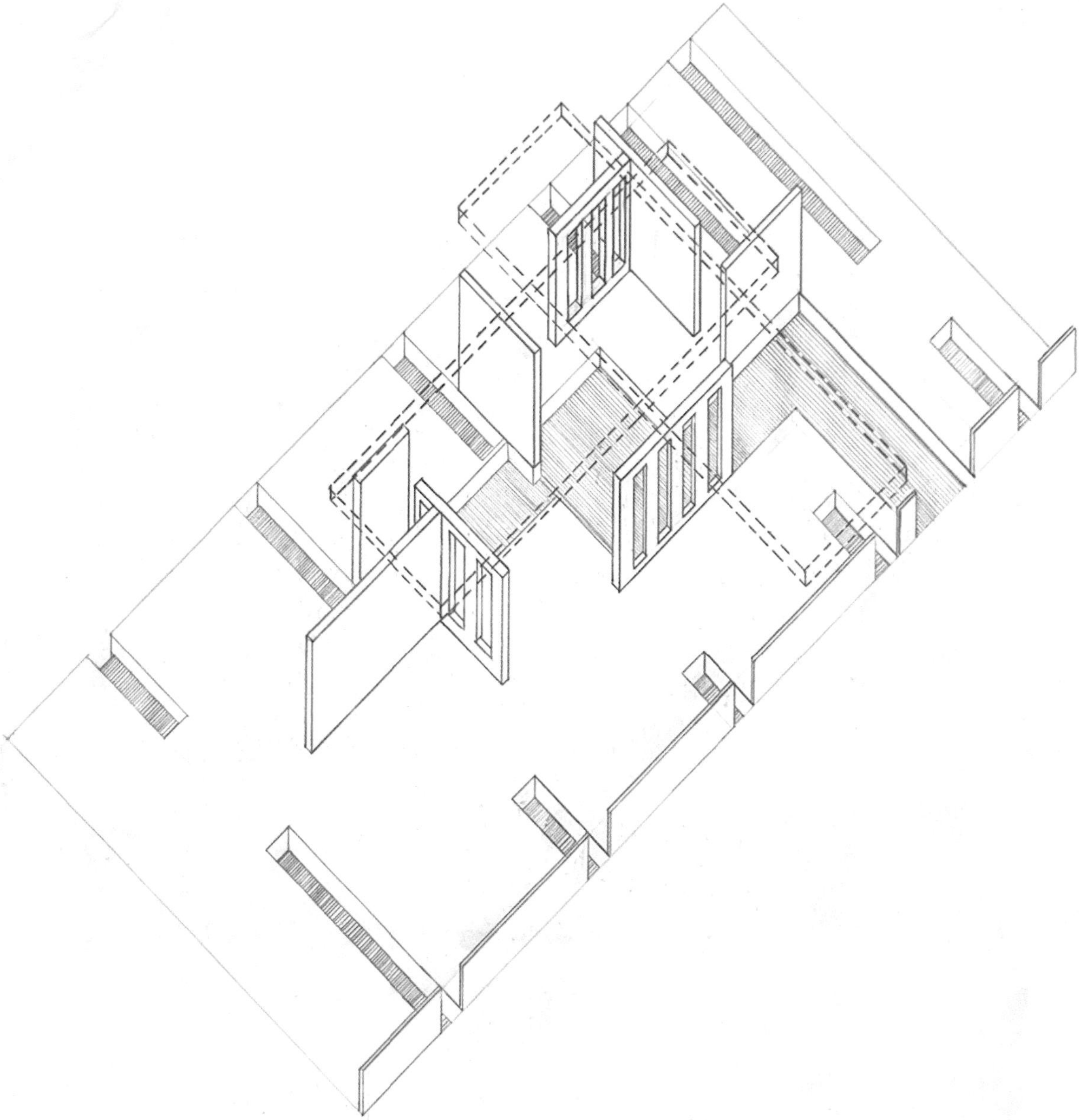

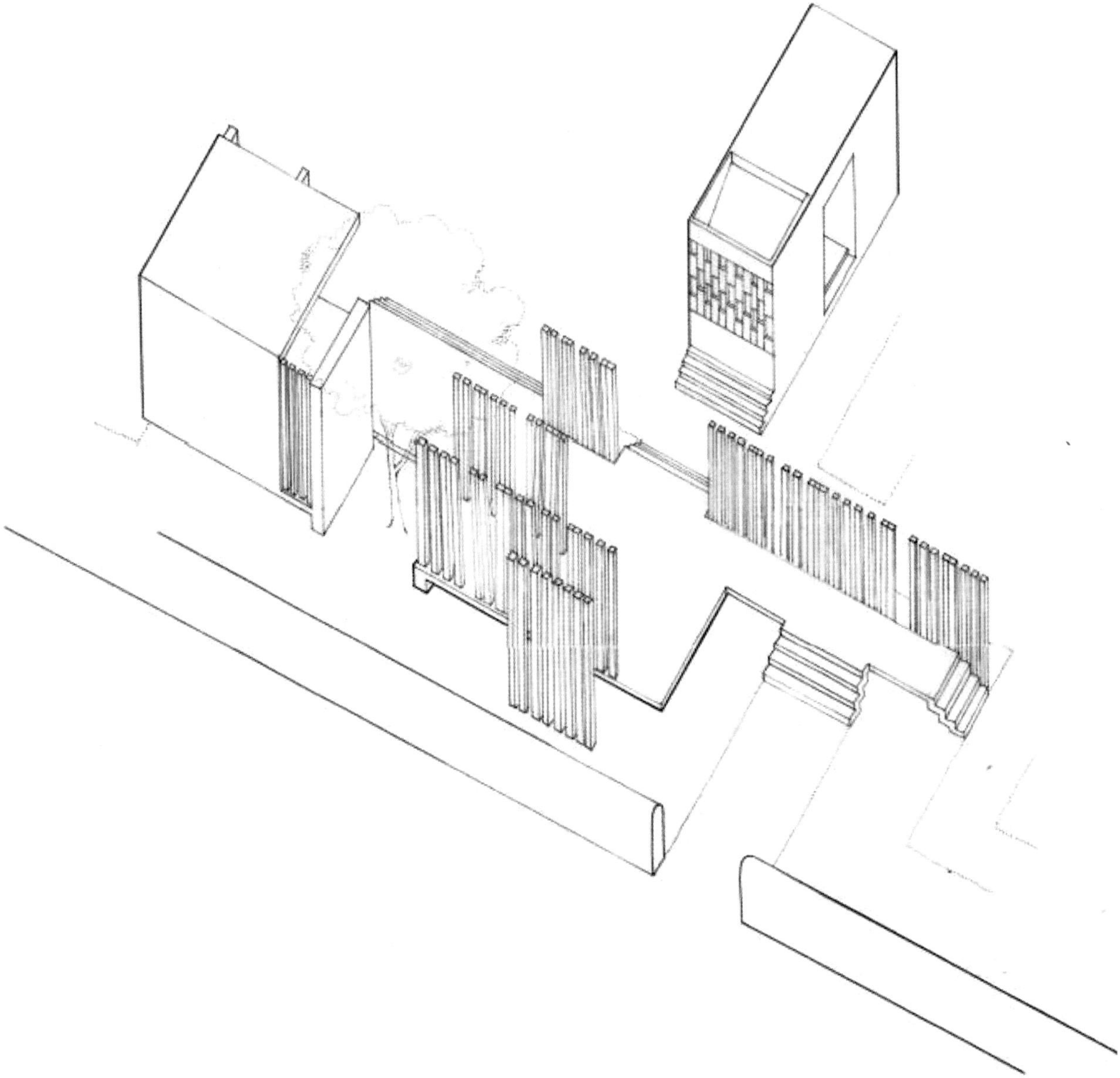

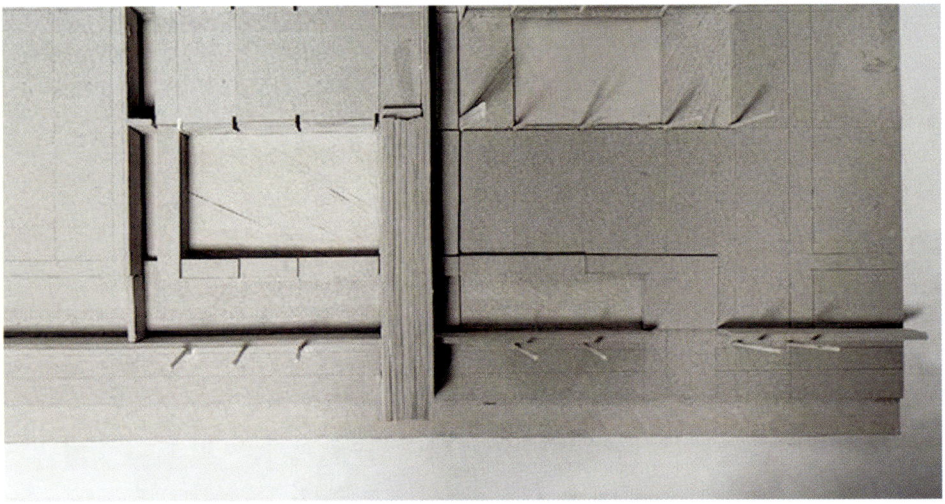

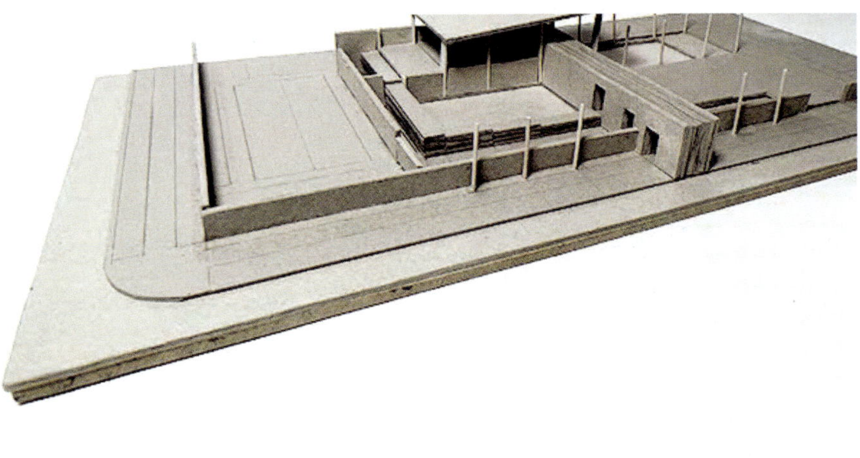

s> Kristina Hoang **i>** Andrew Liles

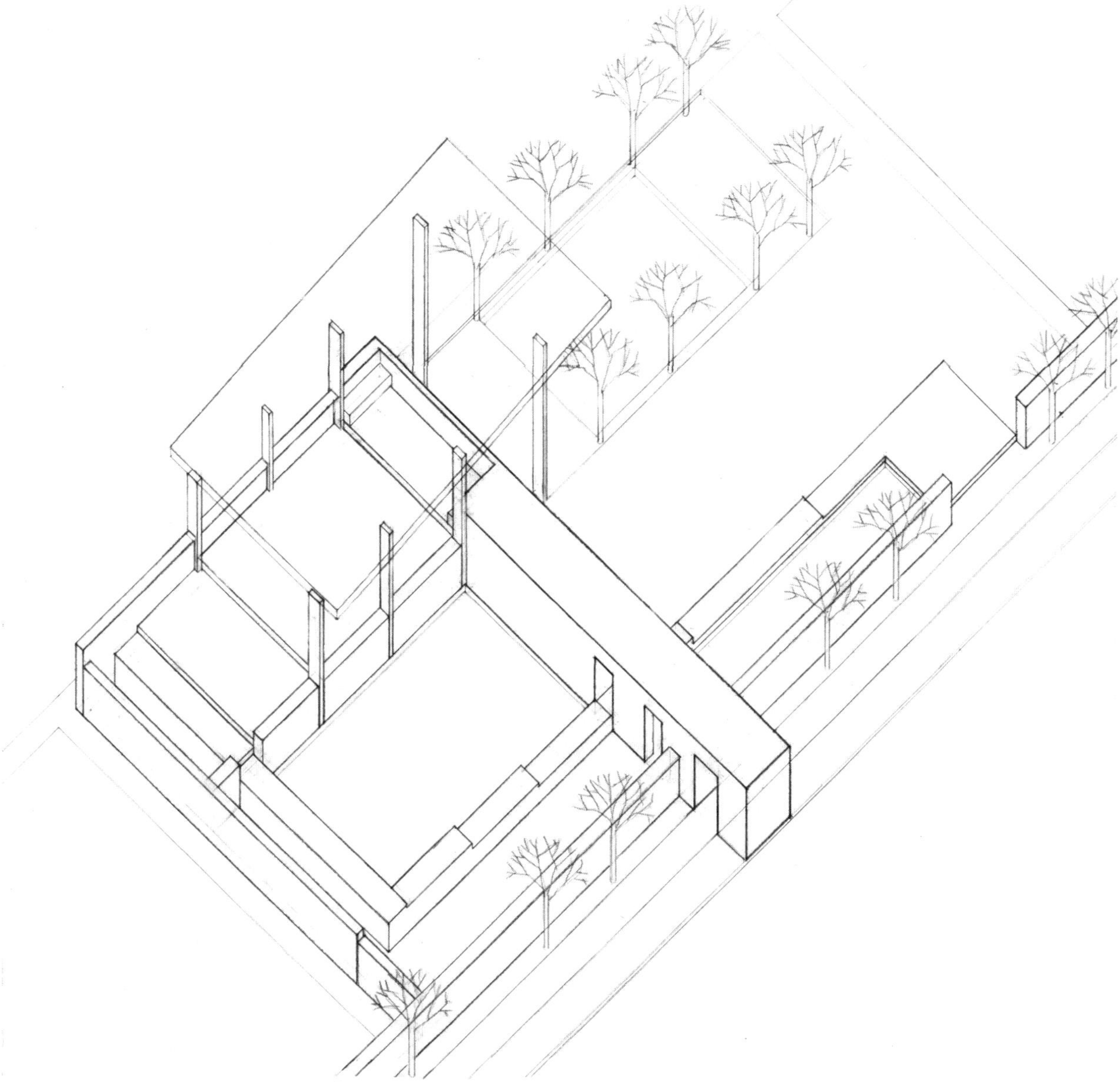

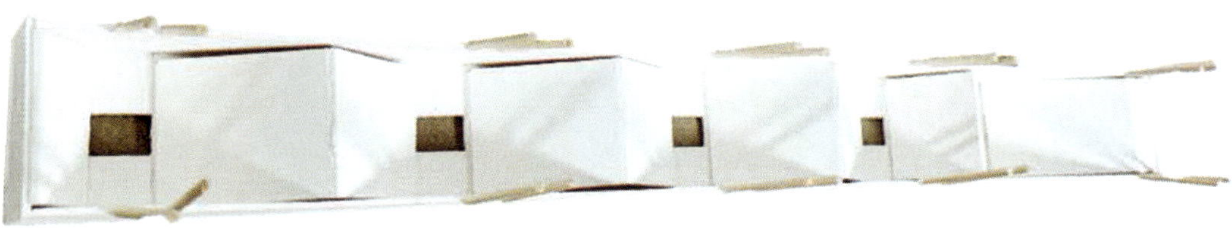

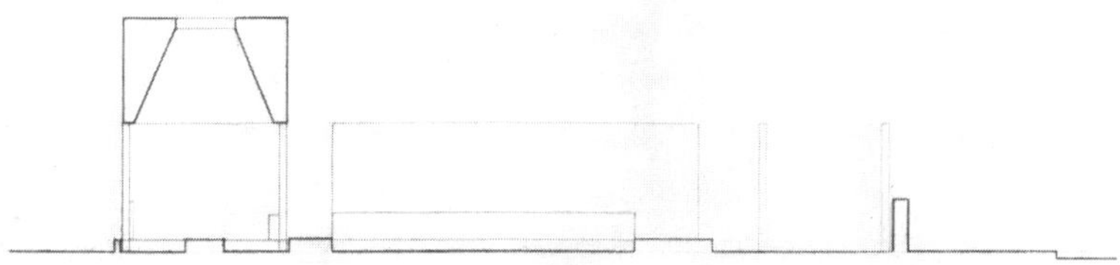

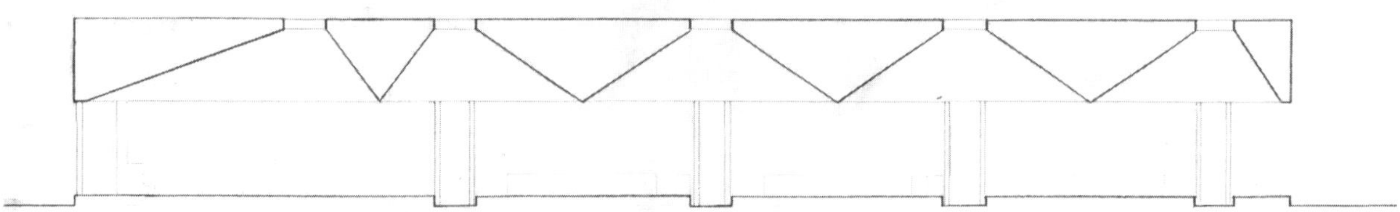

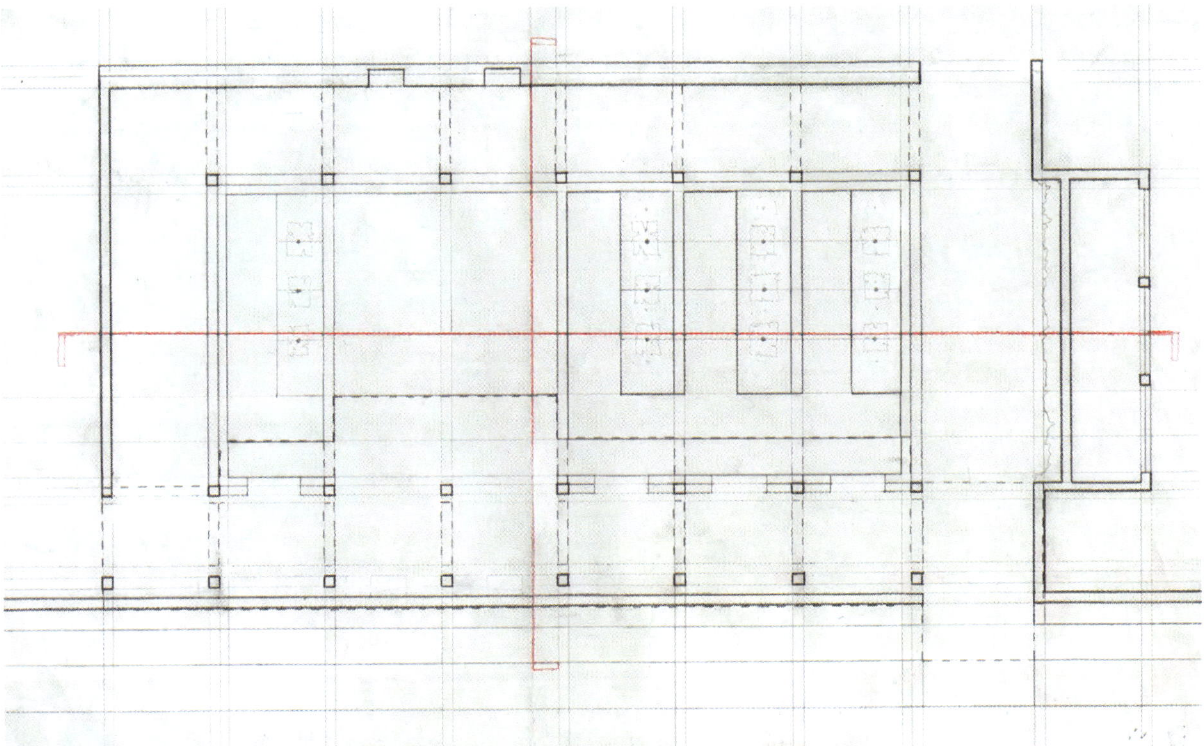

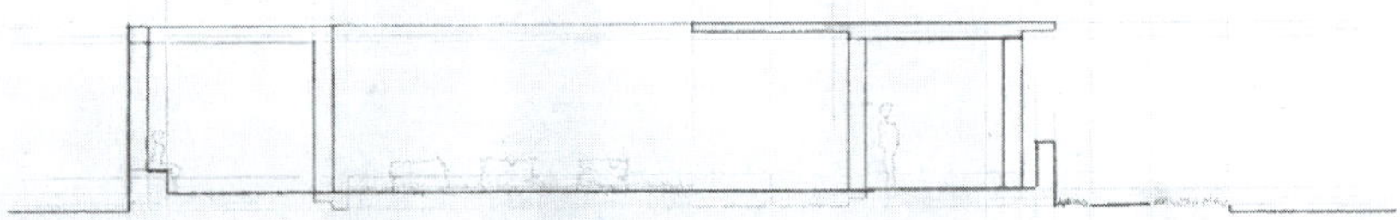

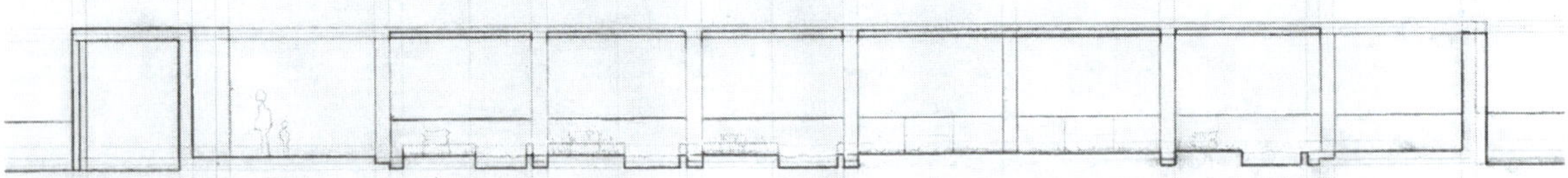

s> Nico Eckstein **i>** Judith Kinnard

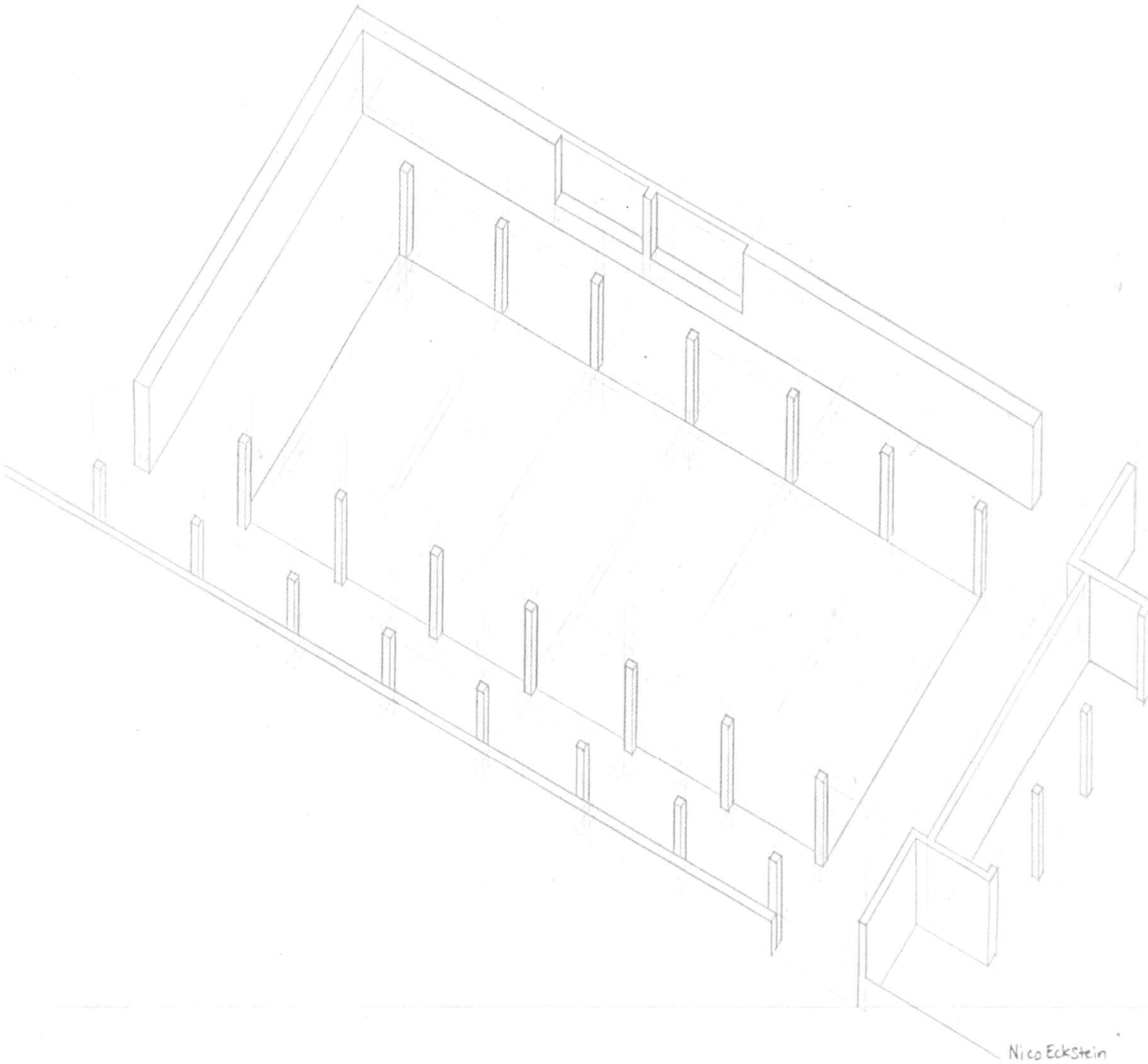

SP22
Adam Modesitt [C]
Omar Ali
Nimet Anwar
Andrew Liles
Wendy Redfield
Edson Cabalfin
Hannah Berryhill

SP23
Wendy Redfield [C]
Nimet Anwar [C]
Ira Concepcion
Charles Jones
Andrew Liles

SPACE

Core studio 1012 explores fundamental methods and principles of architectural design, pushing students to conceive, analyze, and represent architectural form and space. The course introduces various forms of architectural representation and discipline-specific terminology, enabling students to communicate ideas with increasing skill and precision both visually and verbally. Students develop critical thinking skills through project-based learning and experimentation.

The role of form in architecture and design is a key component of the studio. Form is an expansive term that involves shape, configuration, arrangement, manifestation, type, operative sequence, and performance state. In architecture and design, form is closely related to geometry through position, orientation, size, proportion, and pattern. This studio explores the role of form in the construction of objects and space, while also considering form in drawings and images.

The semester pedagogy is intended to advance representational skills, introduce new tools for managing complexity, and expand design vocabularies. Over the course of the semester, projects scale up progressively in size, intricacy, and complexity. "Scaling up" provides an entry point for exploring fundamental architectural problems such as tolerance, fidelity, and resolution.

Computer-aided design (CAD) software is introduced as a new tool for drawing and modeling, with Rhino as the primary tool for almost all work. Greater proficiency with Rhino expands design possibilities, reduces production time, and promotes a deeper understanding of the role of form in architecture and design.

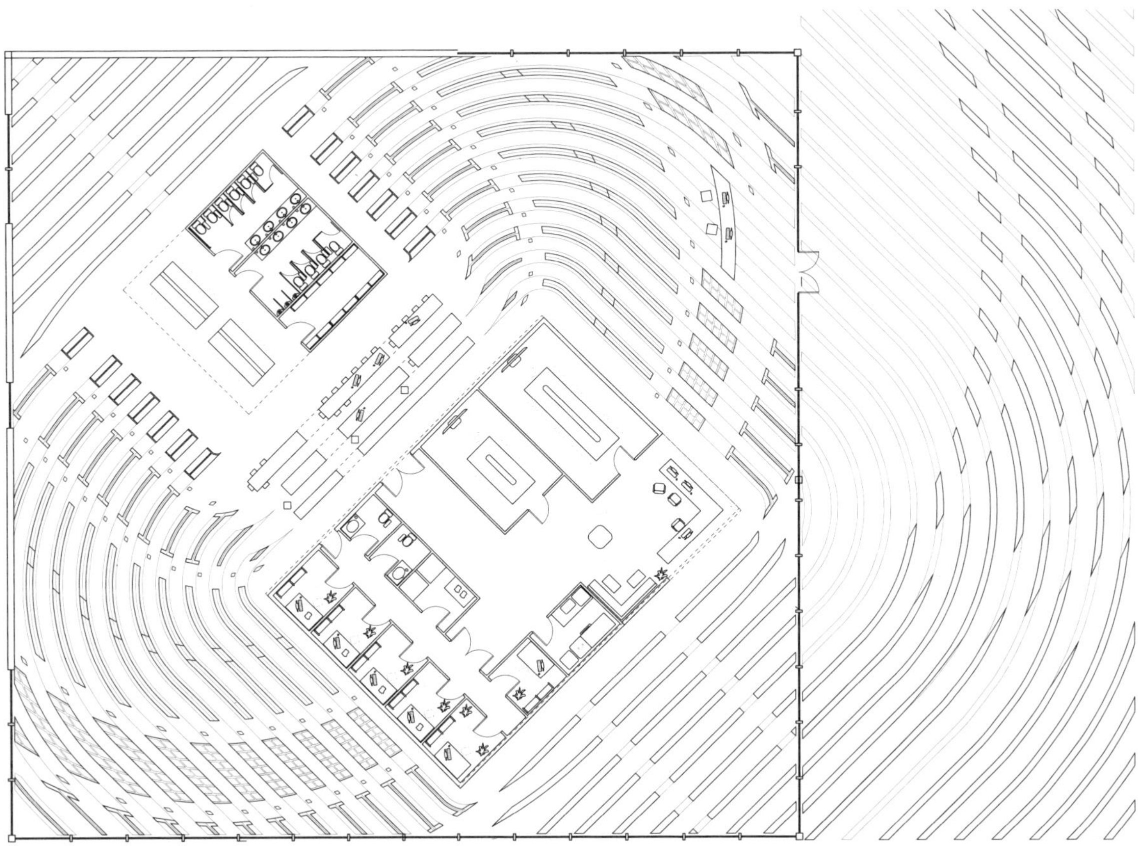

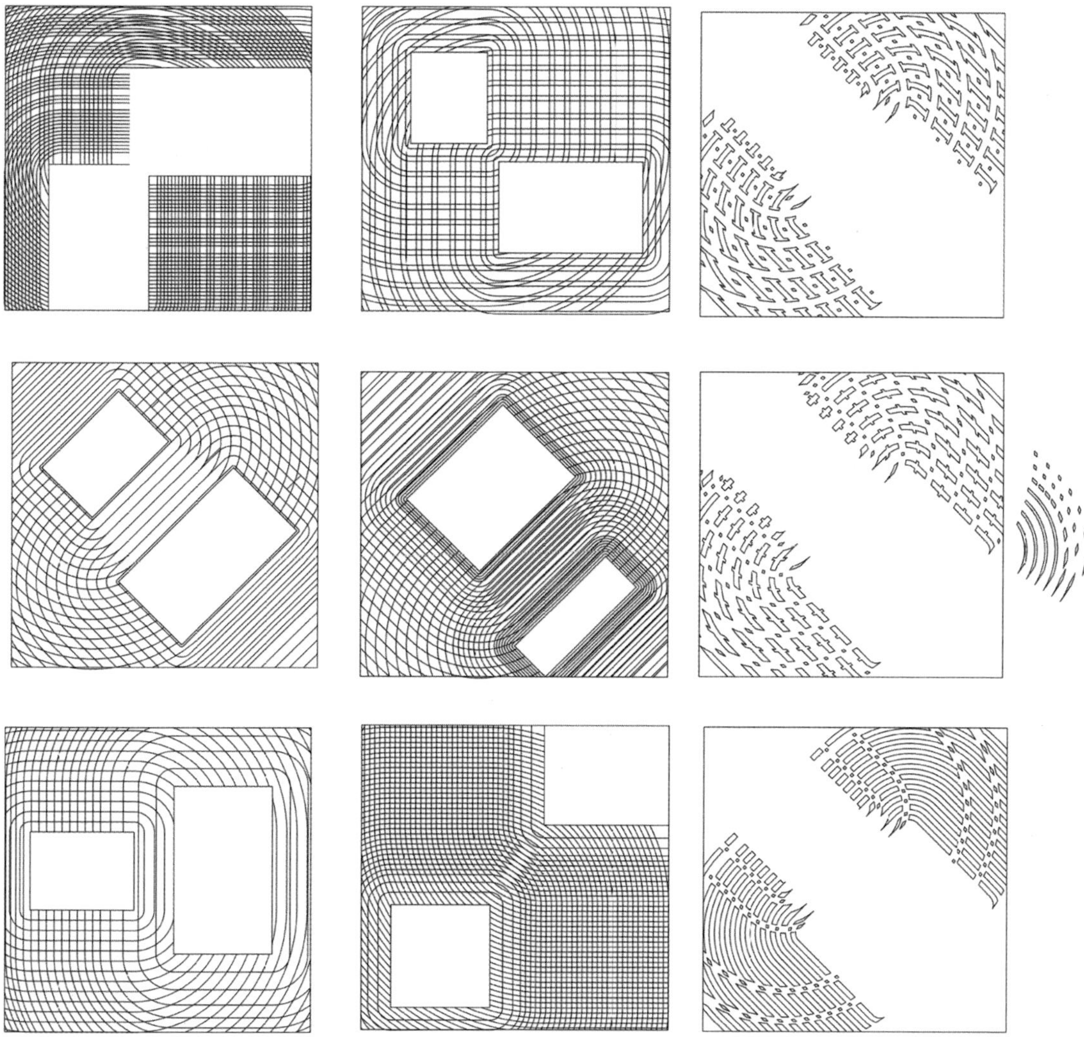

UG

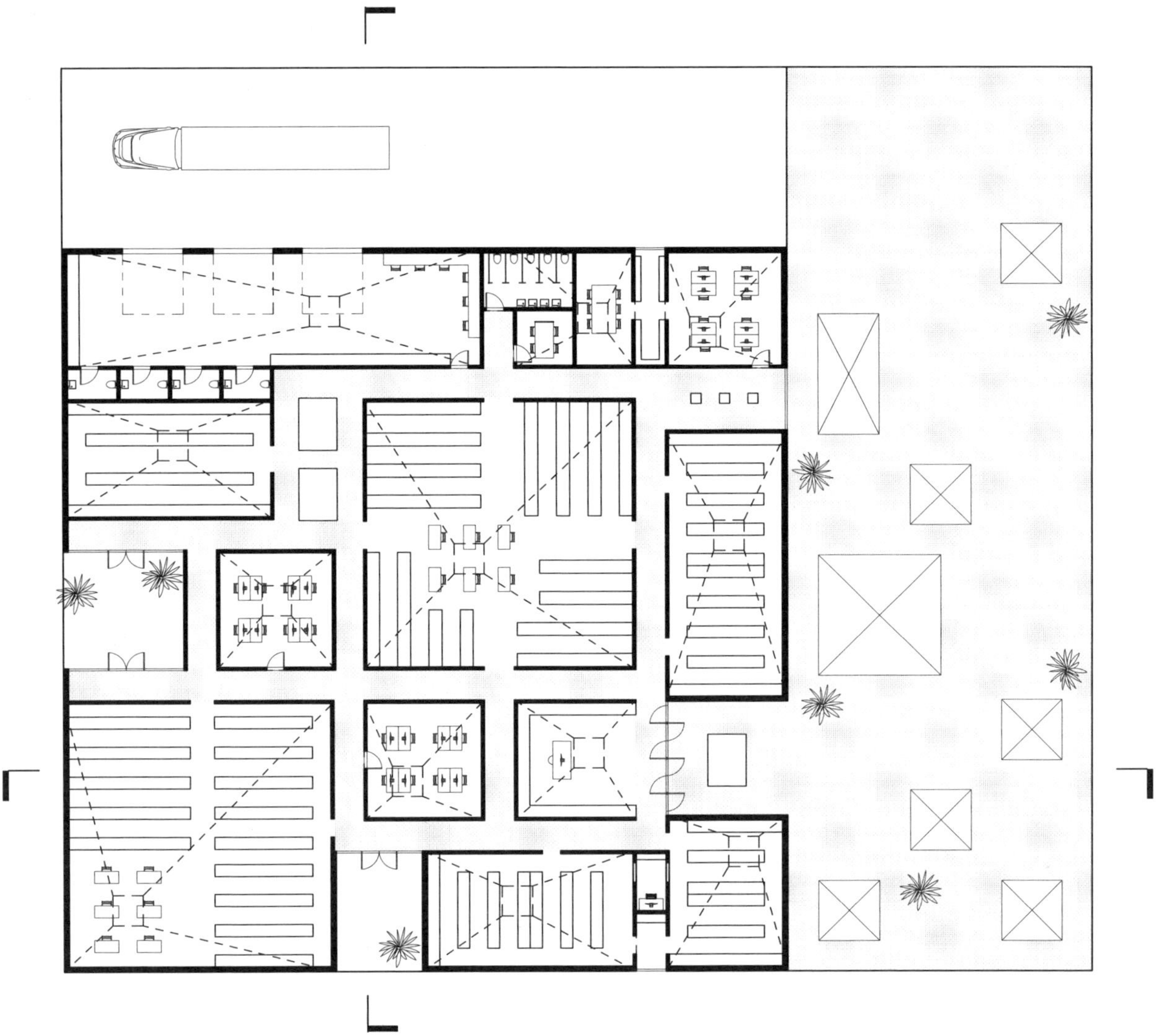

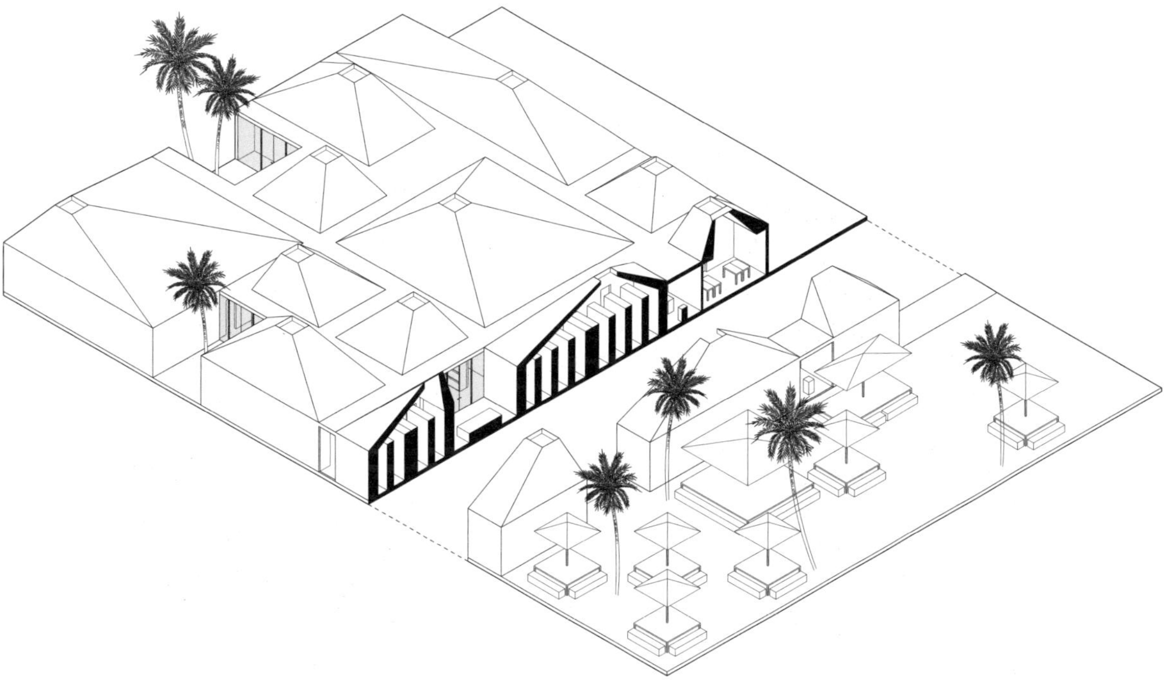

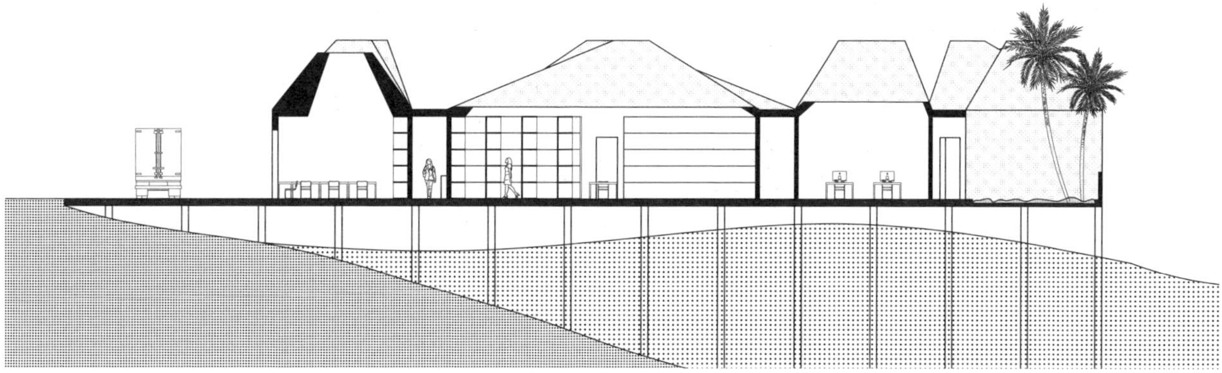

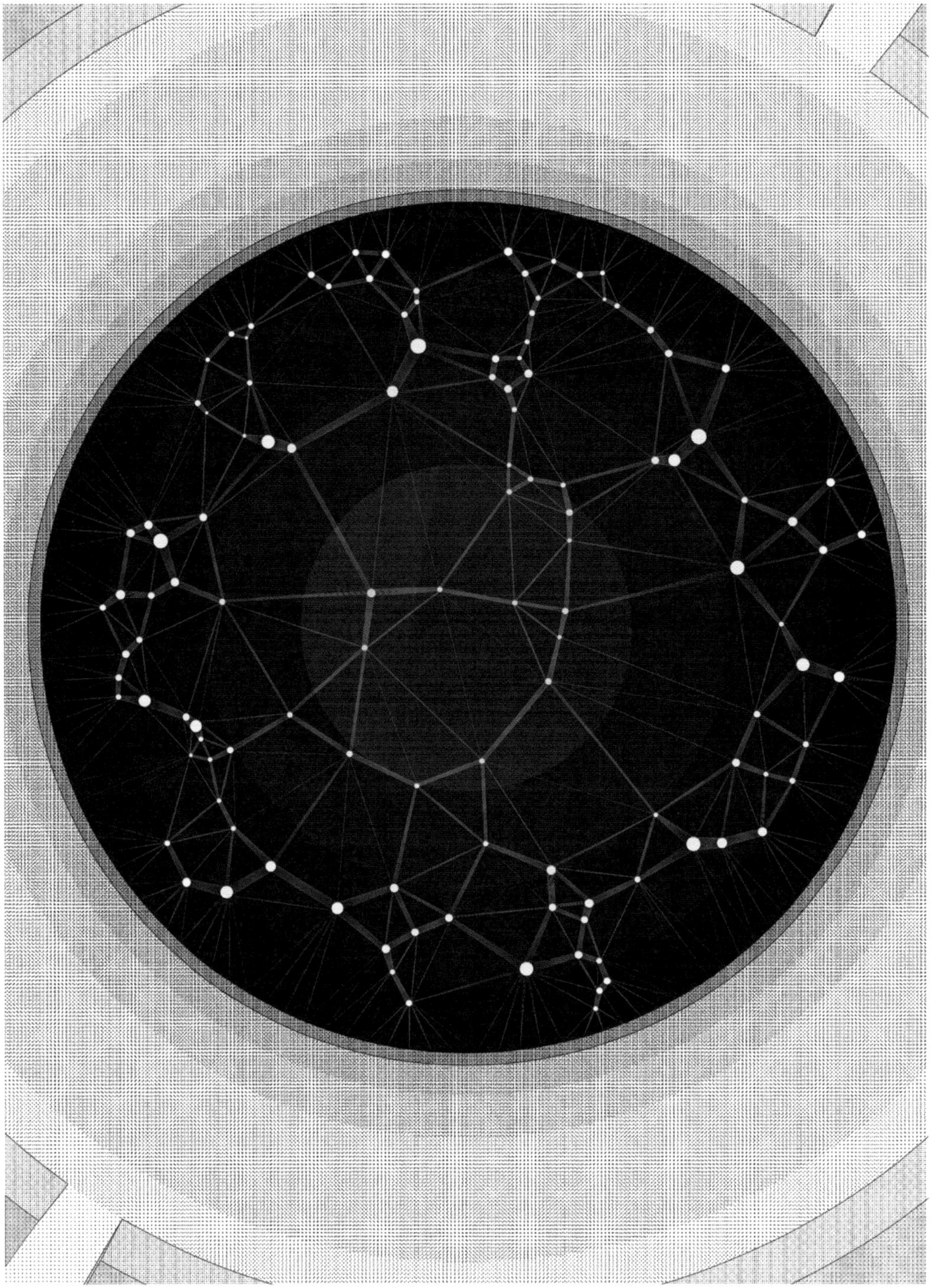

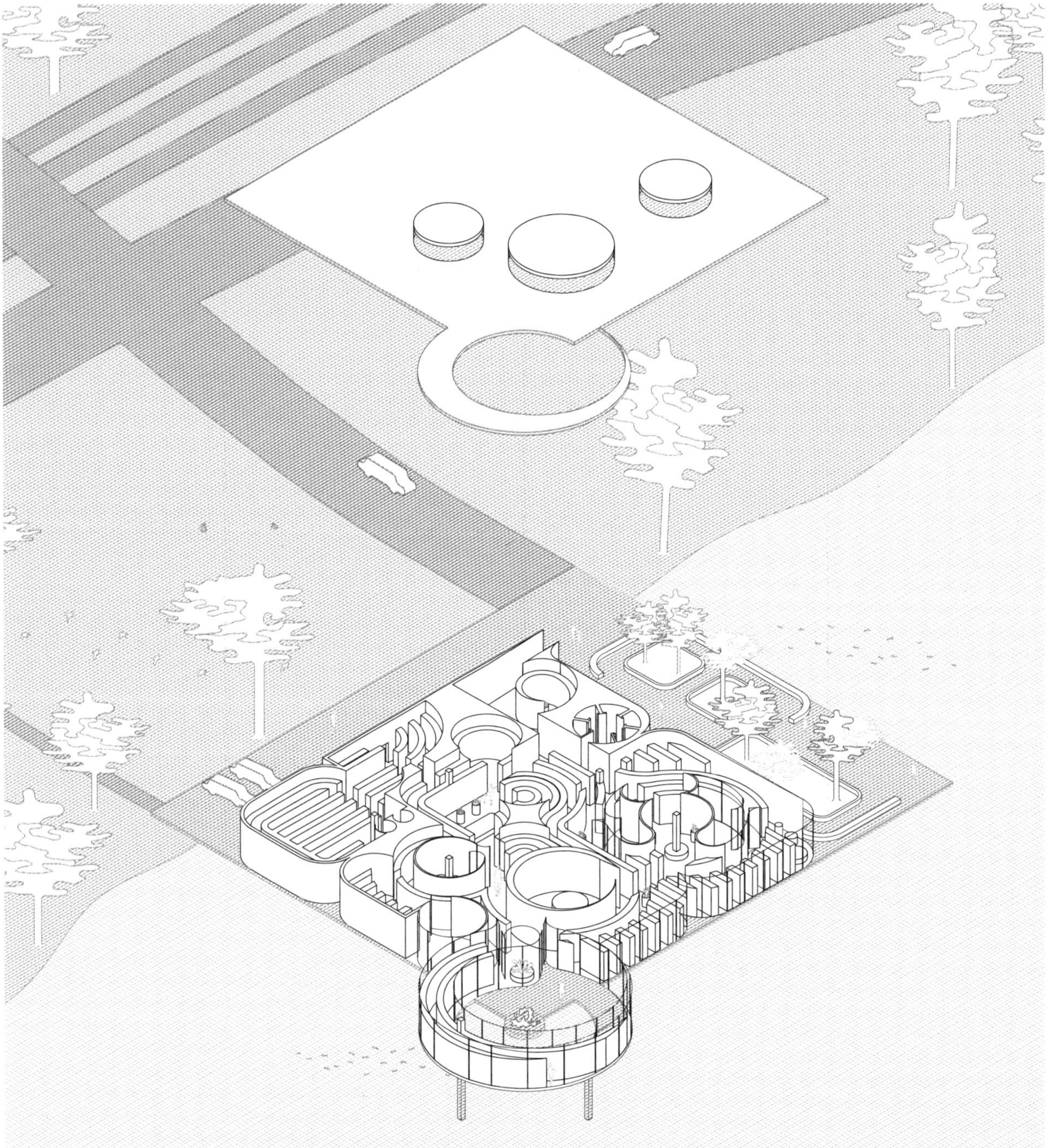

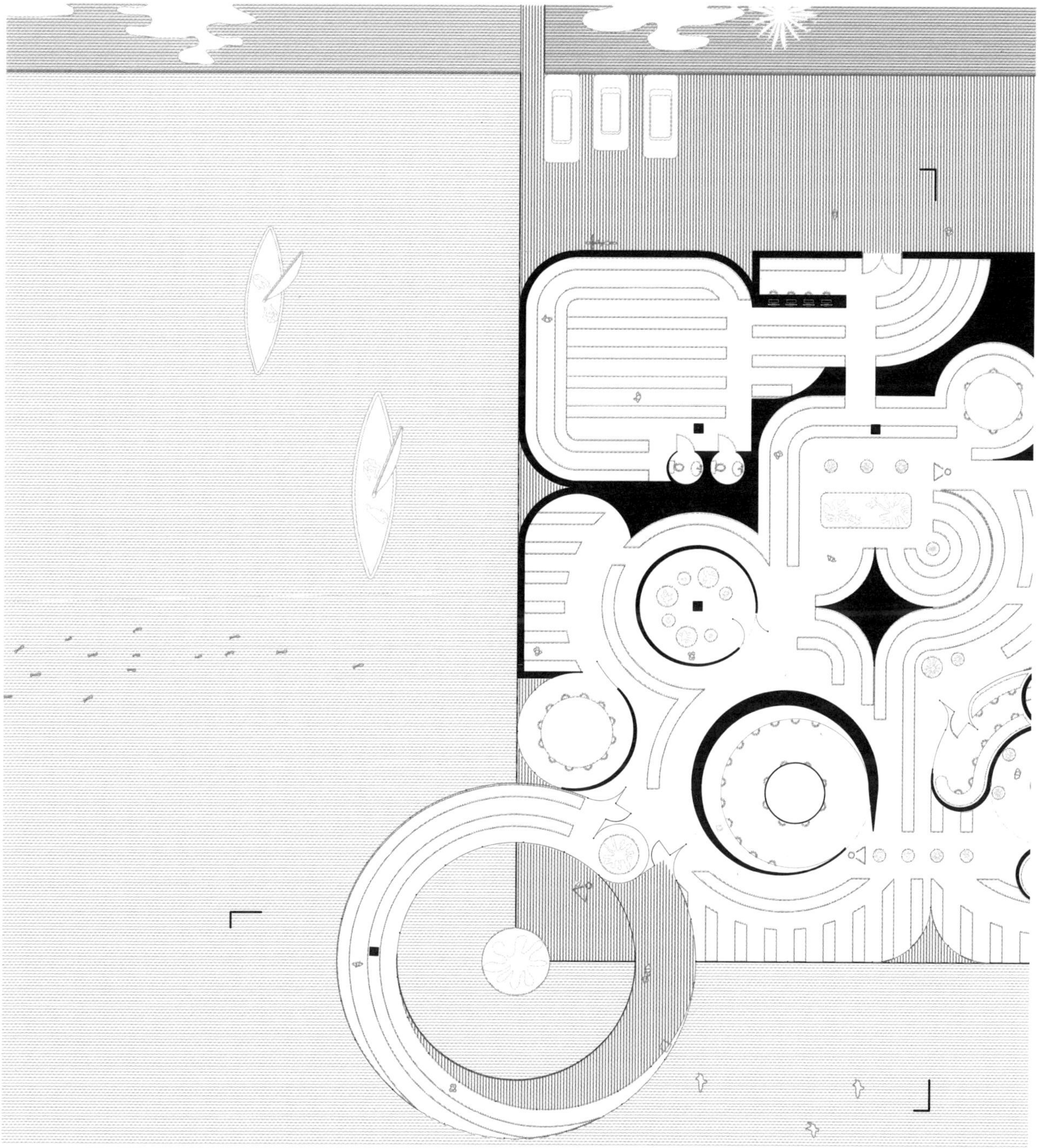

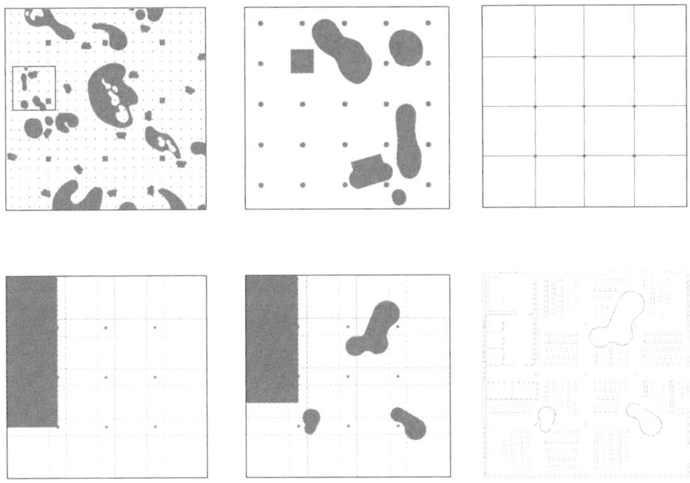

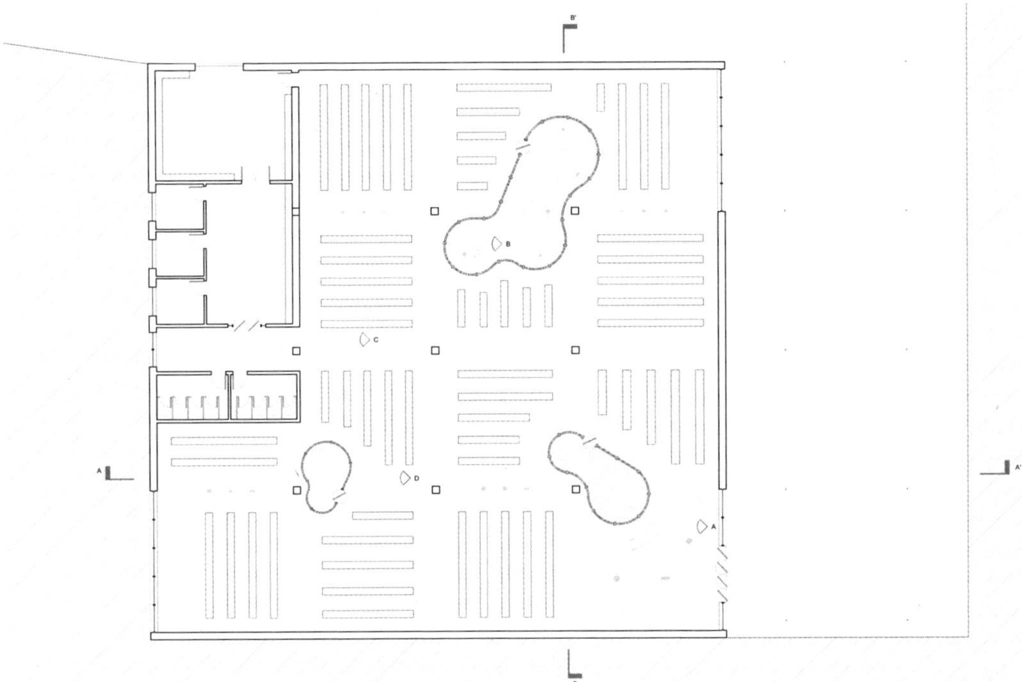

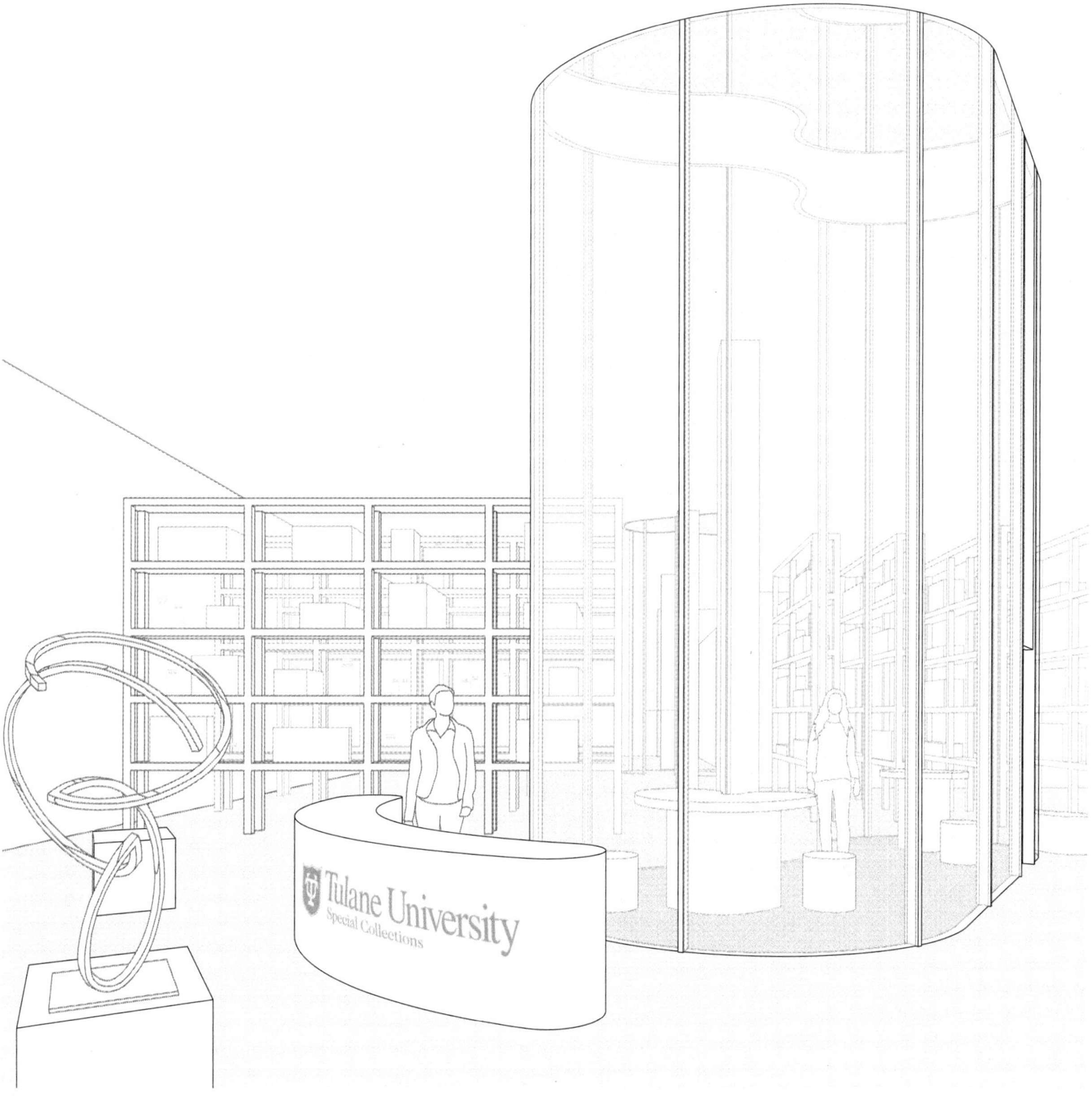
Tulane University
Special Collections

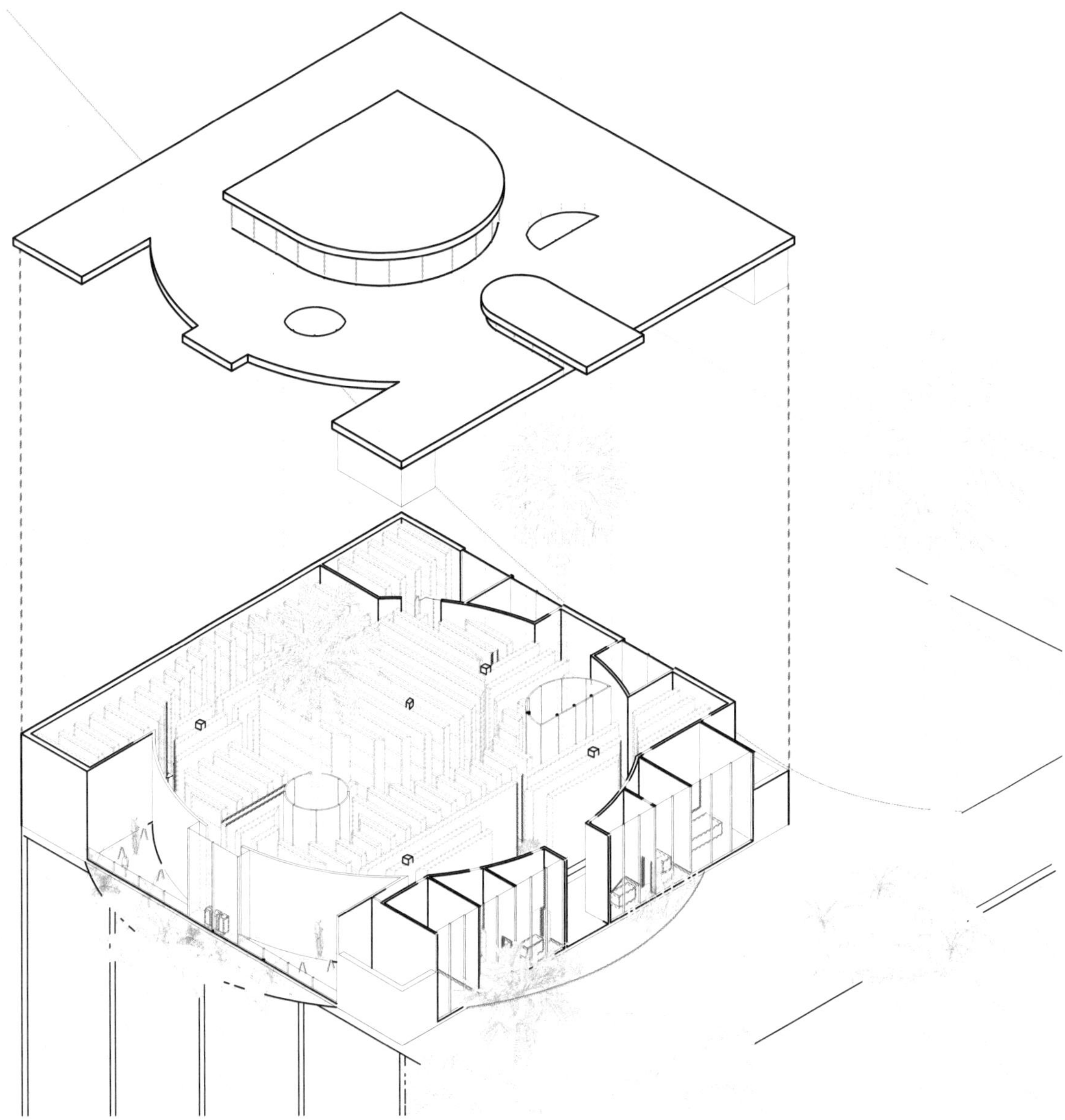

s> Molly Carroll

i> Nimet Anwar

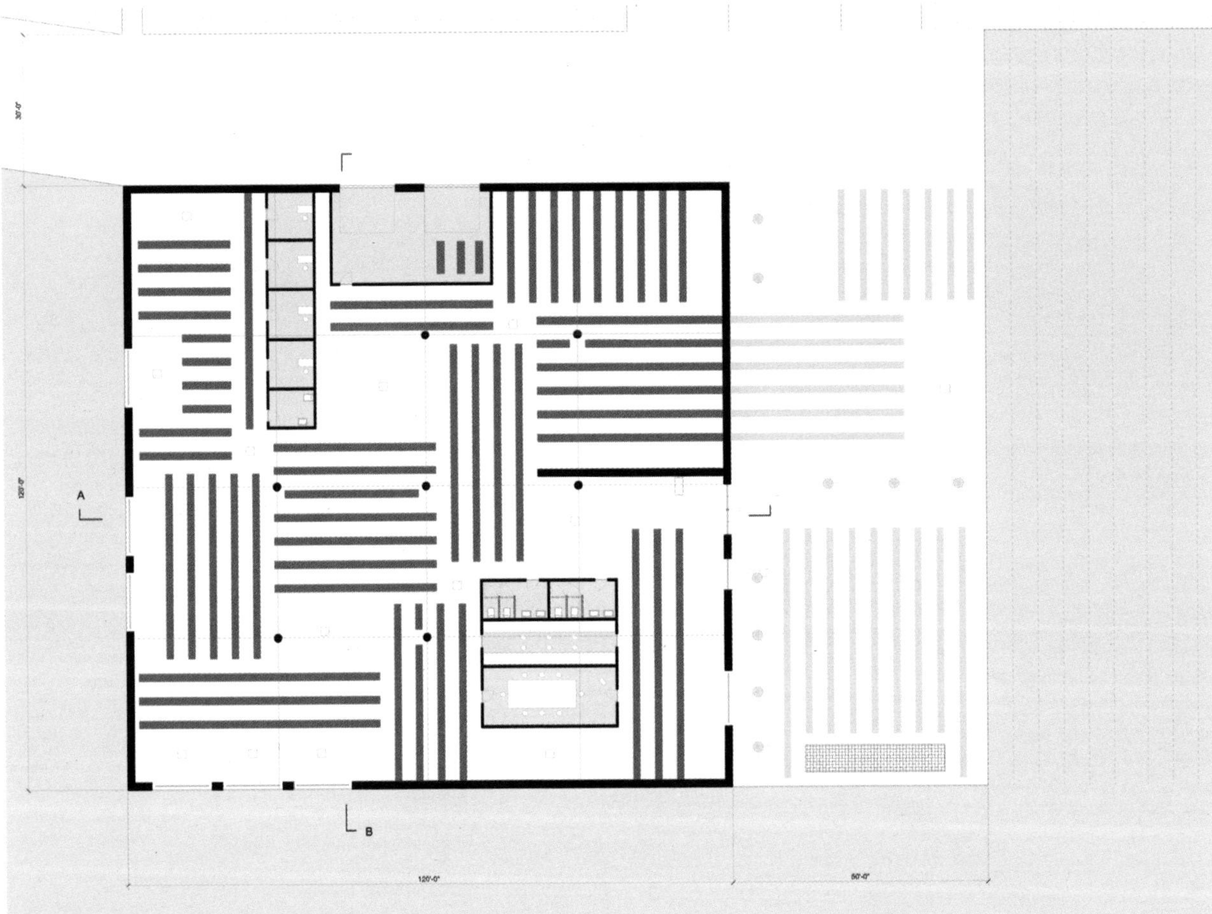
A
B

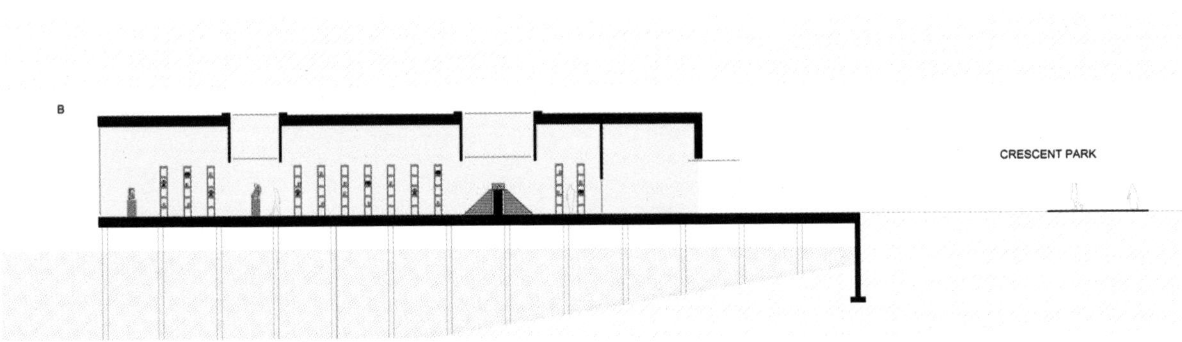
B
CRESCENT PARK

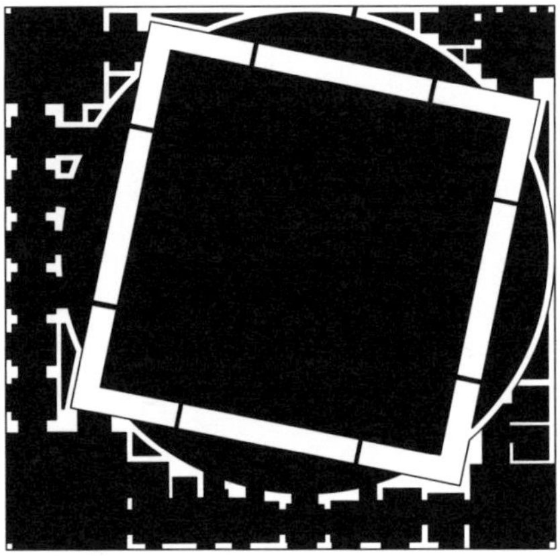

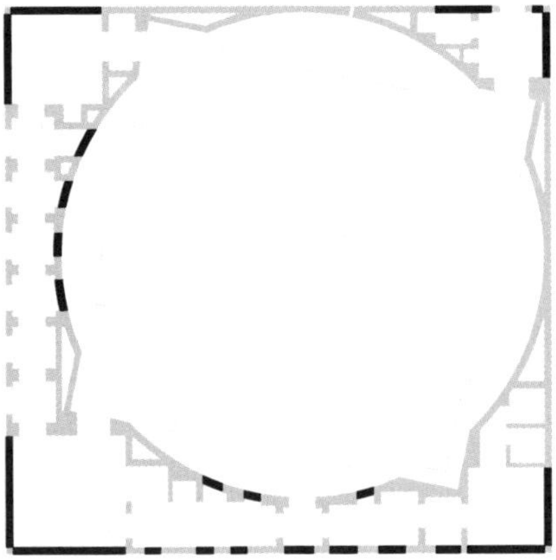

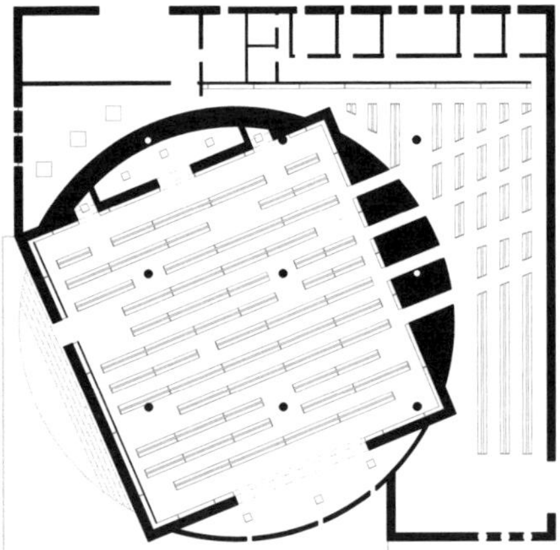

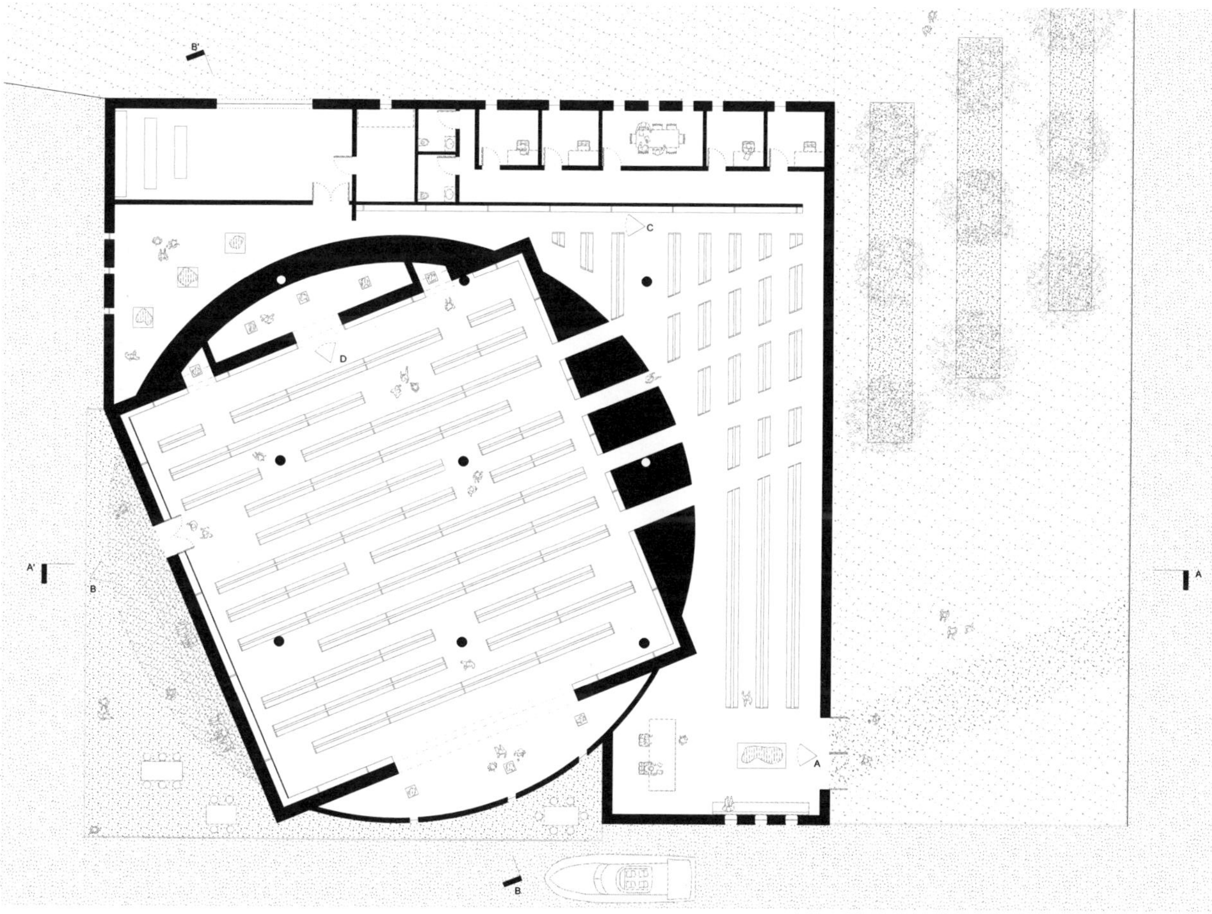
B'
C
D
A'
B
A
A
B

FROM CONCEPTS TO PROGRAMS

Andrea Bardon de Tena [C]

ARCH 6011, Culture, is the first half of a two-studio sequence for new graduate students enrolled in summer courses. The course aims to help new students build foundational skills of the design and graphic process and to learn the nature of studio work. This becomes the first step of the largest goal during their 3.5- year program: to become "visual thinkers" who transform their ideas into spatial organizations. Through the analysis and transformation of a case study, the students explore the relationship between concepts, spatial schemes, and programs. This process helps them understand the importance of clarity on the architectural scheme, by working with the "front of the house" and "back of the house" in different typologies: one public museum or gallery, and one private housing project.

This design studio serves as an introduction to architecture culture at various levels, helping students to acquire the basic working tools to approach the design of architectural proposals. During the first four intensive weeks, the methods of representation, the analysis of existing architectural projects, and the working culture of the studio are addressed. Starting from a case study, each student goes through a process of extraction, abstraction, exploration, relocation and reprogramming, to arrive at their proposal integrated into the urban fabric of New Orleans. The use of diagrams, catalogs, 2D orthogonal representation, axonometric projections and perspectives help them in this process while learning the main representation strategies.

The methodology encourages the class to develop their proposals using both general thinking, to work with clear ideas, and specific thinking, to contextualize them in the real context and needs. The case study transformation work is intended as a design methodology and an introduction to the main architects in history and their works and theory.

During the development of the course, students intersperse studio work with other complementary activities that are considered fundamental for their introduction to Architecture. They attend a digital media workshop to reinforce the learning of new computer programs, including Rhino, Photoshop, and InDesign. Weekly lectures from other university professors, visits to architectural studios, and Tulane's workshops and resources supplement the students' development. The goal of these activities is twofold: to help them integrate into Tulane and make use of the resources available to them, and to show them different career paths that exist in the field.

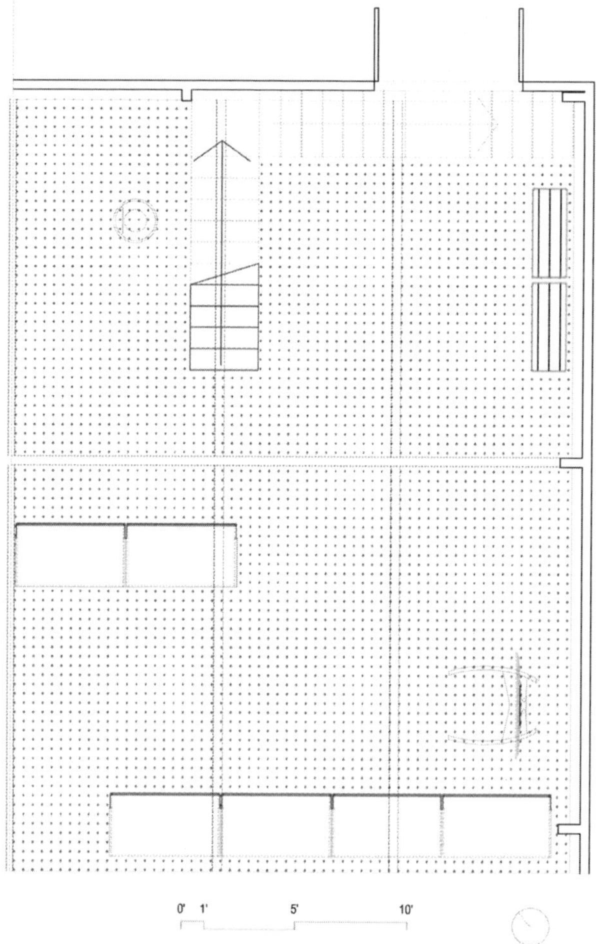
0' 1' 5' 10'

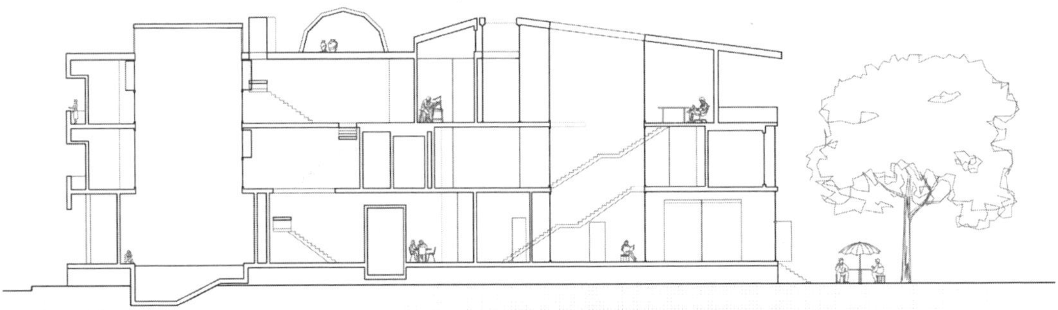

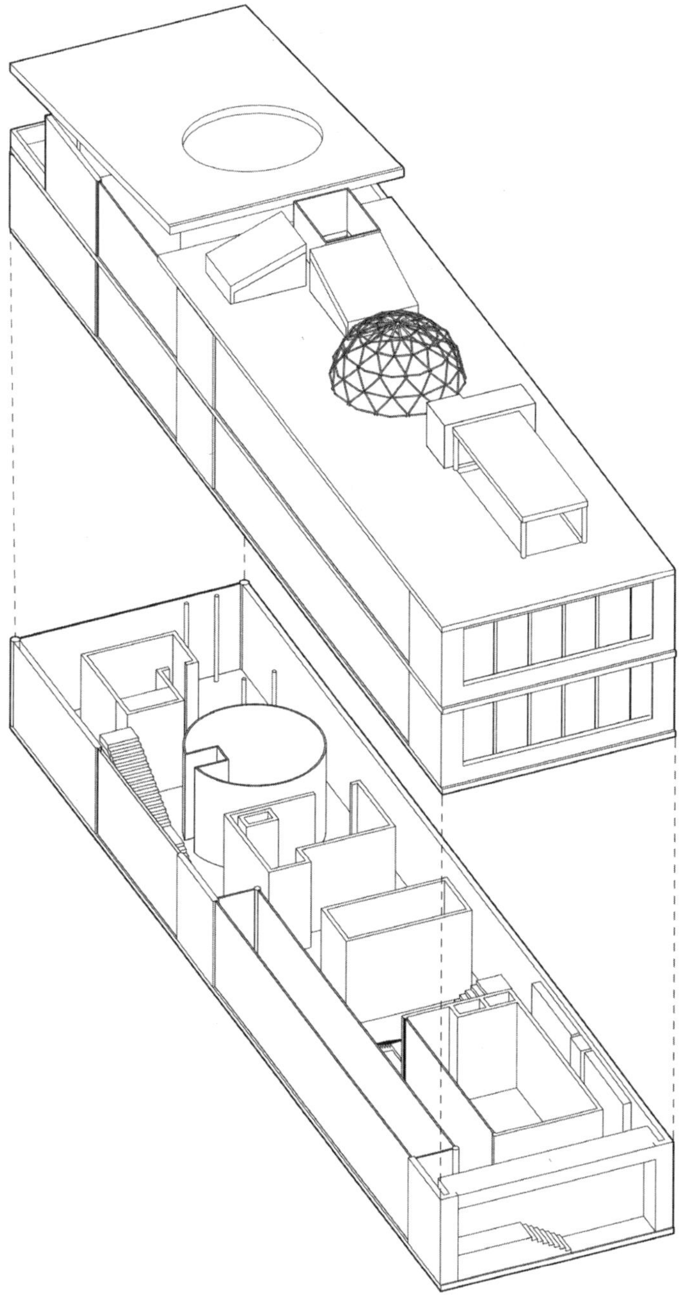

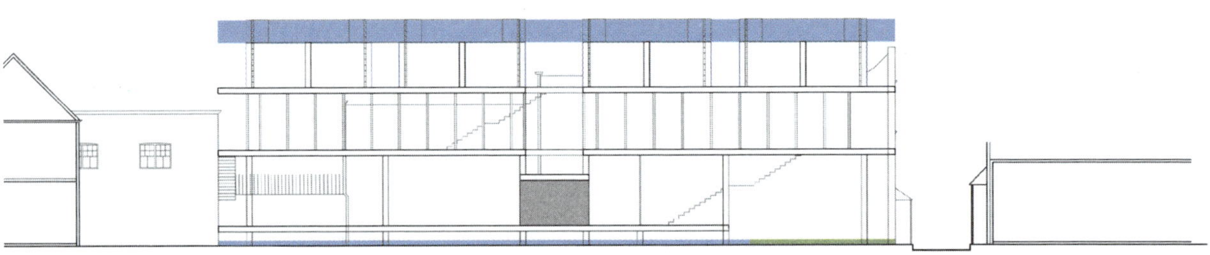

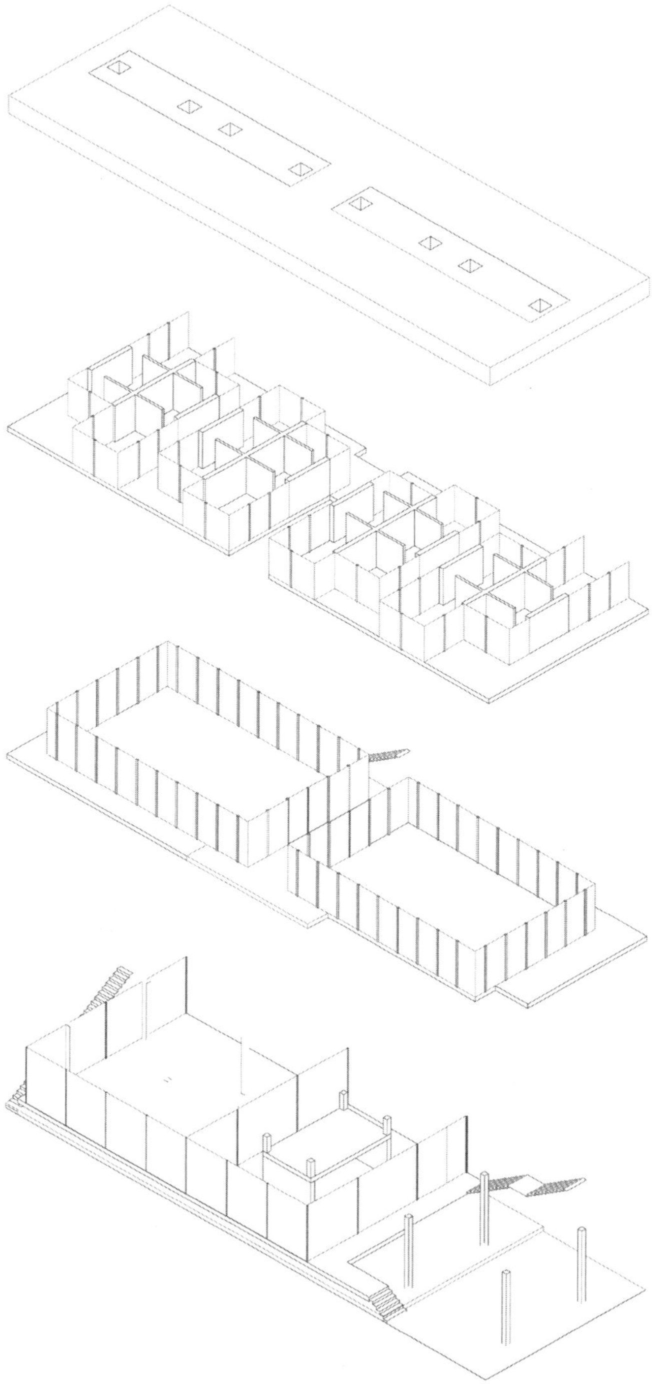

PROTOTYPING SOLUTIONS

Nick Perrin [C]

Prototyping Solutions is an upper-level course that draws upon foundational design skills from studio prerequisites to explore form-making through physical prototypes. Students begin the semester with analyzing an archetype (e.g, stool, table, sawhorse), then work to develop concepts, uses, and formal qualities to be tested on their own design. Students embark upon a semester-long iterative project that includes material studies, physical and digital modelmaking, and constructed fabrication drawings that inform individual projects at full scale.

Understanding design as inseparable from making is central to this course. The final builds are simultaneously studied work and objects for refinement. Throughout the course, the cohort examines a range of approaches to production, to better understand the process from ideation through execution. Students also produce a series of drawings and models to develop ideas and different levels of resolution. A "Portmanteau" serves as the object that the class considers and interprets through making, in three different ways. The work is supported by course readings and case studies, to bolster a research-based approach.

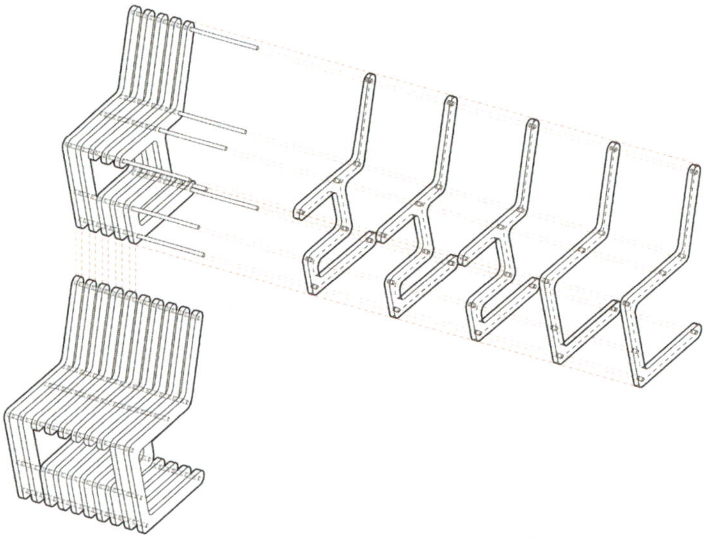

s> Lucy Salas
+ Georgia Cook

i> Nick Perrin

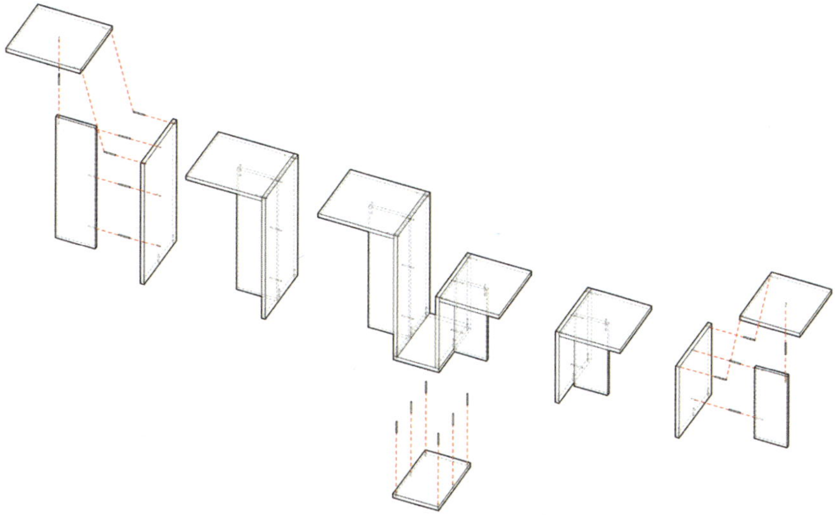

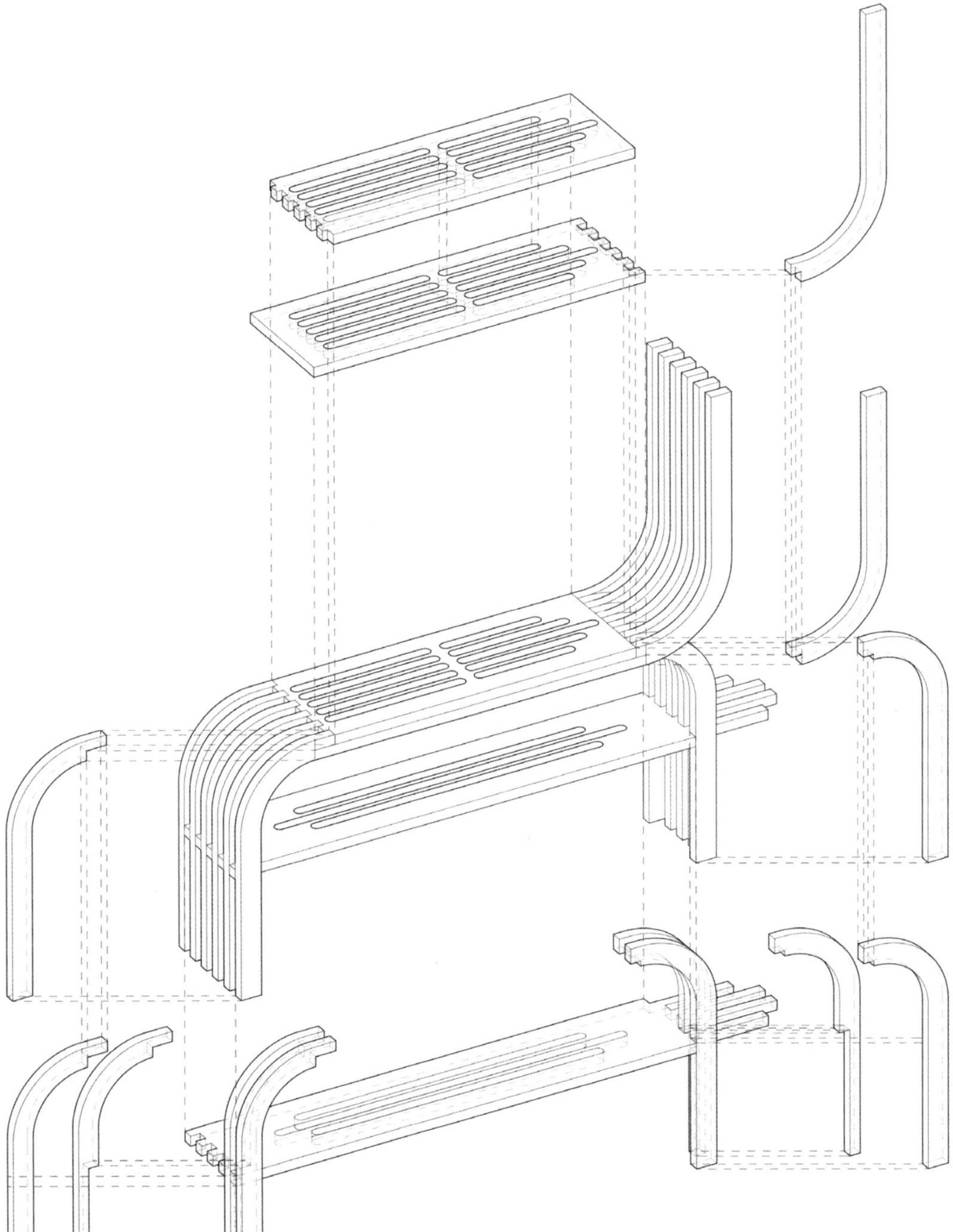

CONTEMPORARY ARCHITECTURE IN HISTORIC CONTEXTS

Ammar Eloueini [C]

This Research Studio focuses on contemporary architectural interventions in historic contexts. For its third year, the studio will focus on the entirety of Magazine Street from the edge of the French Quarter up to Audubon Park. Students will study the historic fabric and the new developments and work on a number of new proposals learning how to be deliberate when making decisions as participants and actors in the evolution of the identity of a place.

The specificity of a place is something that is constant over time, whereas the identity of a place is something that gradually evolves over in time. It is important for architects to be aware and deliberate when making decisions of this distinction as participants and actors in the evolution of the identity of a place.

Change is inevitable. The main question of this Research Studio is: How as architects we can think about the future of cities in a way that will preserve the character of a place while responding to urgent needs for transformation and densification? What are the ways in which architects can work on public spaces and the built environment, take into consideration the cultural heritage of a place while innovating with design?

s > Olivia Georgakopoulos + Camille Kreisel **i >** Ammar Eloueini

s > Olivia Georgakopoulos + Camille Kreisel **i >** Ammar Eloueini

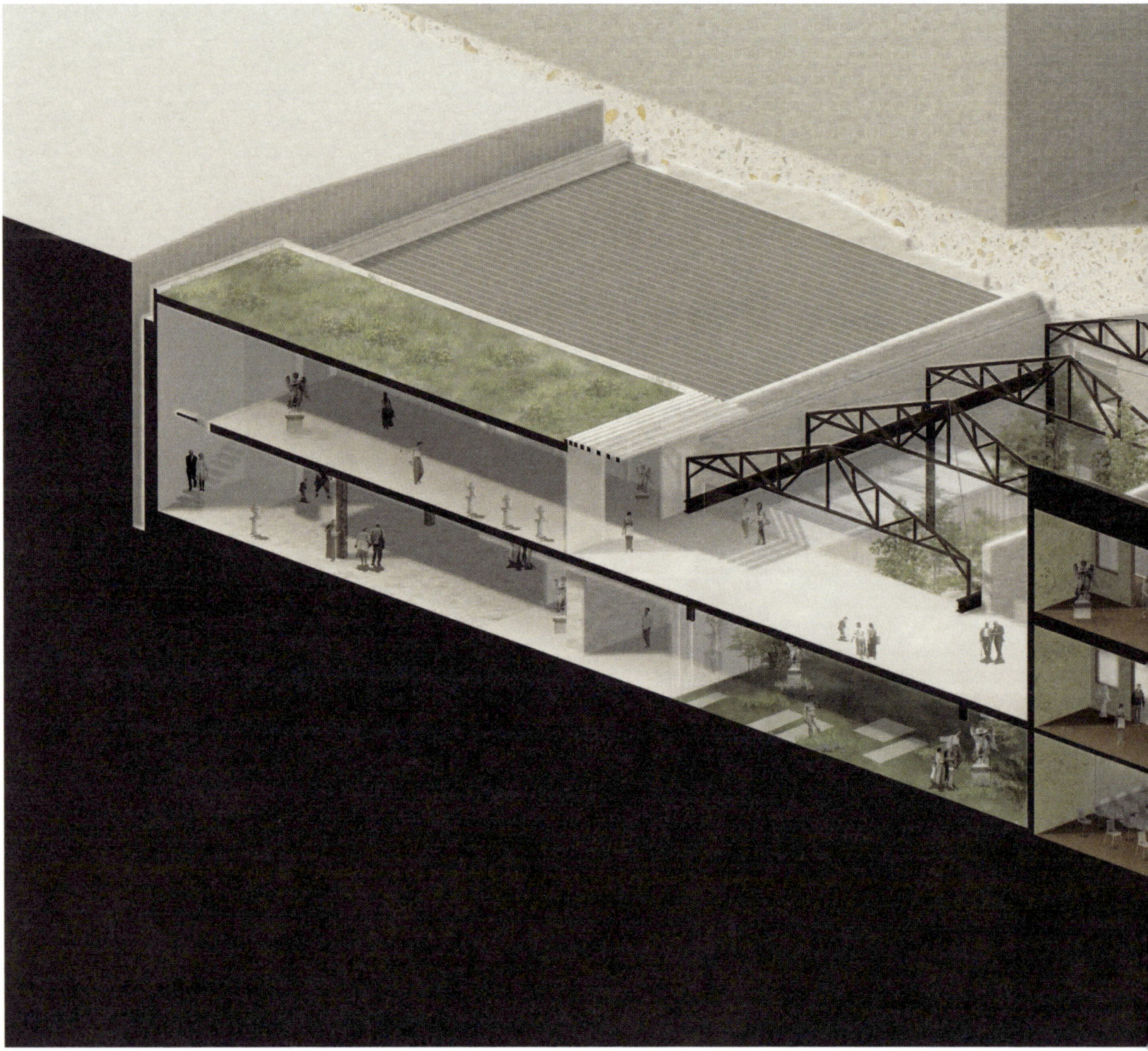

s > Olivia Georgakopoulos + Camille Kreisel   **i >** Ammar Eloueini

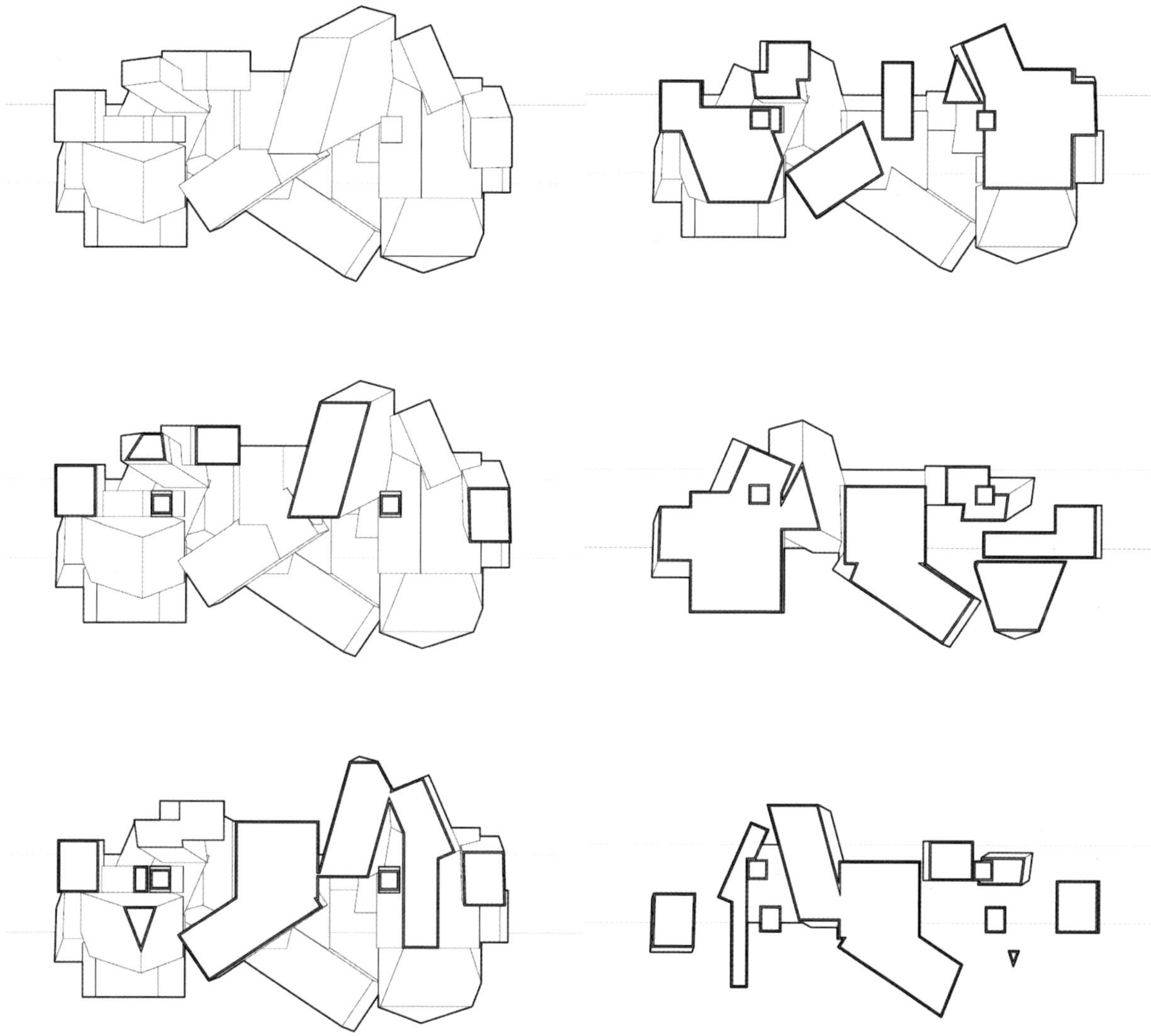

SUSTAINABILITY IN THE TROPICS. FCAT ECUADOR

Edson G. Cabalfin [C]
Anna Yoachim
Nick Jenisch
Emilie Taylor Welty

What is sustainable design? The Mintz Global Research Studio for undergraduate and graduate architecture students aims to consider this question in the context of the tropical Andes in partnership with FCAT (Fundación para la Conservación de los Andes Tropicales), a non-profit based in Ecuador dedicated to the conservation of tropical biodiversity. Through investigating materiality and building function, community input, and the environmental importance of the region, the studio aims to create a site-specific sustainable design framework and applications as well as offer insight into sustainability in the region. The studio was guided by principles of sustainability and questions of community-engaged design within the context of decolonization. Local materials and new technologies were also explored in adapting construction to the tropics.

Throughout the semester, the class worked closely with FCAT staff and stakeholders, researchers and experts. The class asked the question: How might we design different possibilities addressing FCAT Ecuador's needs in the present and the desire to grow in the future? Discussions and feedback sessions were held at multiple points during the semester. By midterm, the class developed multiple options under four building types: housing, field station, commissary, multi-purpose hall/laboratory. During Spring break, the class visited the site of FCAT's forest reservation the Choco area, in Cube, Esmeralda Province as well as the capital city of Quito in Ecuador. By the end of the semester, the class presented consolidated and revised versions of the four building types within a planned masterplan for the site.

The studio explored and developed four building type projects. These are:

1. **Housing** (Amanda Bond, Chelsea Kilgore, Tom O'Brien, William Trotter): The project proposed building additional new housing for the FCATeros or the on-site staff, including the possibility of accommodating guests and visitors.

2. **Commissary** (Katherine Dunn, Lauren Liroff, Naomi Smith): The new two-story commissary facilities is proposed as an expansion of the existing dining and kitchen area close to the dormitory area.

3. **Multi-purpose Hall and Laboratories** (Mario Das, Simone Golder, Kaylan Mitchell). Envisioned as the welcome area for visitors and guests, this new two-story building is being connected to the existing commissary building while providing new social spaces, storage areas, and a library.

4. **Field Station** (Anna Kathryn Becker, Sarie Keller, Olivia Vercruysse): The proposed field station is intended to provide short-term housing facilities or long-term research stations relatively far from the main compound.

s > Housing: Amanda Bond + Chelsea Kilgore + Tom O'Brien + William Trotter

i > Edson Cabalfin

s > Housing: Amanda Bond + Chelsea Kilgore + Tom O'Brien + William Trotter

i > Edson Cabalfin

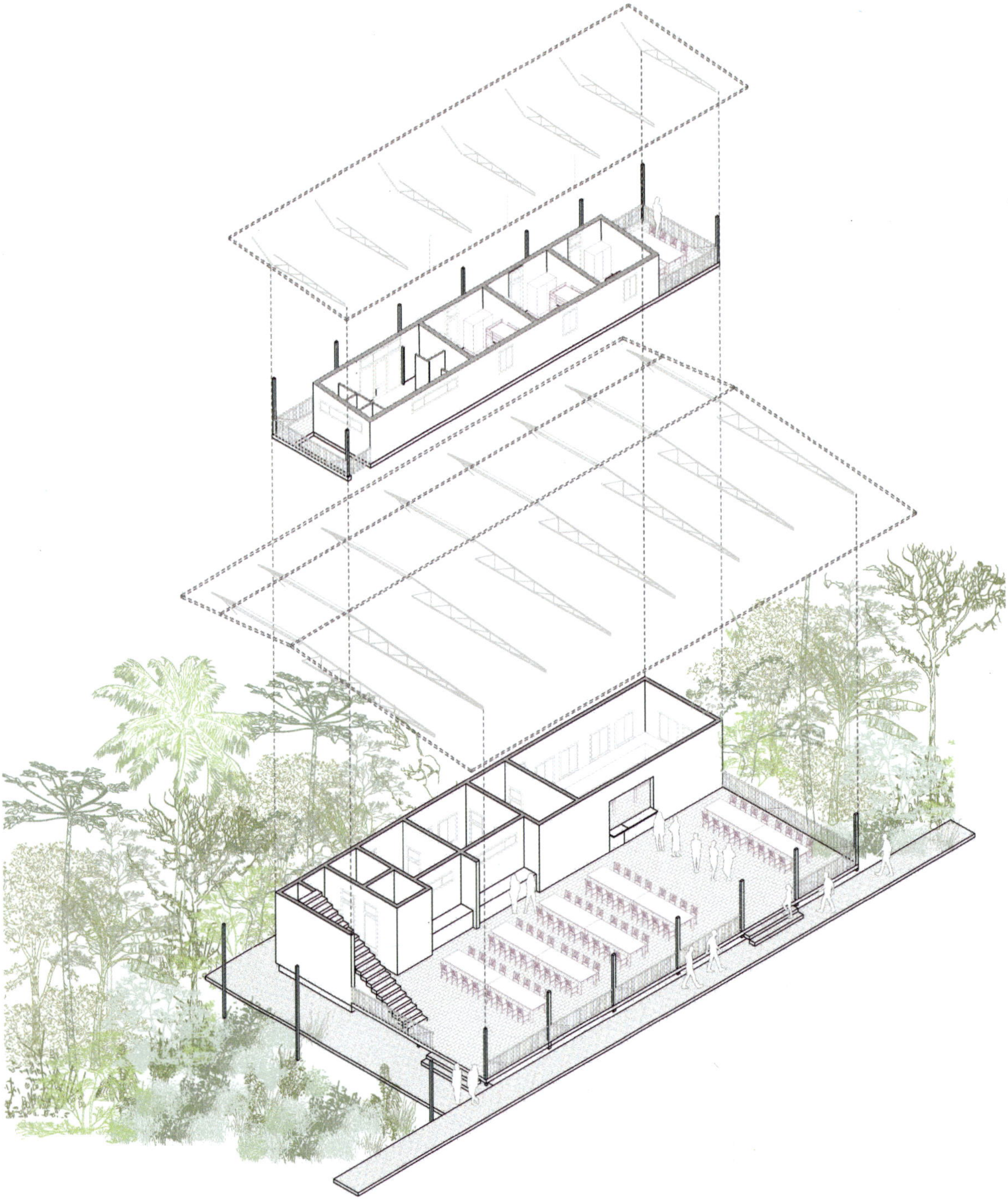

s > Commissary team: Katherine Dunn + Lauren Liroff + Naomi Smith

i > Edson Cabalfin

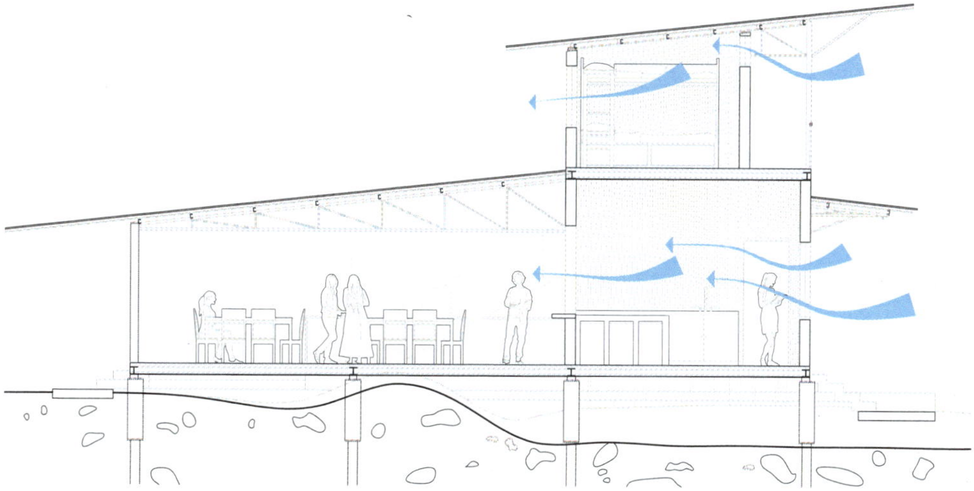

s > Multi-purpose Hall and Laboratories:
Mario Das + Simone Golder + Kaylan Mitchell

i > Edson Cabalfin

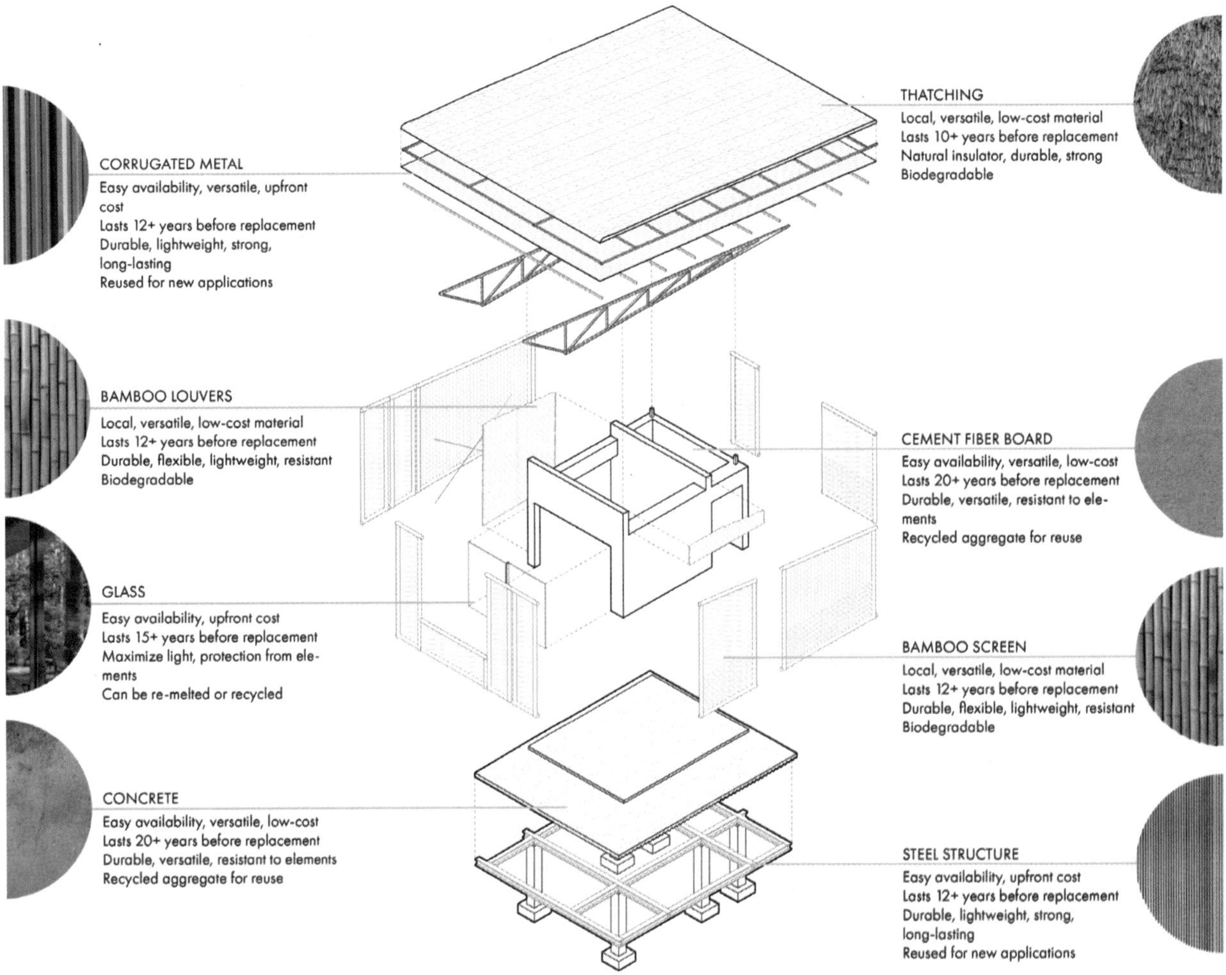
CORRUGATED METAL
Easy availability, versatile, upfront cost
Lasts 12+ years before replacement
Durable, lightweight, strong, long-lasting
Reused for new applications
THATCHING
Local, versatile, low-cost material
Lasts 10+ years before replacement
Natural insulator, durable, strong
Biodegradable
BAMBOO LOUVERS
Local, versatile, low-cost material
Lasts 12+ years before replacement
Durable, flexible, lightweight, resistant
Biodegradable
CEMENT FIBER BOARD
Easy availability, versatile, low-cost
Lasts 20+ years before replacement
Durable, versatile, resistant to elements
Recycled aggregate for reuse
GLASS
Easy availability, upfront cost
Lasts 15+ years before replacement
Maximize light, protection from elements
Can be re-melted or recycled
BAMBOO SCREEN
Local, versatile, low-cost material
Lasts 12+ years before replacement
Durable, flexible, lightweight, resistant
Biodegradable
CONCRETE
Easy availability, versatile, low-cost
Lasts 20+ years before replacement
Durable, versatile, resistant to elements
Recycled aggregate for reuse
STEEL STRUCTURE
Easy availability, upfront cost
Lasts 12+ years before replacement
Durable, lightweight, strong, long-lasting
Reused for new applications

s > Field Station: Anna Kathryn Becker + Sarie Keller + Olivia Vercruysse

i > Edson Cabalfin

trip to

Ecuador

The students of the Research Studio Sustainability in the tropics: Ecuador travel to Ecuador in spring 2023 to visit the place where they develop their proposals and other places in the country. They are both undergraduate and master's degree students.

The project will take place in the FCAT Reserve, located about 30 km West of the city of Quininde (0.3269° N, 79.4648° W). As the FCAT reserve grows and programs shift and expand, critical components of this effort include developing a sustainable design framework that adapts to this growth and visioning and building design that supports FCAT decision making and responds to the site specific context, combining innovative design that uses local materials and knowledge with modern techniques to adapt to the local environment. The site is extremely humid, ecologically diverse, and economically vital, offering a rich opportunity to conduct research to assist FCAT, explore sustainable design in the tropical Andes, and critically consider public interest design in an international context.

Quito and Cube (FCAT Reserve)
Sustainability in the Tropics
Edson Cabalfin
April, 2023

Fig. 1
Model and sketches

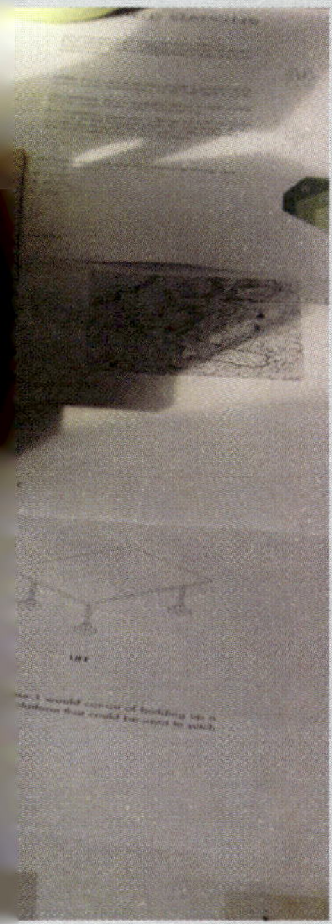

Fig. 2
Studio project site
FCAT, Ecuador.

Fig. 3
Basketball courts.
Quito, Ecuador.

Fig. 4
Students at FCAT Reserve.

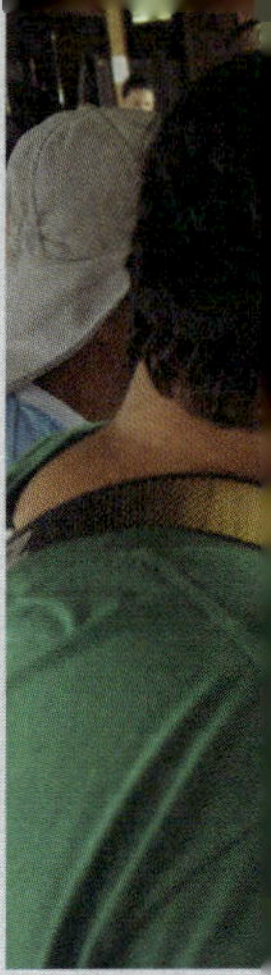

Fig. 5
Residence.

Fig. 6
FCAT Team work.

Fig. 7
Outside view of Basilica del Voto Nacional. Quito, Ecuador.

Fig. 8 and 9
FCAT Team work.

Fig. 10
Bronze doors at Basilica del Voto Nacional
Quito, Ecuador.

Fig. 11
FCAT Reserve.

trip to Barbados

Barbados Research Studio
Brent Fortenberry
April, 2023

The group of students from the research studio ARCH 6042 + PRES 6042 travel to Barbados in the spring of 2023. The course led by Brent R. Fortenberry combines master's students from both the Architecture and Historic Preservation programs at Tulane School of Architecture.

NAKESHA'S
RUSH HOUR
BAR
Banks

STOP RH LITTE
STOP
RH

trip to

Mexico

Tulane School of Architecture graduate students and faculty member Rubén García Rubio explored Mexico City during one of the school's Spring Break trips abroad.

The group visited Teatro Telcel, Museo Jumex, Museo Soumaya, Auditorio Nacional, and other landmarks, and even had the opportunity to admire iconic architecture such as Casa Barragan, Casa Gilardi, Parque Hidrico La Quebradora, and the Museo de Antropología.

Graduate students trip
Ruben Garcia Rubio
April, 2023

Re-New Orleans Exhibition

The Albert and Tina Small Center for Collaborative Design
Curated by Emmanuel Osorno

Re-New Orleans is an architecture exhibition masquerading as a storefront display of New Orleans souvenirs. Its subjects are deliberately kitschy: buildings transformed into everyday objects. Through the use of translation, abstraction, and representation, this exhibition aims to remake the touristic image of the city, proposing a new set of buildings that ought to be elevated to iconic status. Ranging from infrastructural projects to beloved local businesses, the selection of fourteen buildings focused on structures that are often overlooked, but whose history, operation, or program are significant for the functioning and making of New Orleans. Antagonizing the image of exceptionalism and authenticity that typically characterize the portrayals of the city, this series of objects proposes an alternate view, one that reflects the perspective of less fortunate residents of the city. By turning them into souvenirs, the image of these buildings can be mobilized to communicate with a wider public and be used as agents of influence and change.

AFTER
CURTAIN
CALL

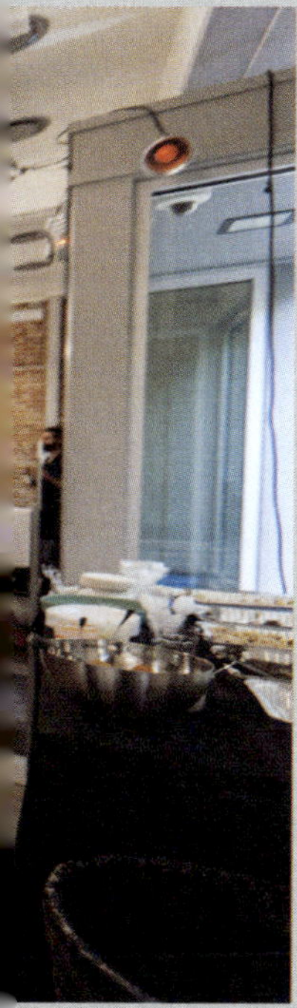

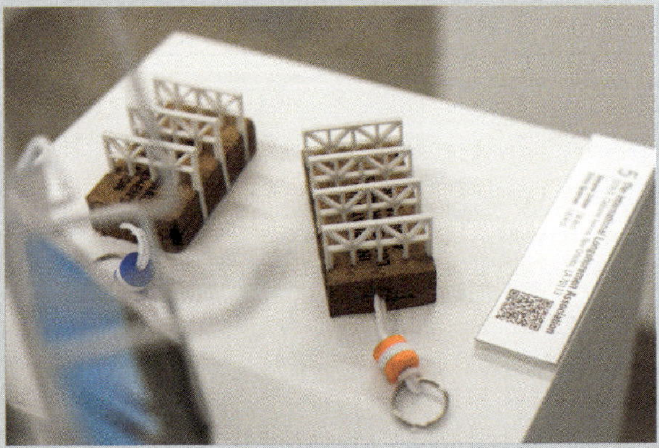

event

Design Symposium

Organized by the Design Program of Tulane School of Architecture

The design symposium, curated and planned by TuSA students, featured invited speaker lectures, panel discussions, and workshops to explore the roles of materiality and representational mediums within the disciplines of architecture and design.

Advancements in representational techniques and material expressions in the design practice have been furthered alongside an increasingly digital world, allowing new methods of functionality and conceptual transference through tactile and digital mediums. This era of evolution sets the stage for designers to shape the future of the built environment and address its challenges through innovative and forward-thinking techniques. The symposium's objectives were to broaden the perspectives and theoretical understanding of materiality and modes of design communication for the TuSA student body and the broader community.

LETTERS
INSTEAD

Extreme Up-Cycling for all of your Extra Fabric

event

Design Showcase

The Design Showcase + Fashion Show is an annual pop-up exhibition at the end of the school year. The inaugural event took place in May 2023 in Kendall Cram Hall at Tulane's Lavin Bernick Center. The event features the work of over 100 students enrolled in 8 Design studio courses at the Tulane School of Architecture. The Showcase highlights the work of our undergraduate students as they learn to use the power of Design to change the world for the better. The event is free and open to the public with food, drinks, and music. Featured projects include topics such as sustainable textile innovation, non-profit organization branding, innovative multi-purpose furniture, nature-inspired light sculptures, and streetcar shelter designs that foster community engagement.

Kendall Cram Hall
at Tulane's Lavin Bernick Center
Curated by Meghan Saas, Hannah Berryhill, and Bo Choi (2023).

culturally

affordable

VISUAL COMM & ADVOCACY

Meghan Saas [C]
Adam Newman
Ray Fontaine

Visual communications & Advocacy course explores graphic design as a tool for communicating information and advocating for various purposes, in both print and digital forms. The course covers topics such as visual analysis and design research methods, informational hierarchy and typographic layout, and how to effectively apply design principles to create impactful visual communication and elicit a meaningful audience response. Live demos of Photoshop, Illustrator, and InDesign help students boost their digital creative skills.

Students enrolled in the BA Design program or pursuing a Design minor are introduced to the course immediately after completing their prerequisite coursework. Visual Communication and Advocacy serves as a gateway to the core Design studio sequence. By encouraging students to think critically about various aspects such as audience, messaging, aesthetic tone, and cultural context, it offers a thought-provoking and engaging learning experience. Through a combination of individual projects and teamwork, students gain valuable insights and hands-on practice in tackling branding and 2D design challenges. The course emphasizes a rigorous and iterative approach, enabling students to refine their skills and creativity while developing a strong foundation in design principles.

The course has two main exercises:

1. Cause Project

Graphic design has the power to act as a megaphone, amplifying messages to reach and resonate with a wider audience. To make a meaningful impact, designers must carefully analyze their audience, refine their message, and skillfully apply a compelling visual tone. By leveraging this power for advocacy, we can steer it towards the greater good. This 10-week project is an immersive journey through the design process—from thorough research and ideation to development, iteration, production, and reflection.

2.Non Sibi Pro Bono Project

In this project, the students engage through service with a local non-profit organization to develop practical materials for the community partner to use. They discuss the role of service and social engagement in the design field, before a site visit to the non-profit organization.

A genuine understanding of the client and their audience is essential to an effective design solution. The students work in a small team of three to propose a design solution. The final review takes the form of a professional design pitch meeting, with each group presenting their final designs directly to the client.

Client: New Orleans Center for Creative Arts
A regional, high-school level, pre-professional arts training center offering intensive instruction in a wide variety of creative areas, while demanding simultaneous academic excellence.

s> Meyers.Malamut.Mesirow **i>** Adam Newman

s> Megan Lauland + Colby Selinger

i> Adam Newman + Meghan Saas

GETLOCALNOLA.

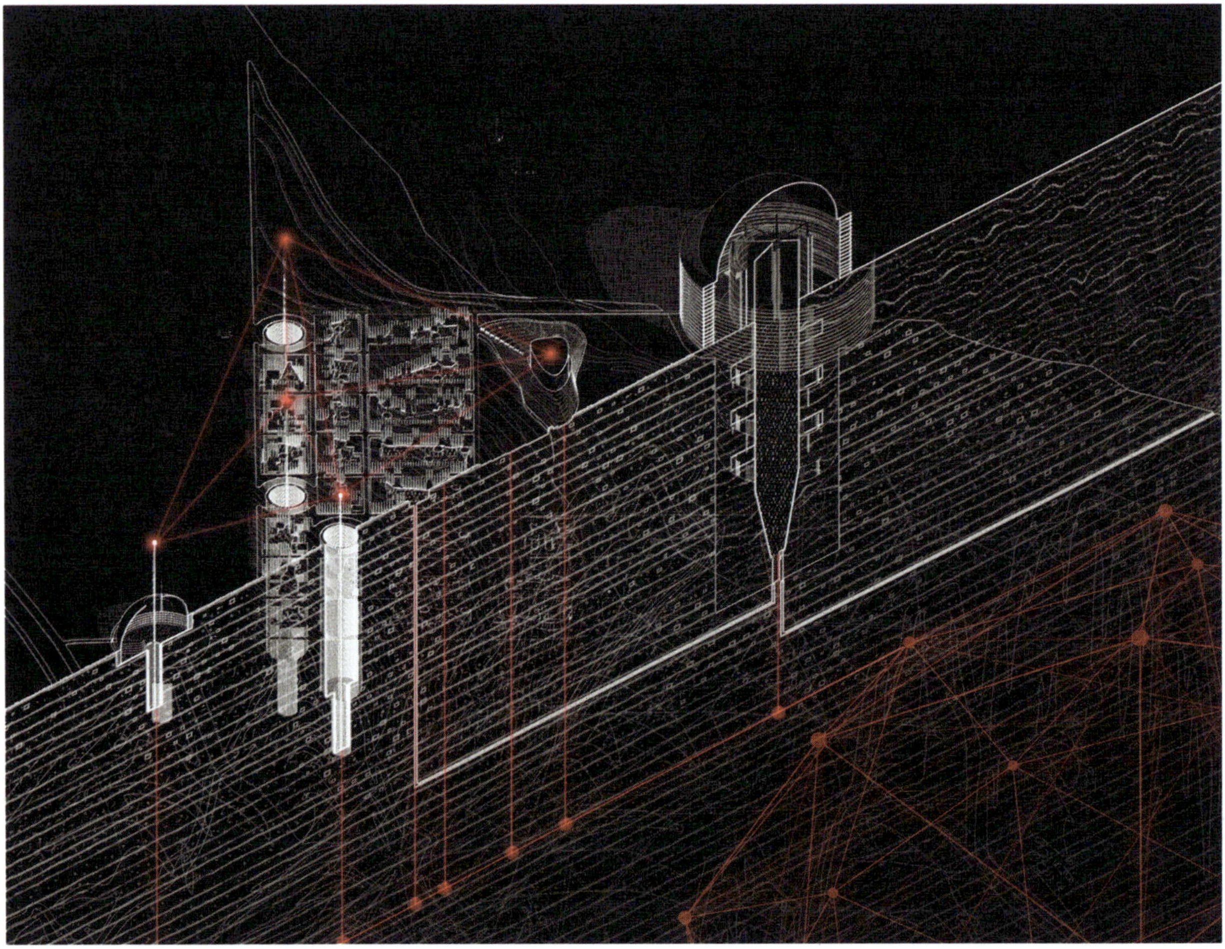

TIGER THEPKANJANA

LIVED CYBERSPACE

A Rehabilitation Center for Media Addiction

ARCH 5990 Instructors
Cordula Roser Gray & Ammar Elouini

s> Tiger Thepkanjana **i>** Cordula Roser Gray & Ammar Elouini

Surveillance, voyeurism and exhibitionism has had a long history. Today, there is the addition of the media as the main means to which we consume information. From this condition, physical spaces have been collapsed into what is inside the screen, all other spaces left unimportant. We all are voluntarily submitting ourselves into a modern digital panopticon, limiting our perception of physical space to the four corners of the screen.

This thesis investigates speculative means to reflect upon this current state of society, establishing changing relationships between architecture and technological advancements. By translating virtual spaces into architecture and the landscape, this thesis attempts to show – through an architectural narrative – how the media have affected our perception of physical spaces, along with dystopian methods to rehabilitate and remediate.

Fig. 1
Experience Through Technology

Fig. 2
Sites of Experience

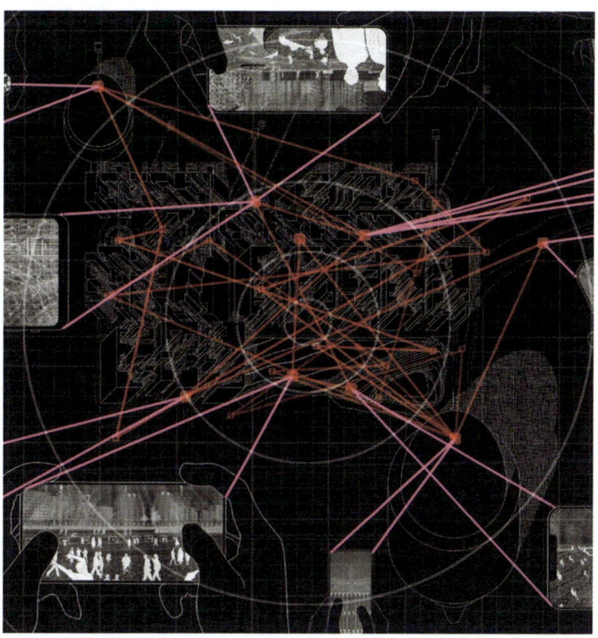

GIULIANA VACCARINO GEARTY

MAPPING MEMORY

Preserving and Restoring the Landscape of Siscily's Tuna Fisheries

ARCH 6990 Advisors
Carol Reese & Iñaki Alday

s> Giuliana Vaccarino Gearty **i>** Carol Reese & Iñaki Alday

This thesis project examines the preservation and adaptive resume of tuna fishery buildings in Sicily. Many of these 17th to 19th century stone buildings, called tonnare in Italian, have been converted from long-abandoned coastal factories to cultural, commercial, and hospitality centers - such as museums, retail districts, and luxury resorts. The investigation is framed around a series of research questions: how do the fisheries in their contemporary forms function within their communities? How can we reckon with the legacies of former production centers via architecture?

While existing scholarship on Sicily's tonnare is rich and active, few historians have examined the preservation and adaptive reuse of these buildings. This considers the fisheries as actively evolving buildings with contested and ever-changing meanings. Focusing on one case study in southeastern Sicily, the Tonnara di Santa Panagia - which is today a ruin- these questions are engaged through exploring how this tonnara might be reinserted back into everyday life in a way that foregrounds its history and terrain.

This project is critical of the numerous fisheries-turned-luxury-developments which inadequately convey the complex material culture and history of the tuna fishing industry in which the tonnara itself is preserved and "reactivated" through minimal programming and any new buildings are located at a distance from the historic site. The landscape becomes one of not only education and history, but also of recreation and ecological restoration: new paths better connect beach-goers to the water and a small "village" of humble apartments lightly set onto the hillside allows visitors to linger. Furthermore, the proposal puts forth a planting strategy in which, over time, a coastal Mediterranean forest will repopulate the landscape.

Fig. 1
Loggia with Roof Addition

Fig. 2
Piazza with Paving and Performance

4

SOCIALLY AFFORDABLE

SOCIALLY AFFORDABLE

Housing

EDSON CABALFIN, Ph.D.
Associate Professor
Associate Dean for Equity, Diversity and Inclusion
Director, Social Innovation and Social Entrepreneurship Program

WHAT IS SOCIALLY AFFORDABLE IN THE BUILT ENVIRONMENT?

Contemplating what is socially affordable in the built environment requires us to interrogate the notion of affordability within the context of sociocultural forces in the built environment. Affordability is not absolute and does not exist in a vacuum. Different societies at various points in history might define affordability in a multitude of ways. For example, social affordability in the middle of the twentieth-century United States might connect to middle-class aspirations during the suburban housing boom brought about by the increase of demand from returning war veterans, the growth of economic prosperity, and the development of automobile culture. [1] In contrast, social affordability within the context of postcolonial Asian nations after World War II might mean the fight for self-determination after colonialism through the rebuilding of cities, the slow recovery of the economy, and the rehabilitation of the national morale. [2]

1 Andres Duany, Elizabeth Plater-Zyberk, Jeff Speck, *Suburban Nation: The Rise of the Sprawl and the Decline of the American Dream,* (New York: North Point Press, 2000)

2 Ryan Bishop, John Phillips, Wei-Wei Yeo, eds. *Postcolonial Urbanism: Southeast Asian Cities and Global Processes,* (London and New York: Routledge, 2003); Anthony King, *Colonial Urban Development: Culture, Social Power and Environment,* (London: Routledge, 1976)

Even in the contemporary context, affordability as it relates to socio-cultural issues, is inevitably diverse and complex depending on the situation and milieu. Social values, or the shared understanding and framework of what is important to a particular social group, are never neutral nor innocent as they are embedded within the historical context. In the built environment, these social values are inevitably manifested in the way people and places are arranged/disassociated, made private or public, segregated/combined, centralized/decentralized, and restricted/allowed to move, among many ways.[3] Inherent to these operations are the dynamics of power relations, or the ways by which certain groups have control or dominance over other groups. Architecture and power mutually inform and reinforce each other. [4]

Associated to this power dynamics in the built environment are issues revolving around social affordability. Some of the issues emerging in today's quest for affordability in architecture and cities center around questions of equitable access and inclusivity. This begs many questions: Who gets to afford? Who are given the opportunities and who are denied? Who has access and who does not have access? Who defines what it means to be affordable?

But these inequities in affordability and accessibility are not just the result of present conditions but are in fact tied to a longer historical trajectory of the systemic oppression and marginalization of certain segments of the population. In the United States, for example, segregation by race was codified in government policies for residential zoning. From 1935 to 1940, the U.S. government, through the Home Owner's Loan Corporation (HOLC), created maps of cities that divided areas into a ranking system based on risk for home mortgage loans.[5] Popularly known as "redlining", areas tagged in red were considered the riskiest and therefore the least likely to be approved for government-insured home loans (**Fig. 1**). However, it was no accident that these red zones coincided in mostly Black neighborhoods. Though since been out of practice after the passing of the Fair Housing Act of 1968, the impact of these racist policies still

3 Carl Nightingale, *Segregation: A Global History of Cities,* (Chicago and London: University of Chicago Press, 2012).

4 Kim Dovey, *Framing Places: Mediating Power in Built Form,* (London and New York: Routledge, 1999)

5 *Mapping Inequality: Redlining in New Deal America*, University of Richmond, accessed through: https://dsl.richmond.edu/panorama/redlining

Fig. 1:
Map of New Orleans by the Home Owner's Loan Corporation (HOLC), 1939. The red zones, labeled as "D: Fourth Grade" are considered high-risk areas, while green zones labeled "A: First Grade" are deemed as the least risk. An example of racialized zoning, the red areas coincided with mostly Black neighborhoods. (Source: National Archives)

MAP OF
GREATER
NEW ORLEANS.
LOUISIANA
COMPILED & DRAWN
-BY-
WM. E. BOESCH.
NEW ORLEANS
U.S.A
COPYRIGHTED AND PUBLISHED BY WM. E BOESCH OF N.O. LA.
RESIDENTIAL SECURITY MAP
LEGEND
A FIRST GRADE
B SECOND GRADE
C THIRD GRADE
D FOURTH GRADE
SPARSELY BUILT UP
INDUSTRIAL
COMMERCIAL
UNDEVELOPED OR FARMLAND
DEVELOPMENT & PARKWAYS
SHUSHAN AIRPORT
SPANISH FORT
MILNEBURG
SWAMP LAND
LOUISVILLE & NASHVILLE YARDS
GENTILLY DEVELOPMENT CO.
ALGIERS NAVAL STATION
AIRPORT
RIVERSIDE SUBDIVISION
ORLEANS SUBDIVISION
ORLEANS PARISH
ST BERNARD
VACANT
MAP OF NEW ORLEANS & VICINITY

reverberates today contributing to the wide racial wealth gap between White and Black homeowners.[6] In New Orleans, this is still evident in the marginalization of Black city residents who live in neighborhoods with high levels of poverty, crime and violence, and hazards affecting physical and mental well-being. [7]

Similarly, we see in many parts of the world, the distinction between the rich and the poor clearly laid out in cities with the side-by-side development of informal settlements with urban growth.[8]High-end commercial centers and luxury residential enclaves are often coupled together with the emergence of slums and squatter settlements. Often denied governmental infrastructure and social services, these informal settlements nevertheless are part and parcel of the urban dynamics providing cheap labor, contributing to its informal economy, and abounding with a strong sense of community.[9] The emergence of informal settlements is not simply the result of the lack of housing but rather a complex web of social inequalities, economic-political structures, and historical cycles of oppression. [10]

Because of the complex nature of social affordability, we can use the concept of "intersectionality" as a lens in which to reimagine a possible solution. Intersectionality was first coined by the legal scholar Kimberle Crenshaw in 1989 to explain how oppression in the legal system works not just as a single issue but rather exists at the overlapping intersection between race, gender, and class.[11] In using intersectionality

---

6 Richard Rothstein, *The Color of Law: A Forgotten History of How Our Government Segregated America*, (New York: Liveright Publishing Corp., 2017); Emily Badger, *How Redlining's Racist Effects Lasted for Decades*, New York Times, August 24, 2017, accessed through: https://www.nytimes.com/2017/08/24/upshot/how-redlinings-racist-effects-lasted-for-decades.html

7 Stacy Seichshnaydre, Robert Collins, Cashauna Hill, Maxwell Ciardullo, *Rigging the Real Estate Market: Segregation, Inequality, and Disaster Risk*, (New Orleans: Data Research Center, 2018), accessed through: https://www.datacenterresearch.org/reports_analysis/rigging-the-real-estate-market-segregation-inequality-and-disaster-risk/

8 United Nations Human Settlements Programme. *Challenge of the Slums: Global Report on Human Settlements 2003*, (London: Earthscan Publications, 2003), accessed through: https://unhabitat.org/the-challenge-of-slums-global-report-on-human-settlements-2003; Mike Davis, Planet of Slums, (London: Verso Books, 2017)

9 Justin McGuirk, *Radical Cities: Across Latin American in Search of a New Architecture*, (London: Verso Books, 2015).

10 Robert Neuwirth, *Shadow Cities: A Billion Squatters, a New Urban World*, (London: Routledge, 2004)

11 Kimberle Crenshaw, *Demarginalizing the Intersection of Race and Sex: A Black Feminist Critique of Antidiscrimination Doctrine, Feminist Theory and Antiracist Politics*, University of Chicago Legal Forum, 1989(1-8), accessed through: http://chicagounbound.uchicago.edu/uclf/vol1989/iss1/8

as a framework for potentially addressing the social ills of the built environment, we need to acknowledge that the operations of marginalization should be thought of as an interconnected ecosystem that affects people in different ways. Oppression affects different people in different ways. In our attempts to solve the social inequalities brought about by architecture and our cities, we cannot simply focus on one aspect of a person's identity (such as race/ethnicity, gender, sexuality, class, education, religion, marital status, etc.) but instead should consider how the variations and combinations affect certain groups disproportionally compared to others. For example, the issue of affordable housing in New Orleans is impacted by the interlaced connection between tourism and entry of short-term rental services (such as Airbnb) in the city, among other issues.[12] This has only benefitted predominantly White, rich, out-of-town, corporations and simultaneously disadvantages mostly lower-income, less-educated, Black and Brown residents with deep historical ties to the city.[13] To solve this issue of social affordability requires a reconceptualization of how the solution requires a multi-dimensional and multi-sectoral method that considers the complexity of the human condition.

Ultimately, addressing social affordability in the built environment necessitates an empathetic, humanity-centered, systemic approach. Empathy, or the conscientious comprehension and sharing of another person's feelings and experiences, forms the basis of creating alternative futures where people care for each other. Through a systemic approach, we can also begin to challenge and dismantle various systems of oppression that have been put in place for many centuries. We need to acknowledge the differences of people while concurrently celebrating the shared humanity of people. We need to practice compassion in the designing and planning of our cities, extending kindness and respect to both humans and non-humans. It is only by practicing compassion that we can create a more just, inclusive, equitable, and sustainable built environment and future.

---

12 Amanda Hoover, *New Orleans Tried to Control Vacation Rentals with a Lottery. It was a Mess*, Wired Magazine, November 21, 2023, accessed through: https://www.wired.com/story/airbnb-short-term-rentals-new-orleans-lottery/; Tom Perkins, "Like a ghost town: how short-term rentals dim New Orleans' legacy", The Guardian, March 13, 2019, accessed through: https://www.theguardian.com/us-news/2019/mar/13/new-orleans-airbnb-treme-short-term-rentals

13 Murray Cox and Kenneth Haar, *Platform Failures: How short-term rental platforms like Airbnb fail to cooperate with cities and the need for strong regulations to protect housing*, Inside Airbnb, accessed through: http://insideairbnb.com/report/platform-failures; Jane Place Neighborhood Sustainability Initiative, "Short Term Rentals, Long Term Impacts 2.0 Report Released", accessed through: https://jpnsi.org/news-events/archive.html/article/2023/01/24/short-term-rentals-long-term-impacts-2-0-report-released

31 AFFORDABLE PUBLIC HOUSING UNITS IN TORRE BARÓ

HOUSING
Barcelona, Spain, 2023

Credits
Architects: aldayjover architecture and landscape & Burgos Garrido Architects
Director of Execution:
Jesus Arcos, aldayjover architecture and landscape

On the slope of Collserolla Hill, two different types of urban fabric converge: informal houses with difficult accessibility among which trees and water slide down; and blocks of social housing and public facilities along the more accessible routes that disrupt the original small and scattered volumes scale. The building is located between both fabrics, with a ground floor plus five levels facing the main street, and only three floors, fragmented, facing the secondary street and the single-family neighbors.

The volume opens up on this side to allow sunlight into the courtyard, dividing the length of the façade. The courtyard is a stepped void that connects both streets and serves as an access and meeting space for the 31 rental homes and the commercial unit.

The tight budget is invested in materials that are simple but of high quality. The highest level of energy rating has been achieved through passive systems and common facilities, which allows for the exempt for parking spaces and will reduce maintenance and operating costs for tenants.

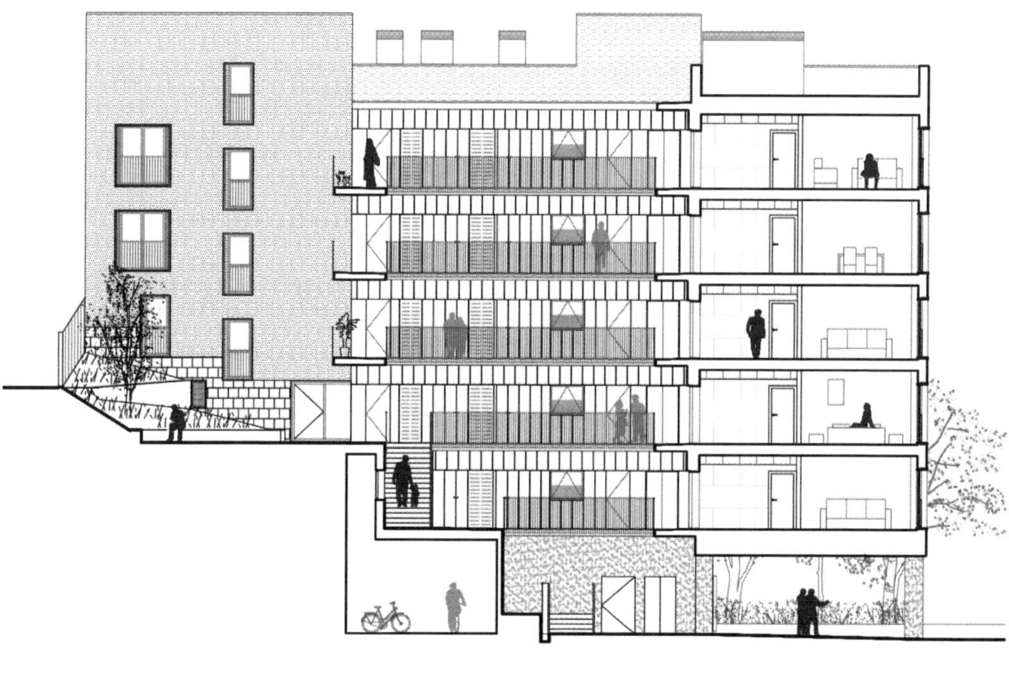

Fig.1
Section

Fig. 2 Plan.
Environmental functioning.

Fig.3 Inside view.
Photo.

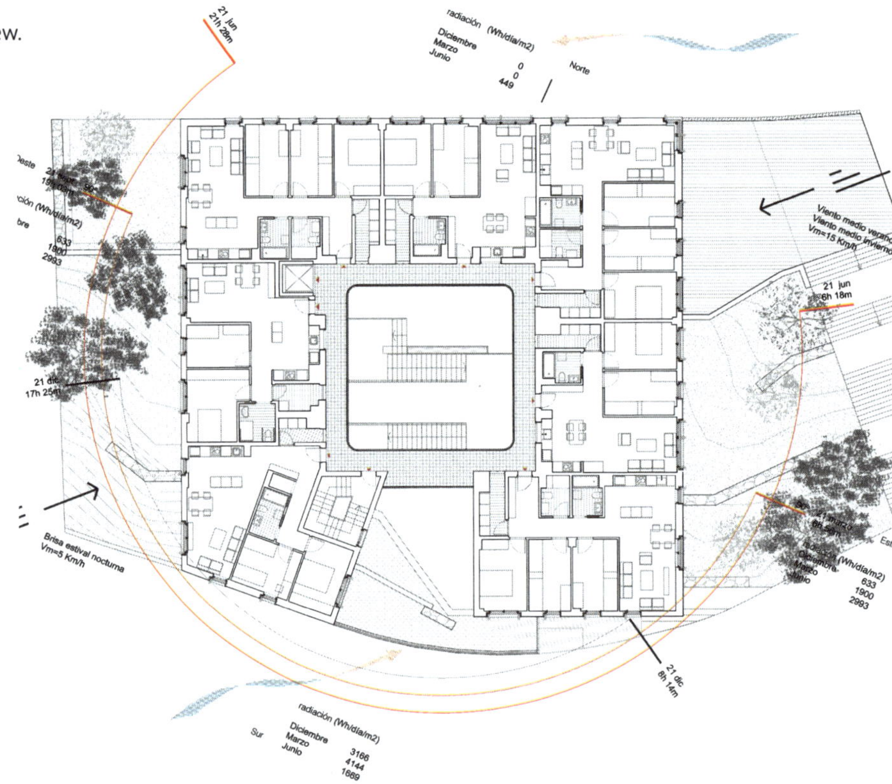

Fig. 4
Central courtyard view
Photo.

URBAN HOSTEL

Irene Keil [C]
Andrea Bardon de Tena
Cynthia Dubberley
Juan Medina
Ross Karsen

The Core 3032-6032 Integrated studio focuses on the development of an architectural project with regards to site relationships, historic context, structure, systems, materiality, and building codes, meeting the NAAB accreditation requirements of the integrated studio. "Integration" encompasses a coordination of building systems during the project development, and reciprocal feedback between intent and material manifestation. The project demands a careful calibration of intent, form, system, material, and technology where relationships have inherent potentials and unanticipated possibilities.

The pedagogical objective of this studio is to lead the student through a design process where architectural ideas, spatial planning, and building technologies are integrated. Students learn to comply with all applicable building code regulations, such as life safety and egress. The content addresses and explores the implications of construction techniques, building technologies, and systems.

The typical project of an integrated studio addresses the design and resolution of a normative building program and building type – a mid-scale building in an urban context. In this course students explore a hostel, a form of low-cost, short-term, shared sociable lodging where guests rent a bed – typically filled with bunk beds and few additional private rooms. All travelers have communal use of lounge(s) and kitchen areas and a variety of other social spaces determined by each student. The building is developed over a more public ground level with hostel functions and adjacent outdoor space. Students work individually, integrating group constellations for components of the design process.

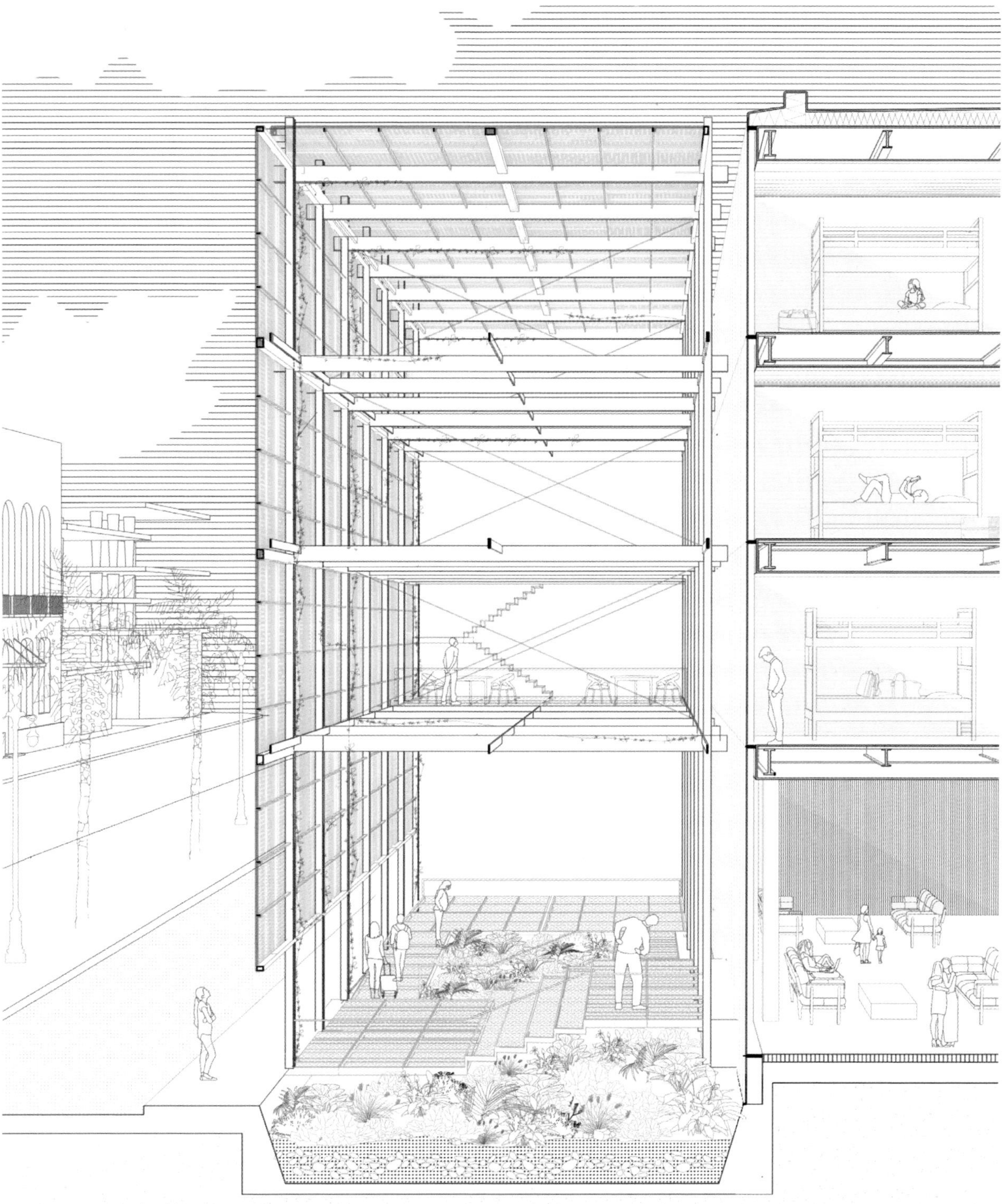

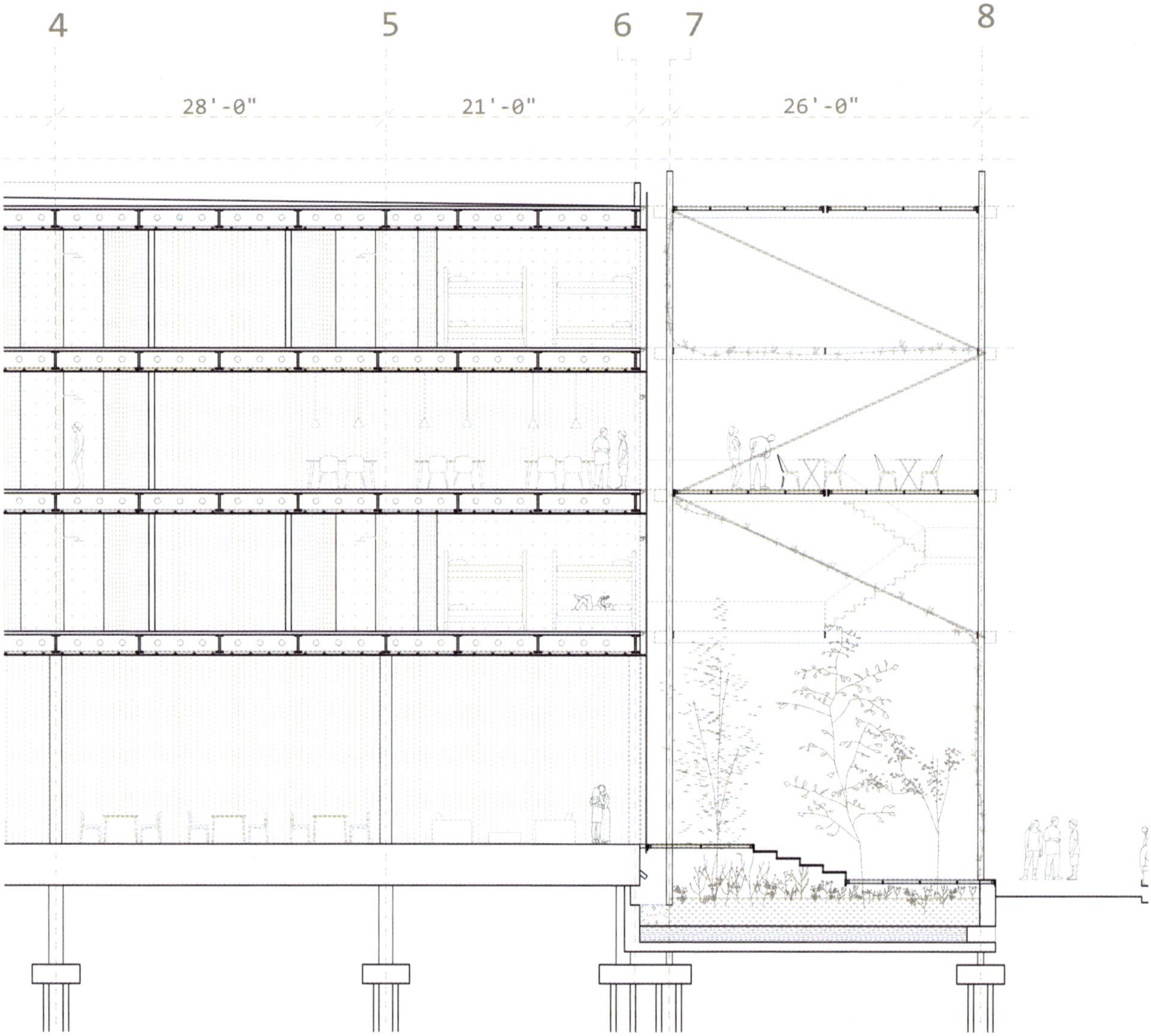
4
5
6
7
8
28'-0"
21'-0"
26'-0"

s> Zaynab Eltaib

i> Juan Medina

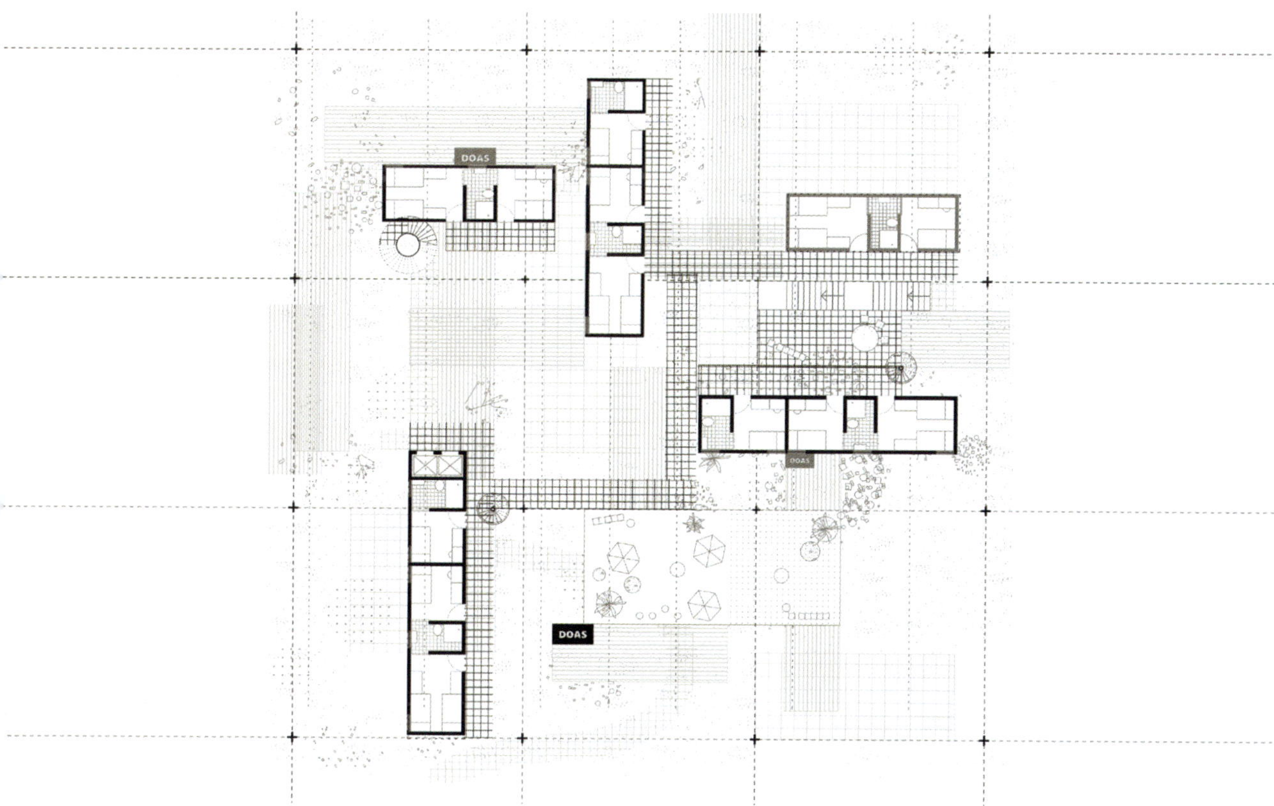

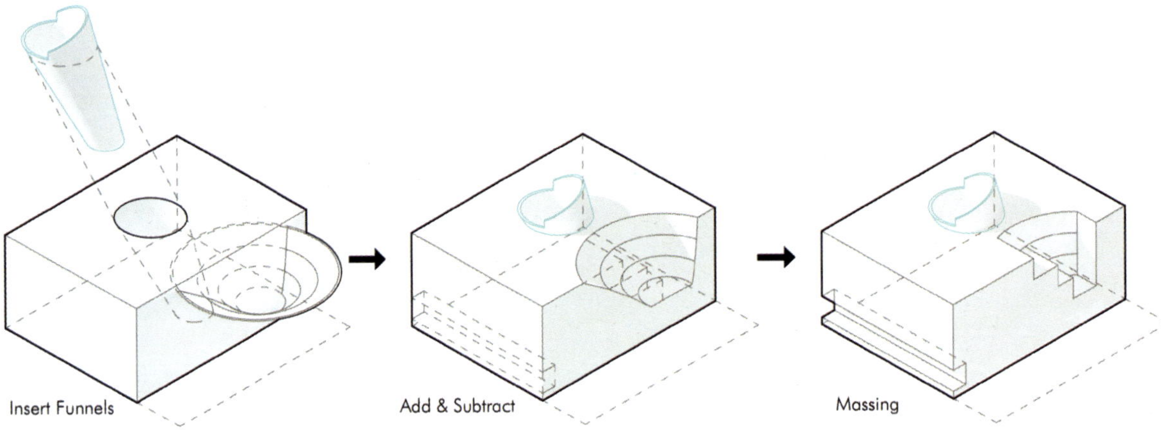
Insert Funnels
Add & Subtract
Massing

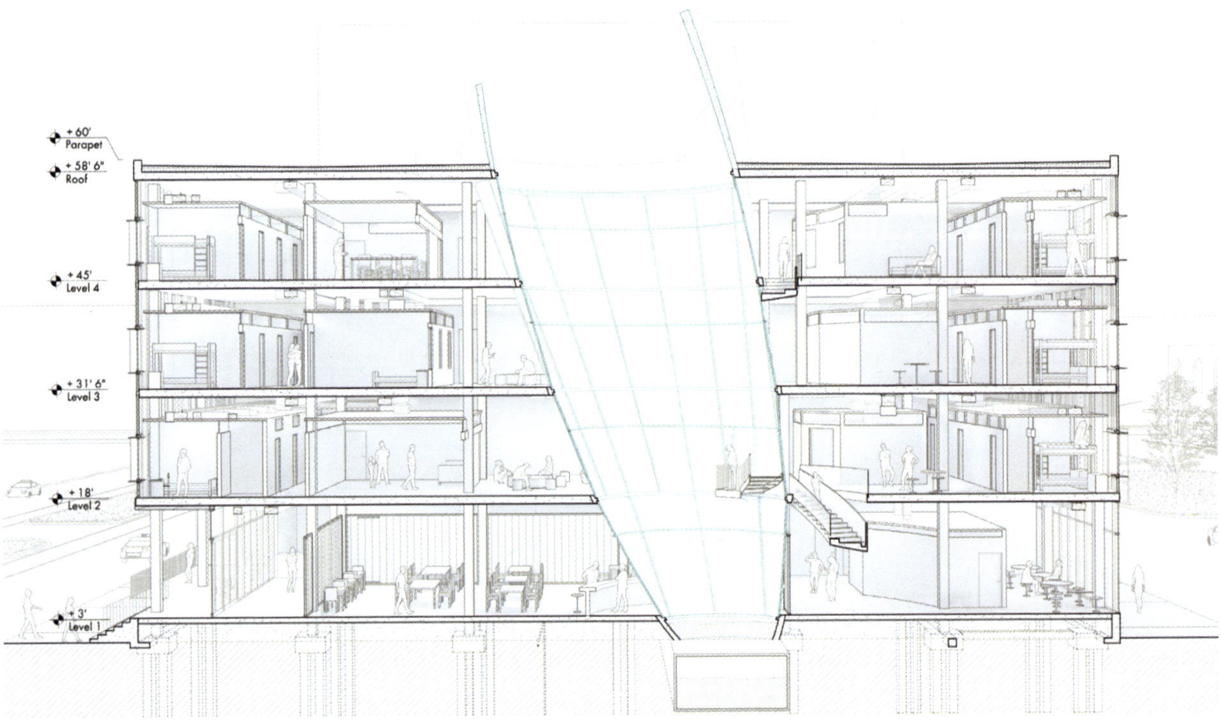
+ 60' Parapet
+ 58' 6" Roof
+ 45' Level 4
+ 31' 6" Level 3
+ 18' Level 2
+ 3' Level 1

s> Joey Tomshe **i>** Cynthia Dubberly

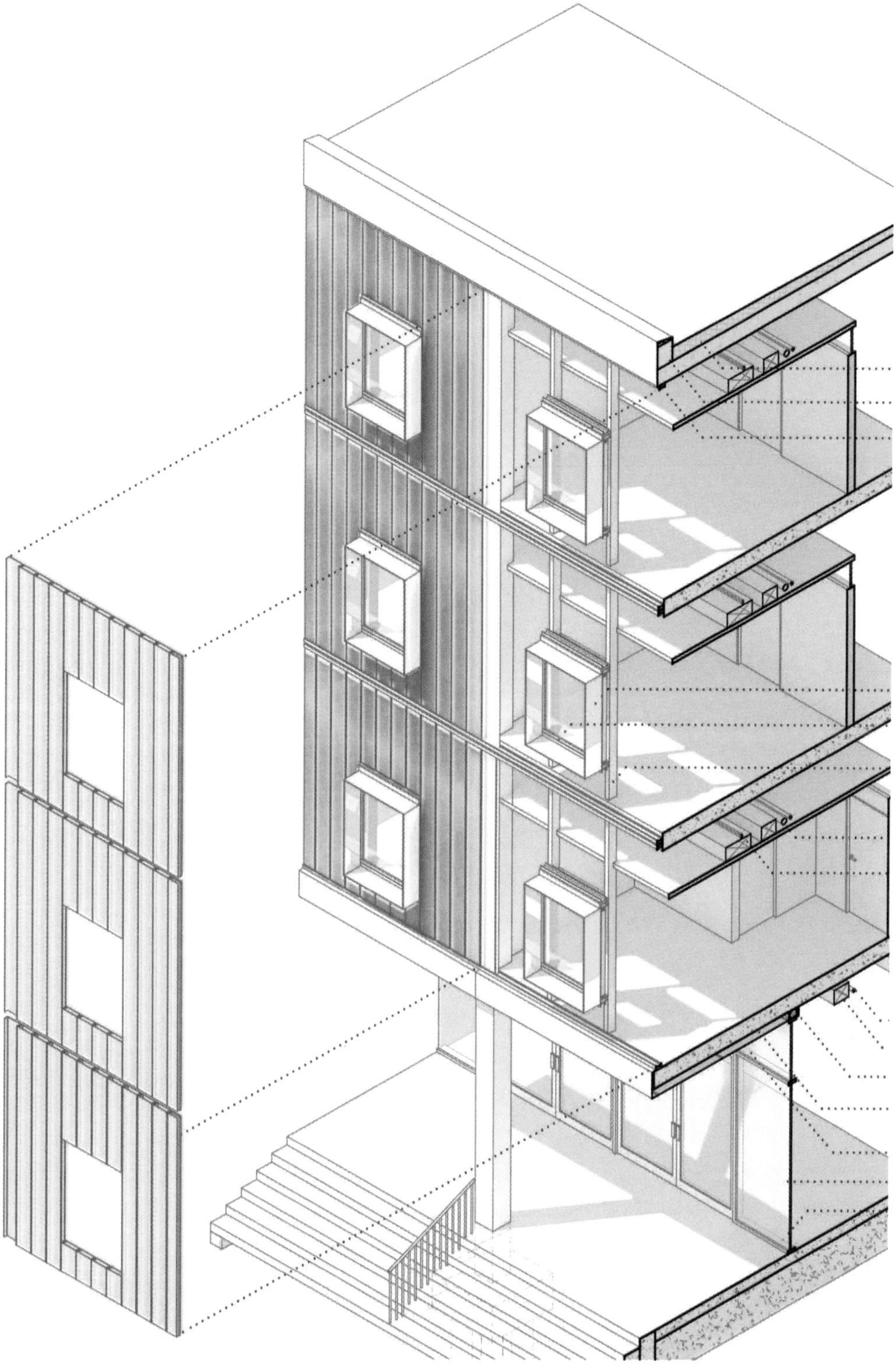

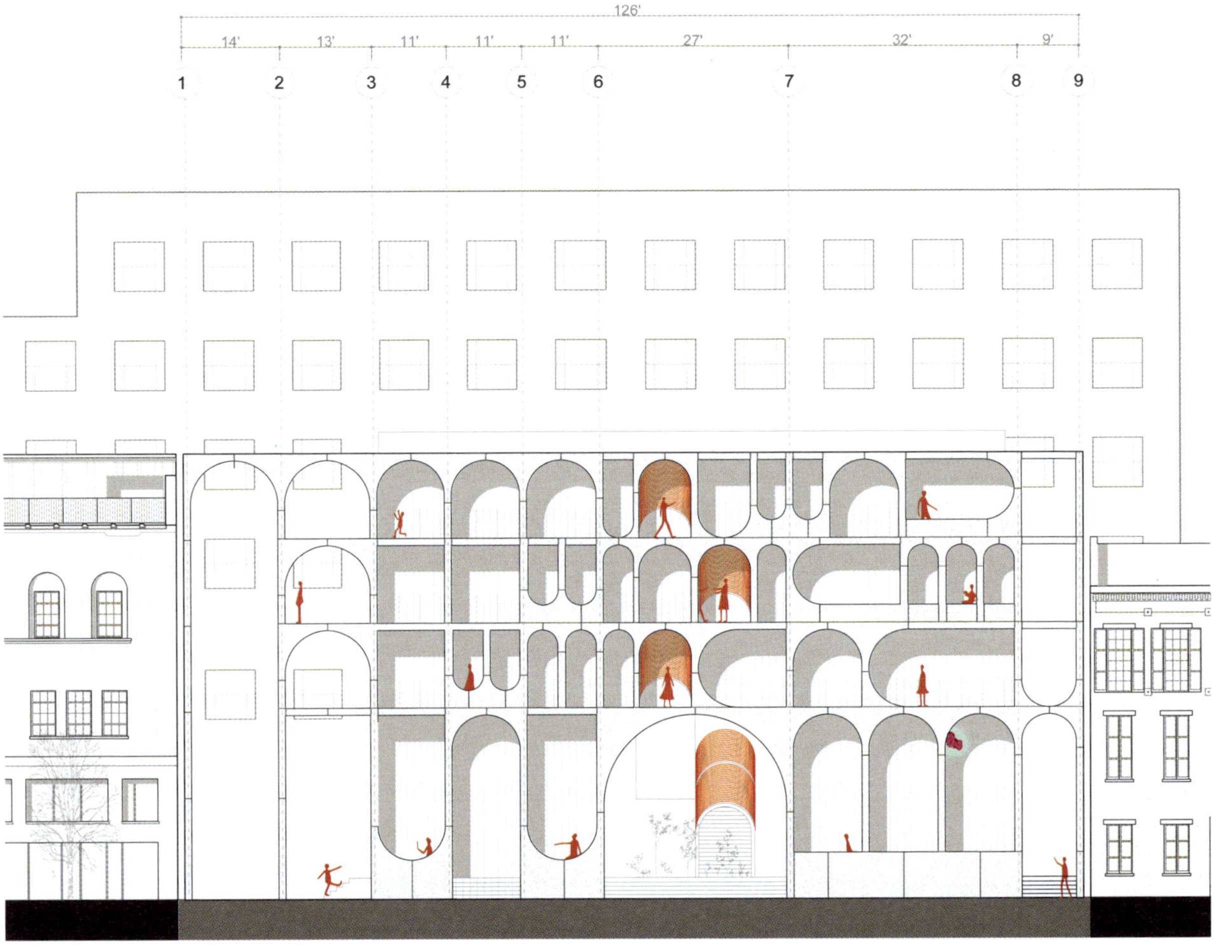
126'
14'
13'
11'
11'
11'
27'
32'
9'
1
2
3
4
5
6
7
8
9

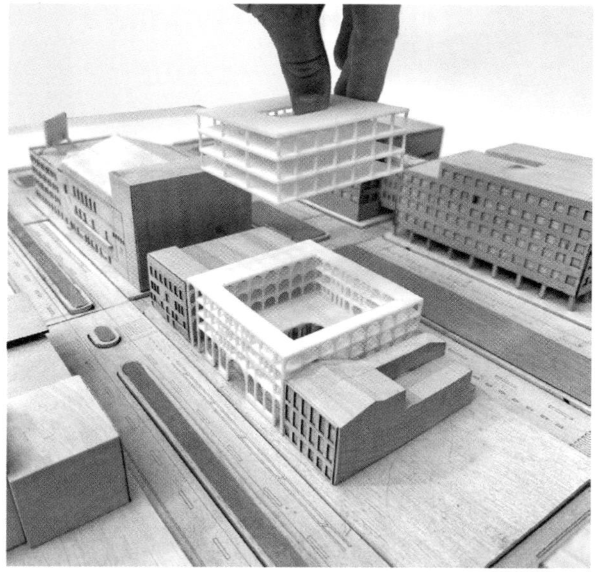

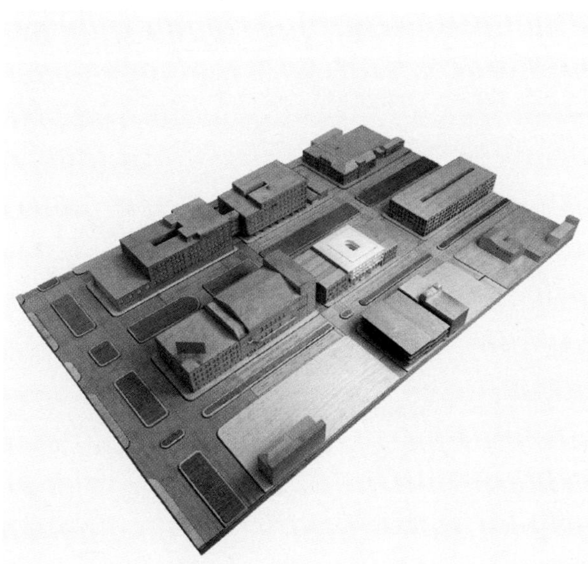

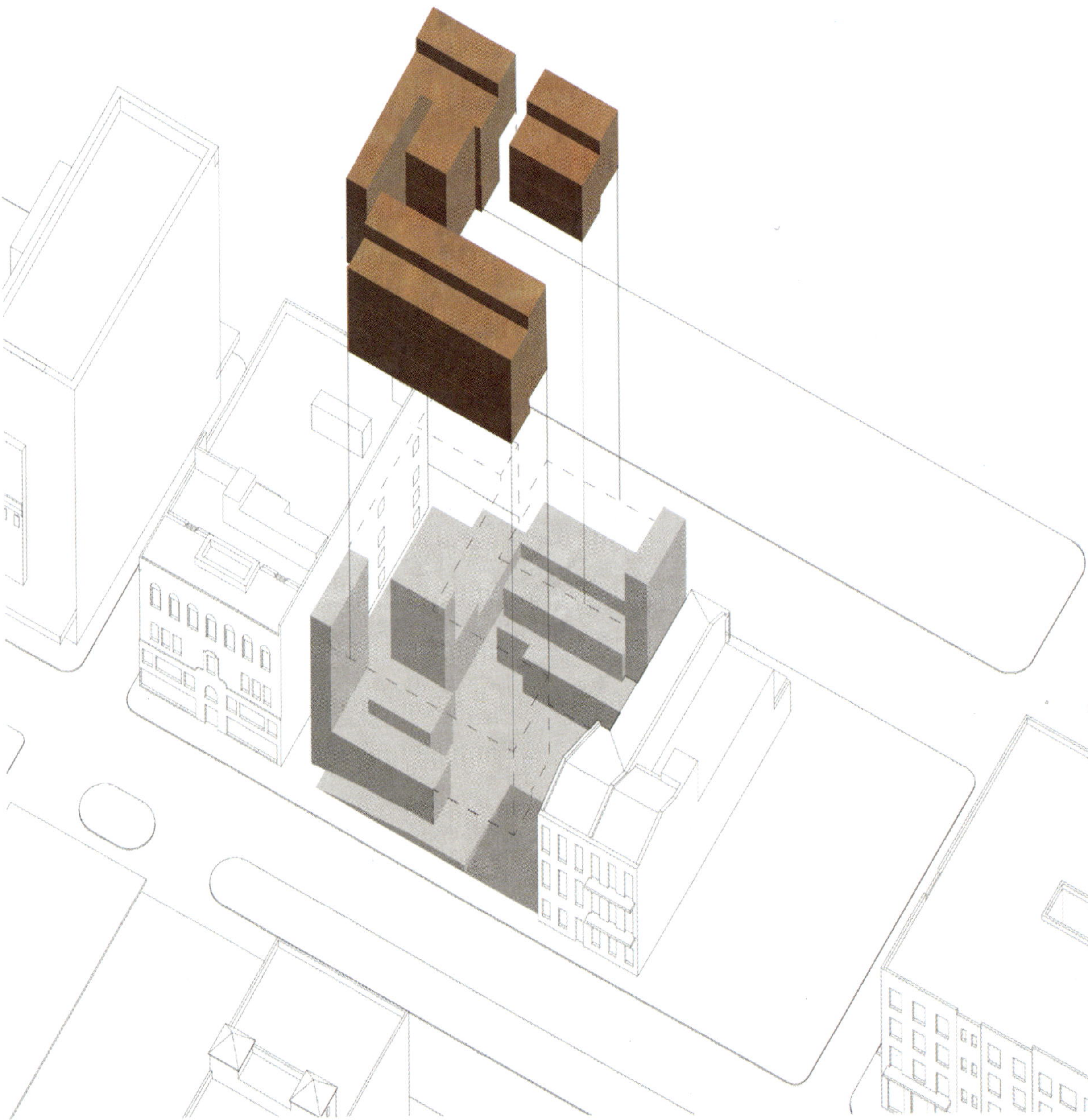

s> Brooke Menhey

i> Andrea Bardon de Tena

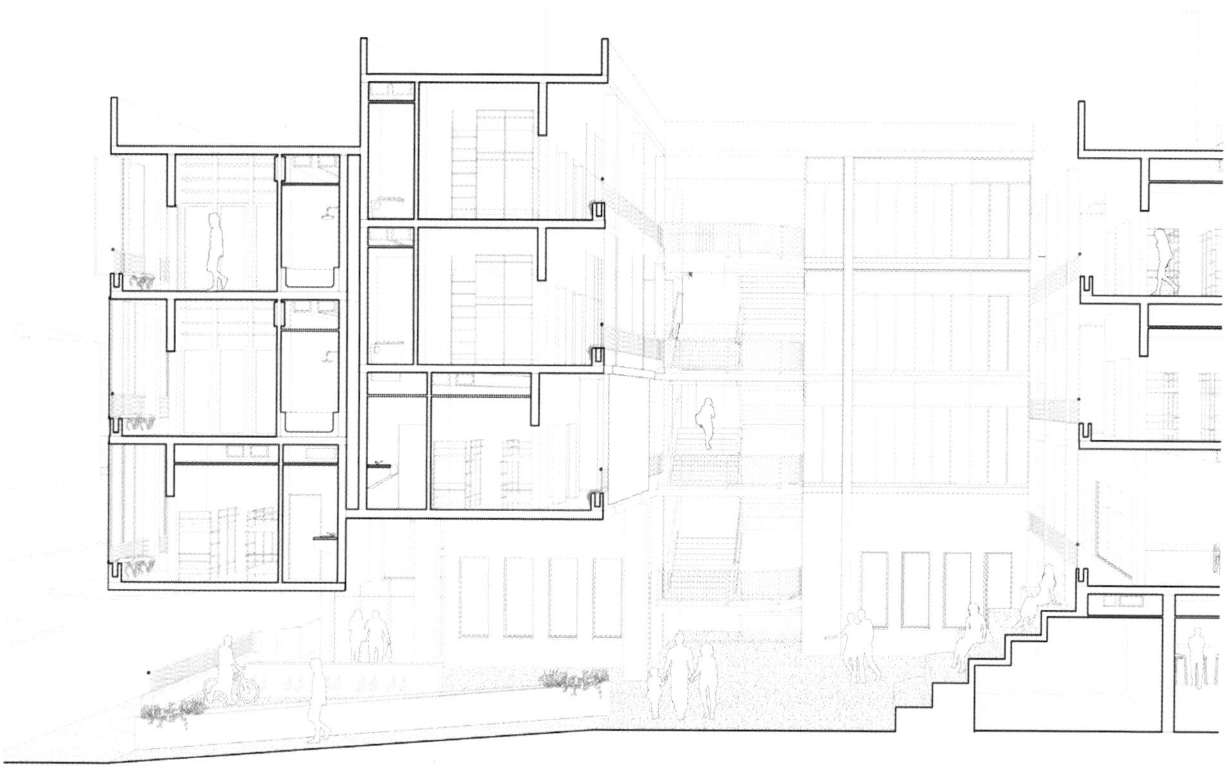

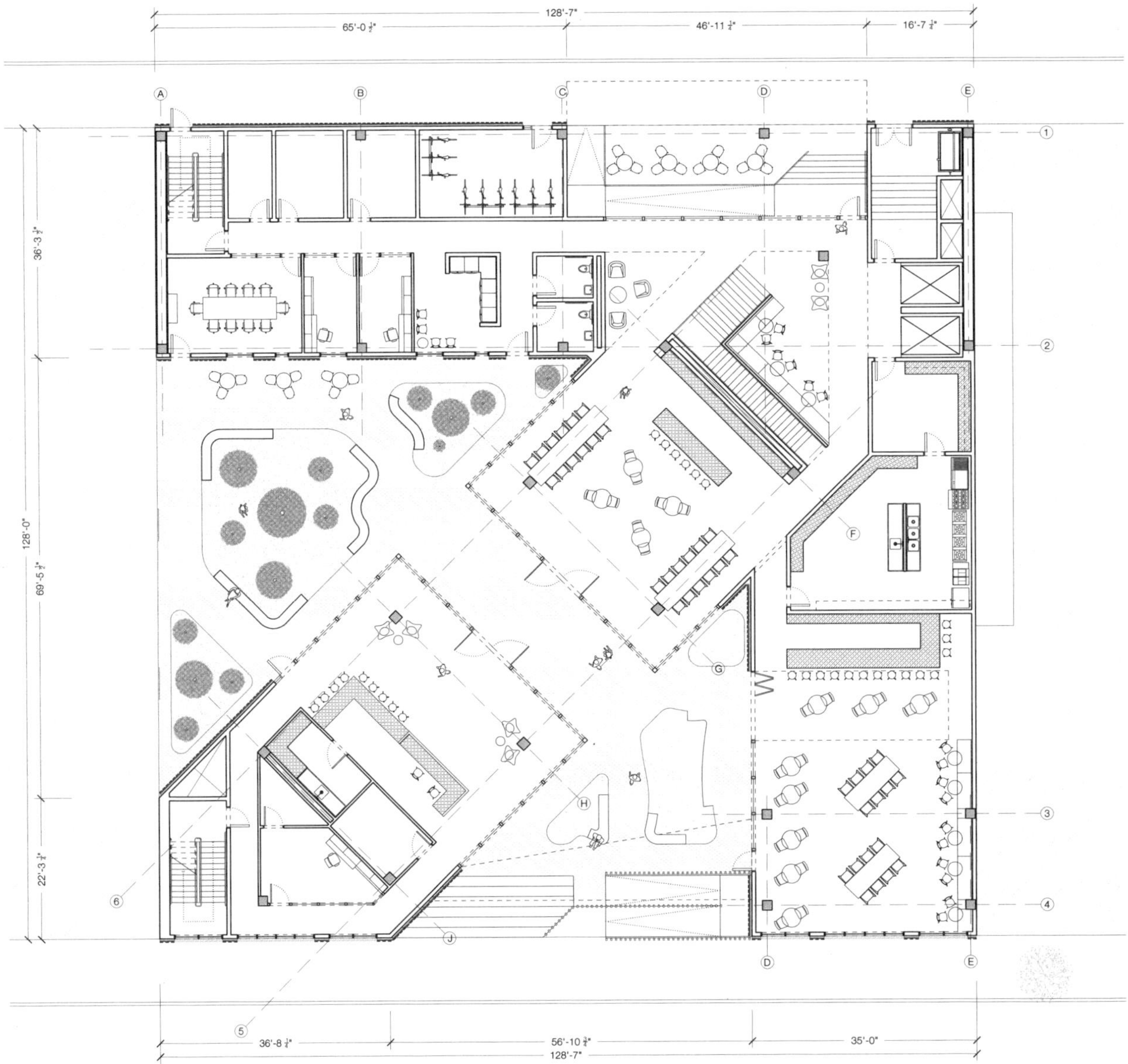
128'-7"
65'-0 ½"
46'-11 ½"
16'-7 ½"
36'-3 ½"
128'-0"
69'-5 ½"
22'-3 ½"
36'-8 ¼"
56'-10 ¾"
35'-0"
128'-7"

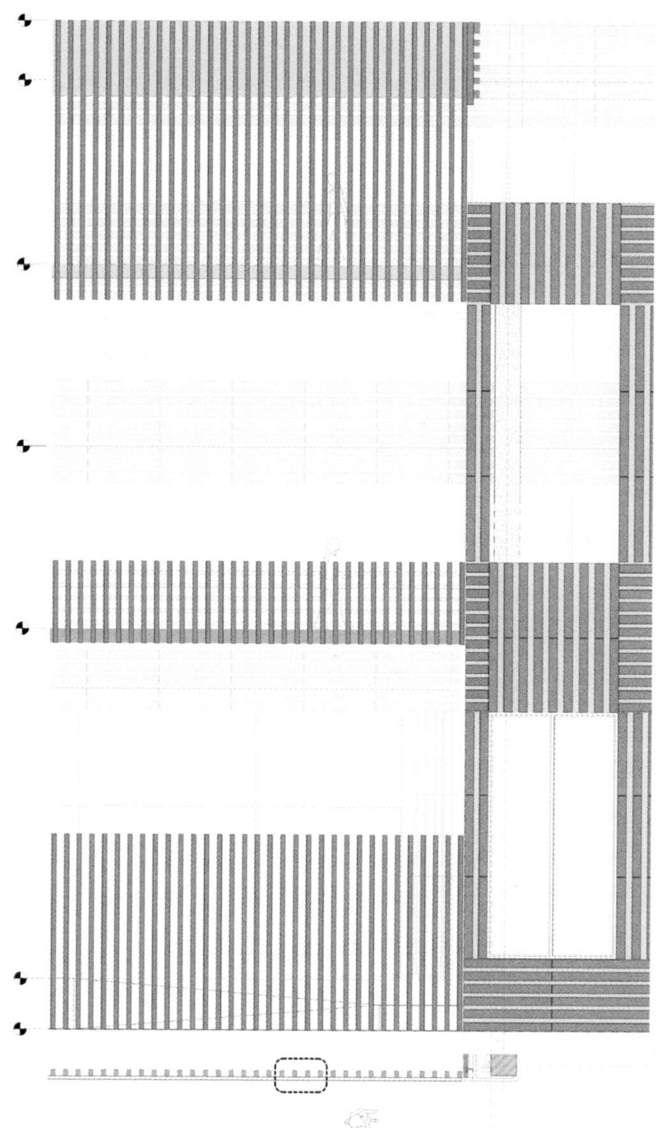

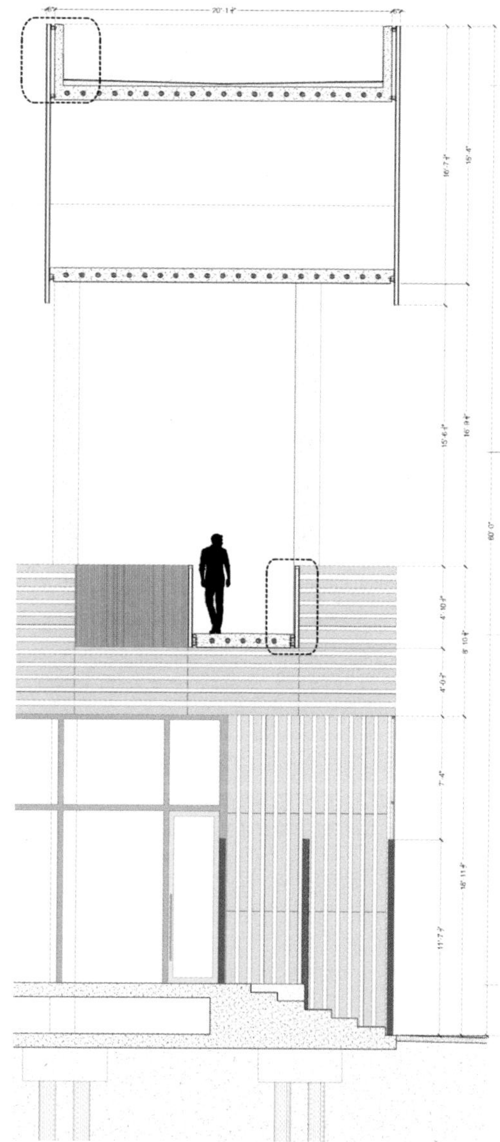

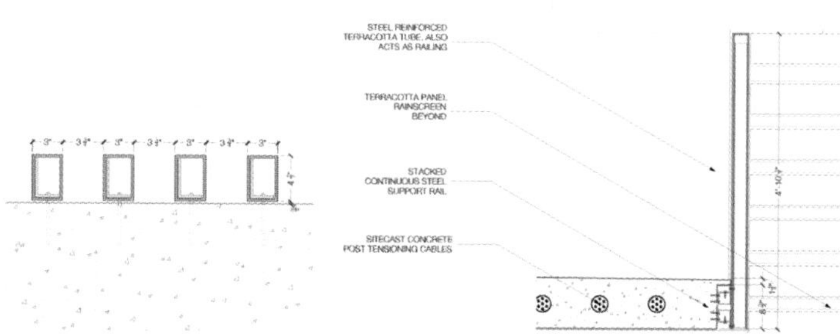
STEEL REINFORCED TERRACOTTA TUBE, ALSO ACTS AS RAILING
TERRACOTTA PANEL RAINSCREEN BEYOND
STACKED CONTINUOUS STEEL SUPPORT RAIL
SITECAST CONCRETE POST TENSIONING CABLES

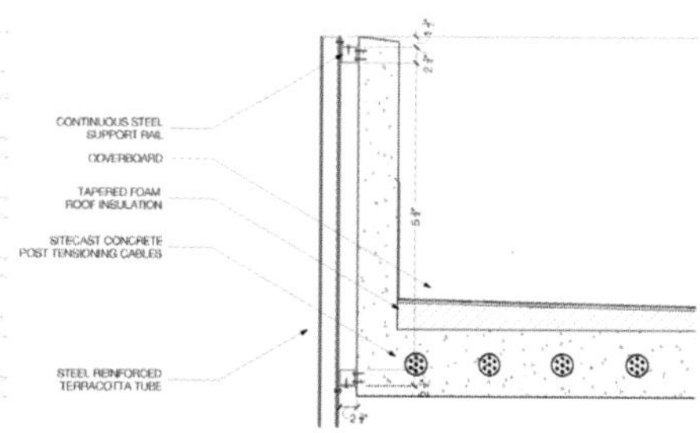
CONTINUOUS STEEL SUPPORT RAIL
COVERBOARD
TAPERED FOAM ROOF INSULATION
SITECAST CONCRETE POST TENSIONING CABLES
STEEL REINFORCED TERRACOTTA TUBE

COLLECTIVE HOUSING

Margarita Jover [C]
Omar Ali
Sergio Alvarez
Patricia Mintguito
Emmanual Osorno

The course is organized into two parts. The first 7 weeks are dedicated to designing 4 small quick projects: horizontal, vertical, sectional homes and the inhabited stair. The second part of the course is dedicated to the design of the collective housing project that combines the previous designs.

The project is the design of collective housing with shared amenities and collective open spaces. it occurs densifying an existing urban block of 300 feet by 300 feet (100meters). The cultural challenge is to keep the 'suburban' lifestyle but with mixed-use, mixed-income and ecologically minded strategies. To that end the notion of the meta-home is put forward for exploration.

This core studio works with the hypothesis that the site has 21 owners. all except 2 agree to redevelop together. One of them decides to participate in the redevelopment and relocate its business. Owners are determined to redevelop themselves self-organizing as an 'intentional community'. They hire an architect to study the transformation from a group of homes to a meta-home that must generate collective usable space, hypothetical profits, and ecological benefits.

The three spheres of knowledge students get in this course are:
1. Visualization that includes the consolidation of the universal language of 'orthogonal projection'.
2. The architectural theme of collective housing with its set of constraints; the methodology of the project as the main field for architectural thinking.
3. Critical thinking in relation to the way we live today and can live tomorrow if we were to do it in a more sustainable way.

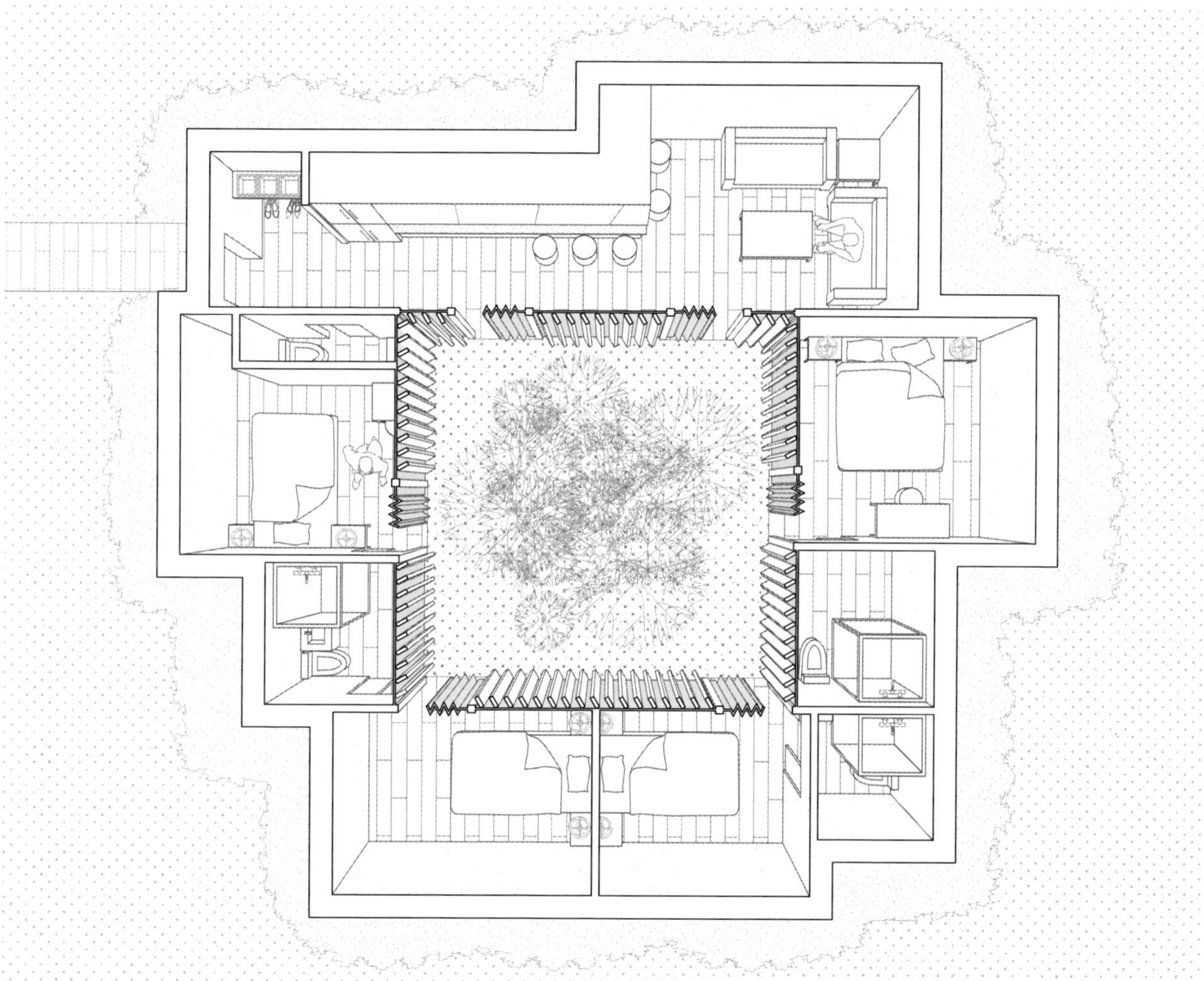

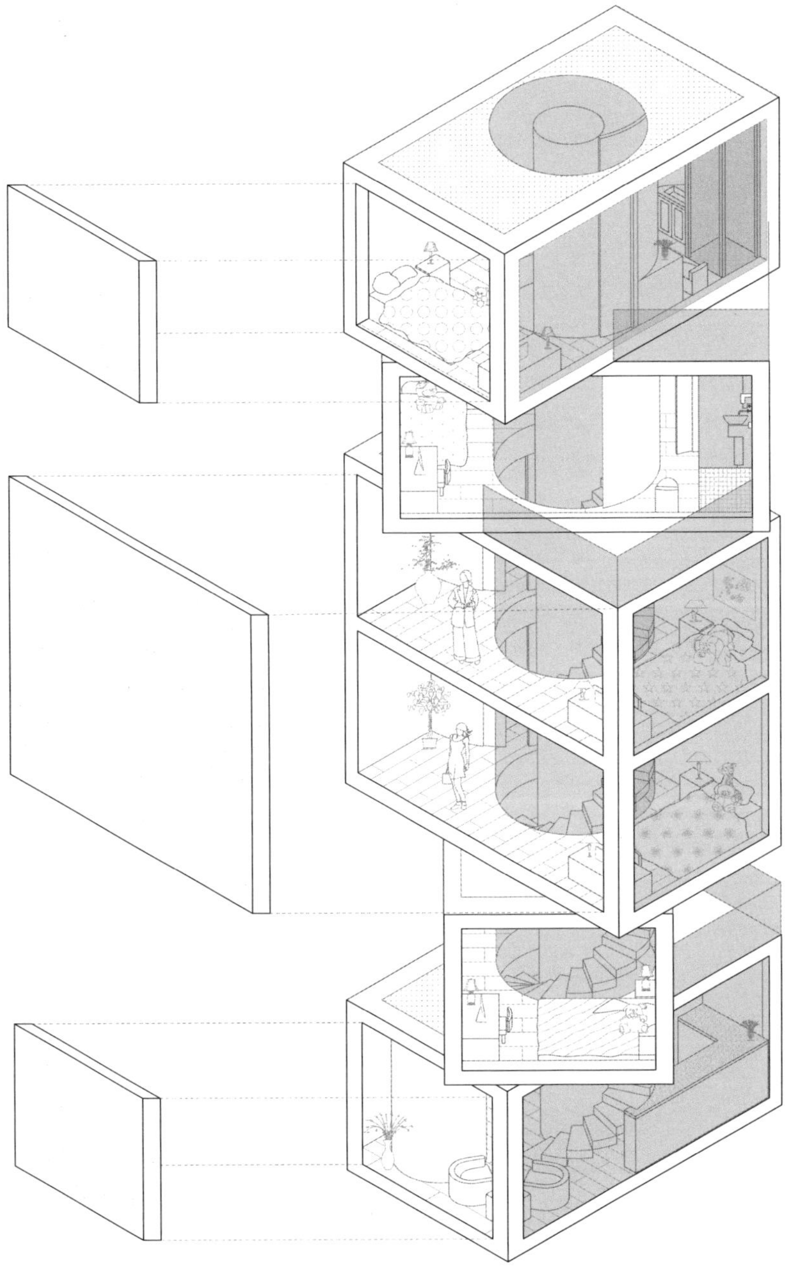

s> Ariel Levine

i> Patricia Minguito

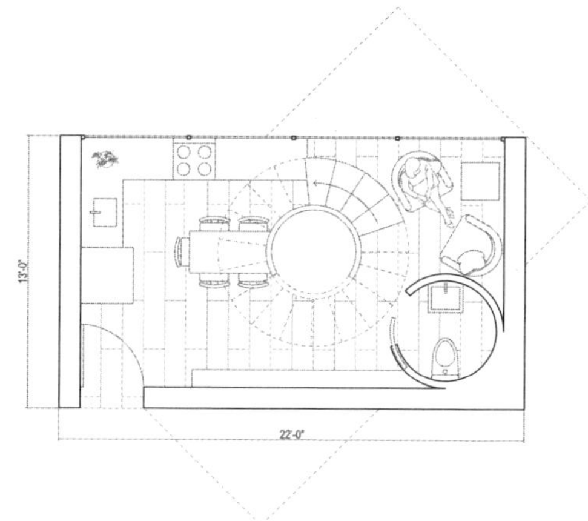

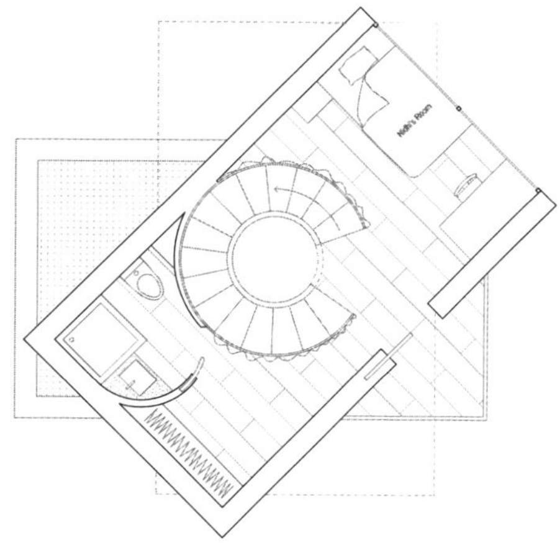

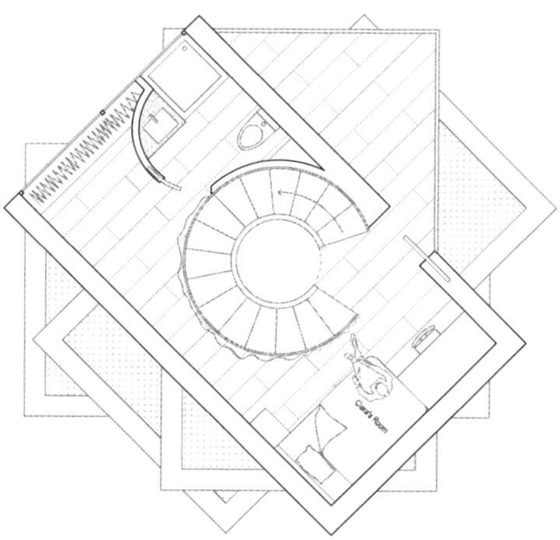

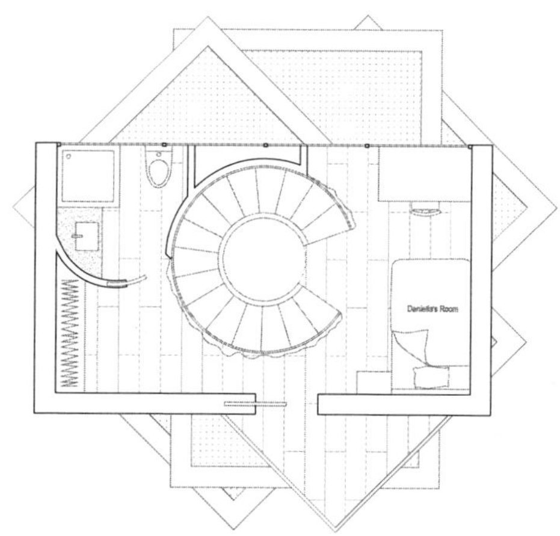

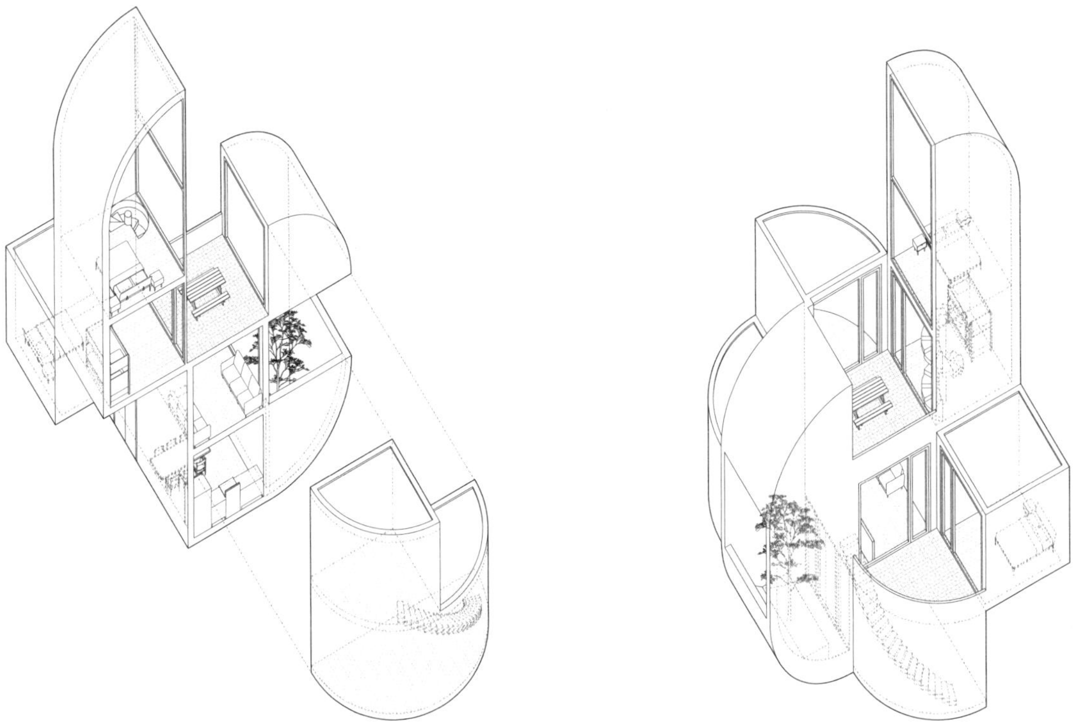

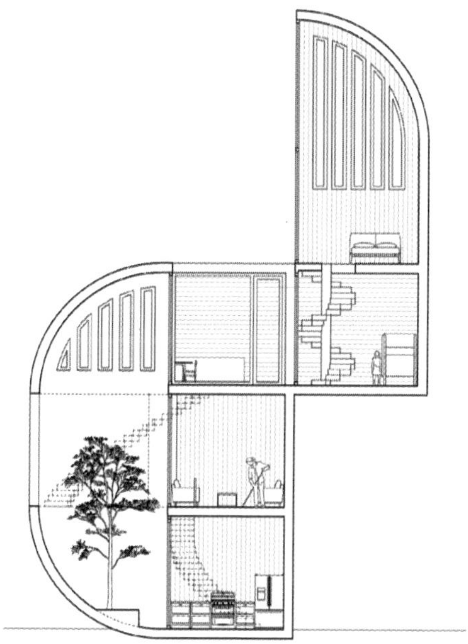

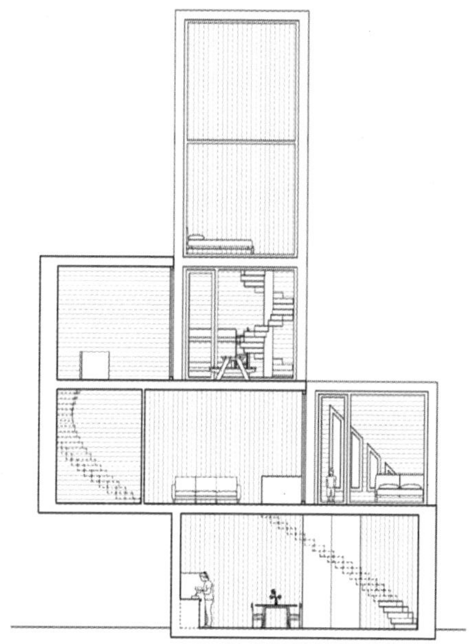

Annunciation St
Arabella St
Joseph St

s> Francesca Lubega **i>** Emmanuel Osorno

trip to

Barcelona

Barcelona is taken as a case study, due to its internationally recognized relevance in architectural and urbanistic culture. This City Port has a long trajectory of notable architecture in different stylistic languages ranging from Roman, Gothic, Historic Eclecticism, Art Nouveau, Modernist, Regionalism, Deconstructivism and more formally constrained recent times. In the early 2000s, the city received the 'RIBA Gold Medal of Architecture' for its urban transformation.

Barcelona Housing Travel
ARCH 6930-4930
Margarita Jover
May-June, 2023

This exceptional urban phenomenon occurred in the period from 1980s to 1992-Barcelona Olympic Games. The country had recovered its democracy in 1978 after a long dictatorship (1939-1975) preceded by a Civil War (1936- 1939).

The selection of projects responds to an interest of the course to emphasize the relationship between cultural, social, and political periods, and the production of housing types.

socially affordable

what is

Home, Sweet Home

The Albert and Tina Small Center for Collaborative Design
Curated by Omar Ali and Nimet Anwar

Home, Sweet Home is a group exhibition by Tulane School of Architecture faculty that collects work from a special topics seminar taught by Architecture and Social Innovation Fellow Emmanuel Osorno and a design-research project led by Architecture and Urbanism Fellow Omar Ali and Visiting Assistant Professor Nimet Anwar.

Evolutive Housing, curated by Omar Ali and Nimet Anwar, is a design-research project. The project considers the effects of gentrification and displacement through the lens of housing in the city of Houston. Housing availability in the city is not unlike the typical offerings of a suburb: single-family homes and multi-family apartment complexes of various sizes and scales. As housing needs grow, Houston is increasingly looking to develop middle-scale housing types.

Cities across the country deal with housing availability issues due to restrictive zoning laws that hinder the application of multi-family housing at the middle scale. Houston is at the forefront of progressive strategies for filling this absence through its relaxed approach to zoning as well as the excess of spec middle-scale housing projects led by developers. This approach has shown to be useful but is keeping housing ownership out of reach to a large percentage of Houstonians. Houston serves as a critical case study that can be applied to cities throughout the country that are amid addressing a scarcity of affordable housing. Architects can lead this conversation by partnering with Community Land Trusts and producing new strategies for socially-minded housing for all.

EVOLUTIVE HOUSING
Common Property and Addressing the Middle Scale in Houston, Texas
NO OFFICE

THIRD WARD

URBANbuild 18

Byron Mouton [C]

Two years ago, as the fabrication of URBANbuild16 was underway, a relationship was established with members of the neighboring Bethlehem Lutheran Church – the oldest historically black Evangelical Lutheran Church in the continental United States. The size of the congregation has waned since hurricane Katrina, but Pastor Ben Groth is working to increase membership and spearhead a successful community outreach initiative, Community Table. Pastor Groth pursued his PhD at Tulane's Department of History, researching the relationship between white supremacy and Christianity. In addition to URBANbuild's focus on the topic of housing, his continued involvement with this studio and the resulting conversations addresses challenges related to broader issues of community engagement and reconciliation.

Bethlehem presently owns two properties in Central City. One is located at 1823 Washington Avenue, and a portion of that lot holds their current Church building. The other property is a large empty "Key" lot directly across the street at 1822 Washington Avenue. Last year's UB17 research team suggested long term coordinated uses for both sites. Contextual analysis, both physical and cultural, combined with interpretations of the New Orleans Comprehensive Zoning Ordinance (CZO) fueled the first steps in the development of master planning options. That research influenced New Orleans' City Council to approve a text amendment to the CZO authorizing the City Planning Commission to allow increased density in residential districts for use as small multi-family affordable developments, if certain criteria are met. This phased URBANbuild project is the first in the city to take advantage of that amendment.

Developments focus on big questions related to housing, complimented by material consideration and resolution at the scale of detail. UB18 occupies Phase 2 of a phased four-year strategy aiming to achieve a "Model" master plan through a series of sequential design build opportunities. Research builds upon the program's past inquiries related to topics of efficiency, density, replication, and affordability - while also aiming to provide physical outcomes of quality thought and design.

Students:

Maria Arrieta
Andrew Balestrary
Malia Bavuso
Mike Brady
Eric Cheesebrough
Jenna DeSmidt
Ashley Dwyer
Anthony Gagliano
Clara Gardner
Lauren Lake
Bryan Montenegro
Ryan Russel
Helen Samay
Katherine Schwab
Ian Shaw
Esther Zulem

s > URBANbuild team **i >** Byron Mouton

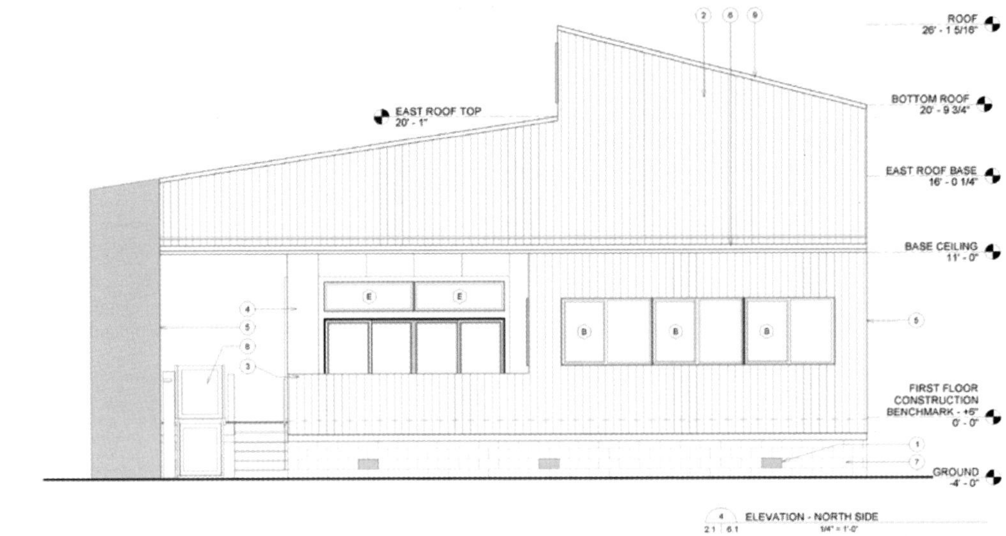
ROOF
26' - 1 5/16"
BOTTOM ROOF
20' - 9 3/4"
EAST ROOF TOP
20' - 1"
EAST ROOF BASE
16' - 0 1/4"
BASE CEILING
11' - 0"
FIRST FLOOR CONSTRUCTION BENCHMARK - +6"
0' - 0"
GROUND
-4' - 0"
ELEVATION - NORTH SIDE
1/4" = 1'-0"

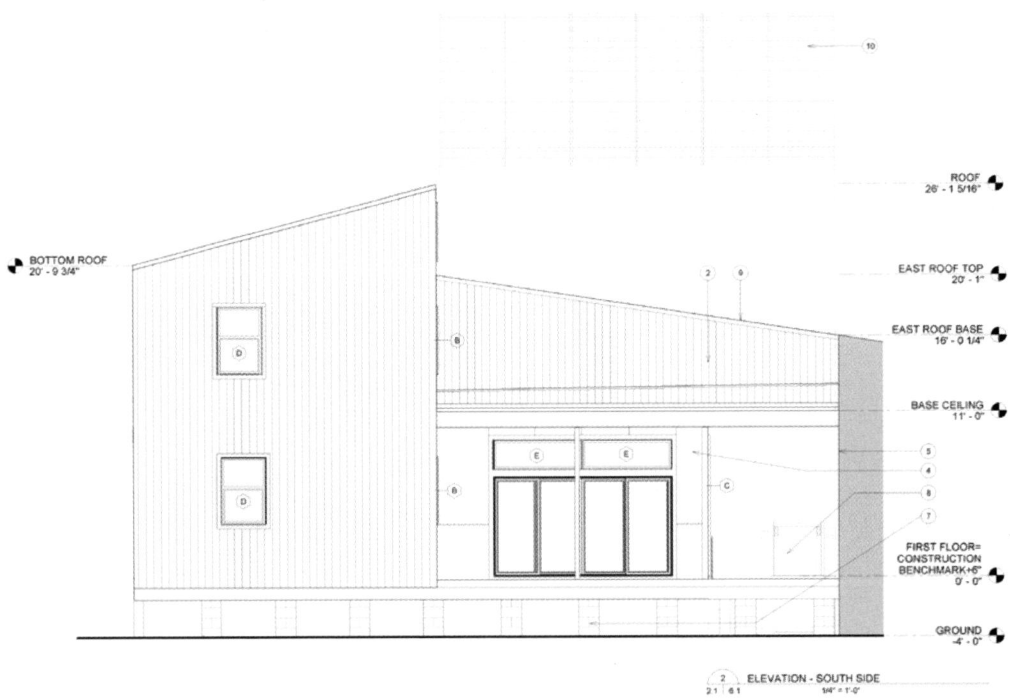
ROOF
26' - 1 5/16"
BOTTOM ROOF
20' - 9 3/4"
EAST ROOF TOP
20' - 1"
EAST ROOF BASE
16' - 0 1/4"
BASE CEILING
11' - 0"
FIRST FLOOR= CONSTRUCTION BENCHMARK+6"
0' - 0"
GROUND
-4' - 0"
ELEVATION - SOUTH SIDE
1/4" = 1'-0"

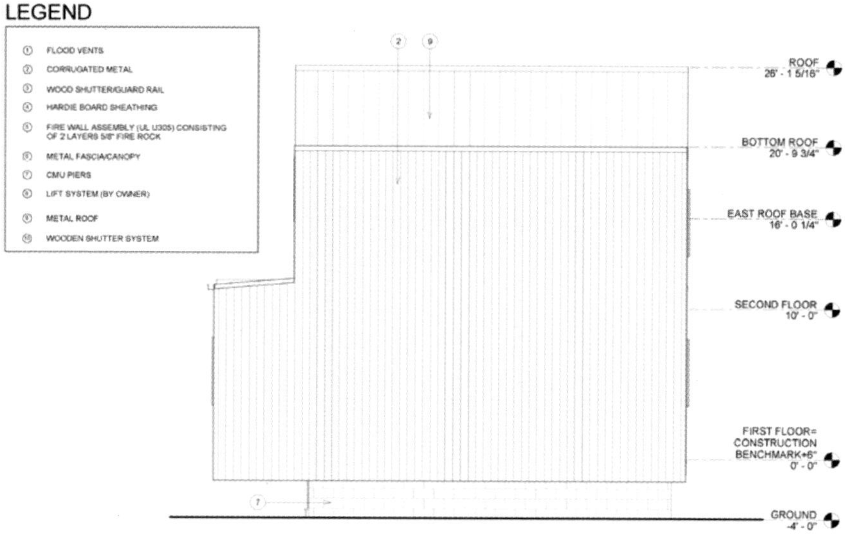

3 ELEVATION - WEST SIDE 1/4" = 1'-0"

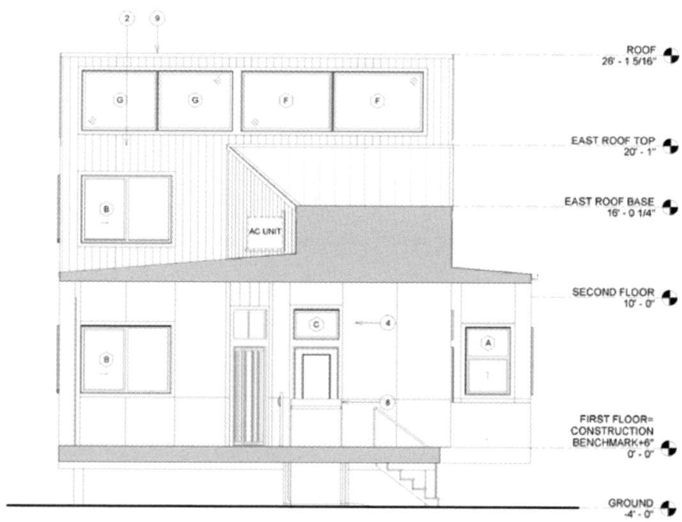

1 ELEVATION - EAST SIDE 1/4" = 1'-0"

Home
Tyve
HomeWr

ARCH 4052-5052 6042 **Research Studio**

UG + GR

FA22 + SP23

s > URBANbuild team

i > Byron Mouton

ARCHITECTURE THROUGH COMPETITIONS

Juan Medina Revilla [C]

Architecture Through Competitions. A Process of Synthesis aims to broaden the skills of design and complex concept development within a competition process and delivery format. Visual representation, scale, architectural language, layout, and hierarchy each play an essential role in the studio, with three real competitions at different scales and as the basis of investigation.

From the research phase and the identification of precedents, through the drawing stage, and ending with presentations of work, the parameters and hierarchies are altered by external statements. Interpreting and utilizing these statements are fundamental to the studio' structure. This process helps students evaluate the methods of representation, which includes choice of scales, colors, and other decisions that best enhance the idea.

The course is facilitated through three competitions at different scales: small, medium, and large. The small competition is individual. The medium and large projects are organized for group work, to emulate real-world scenarios where work distribution and understanding personal strengths are critical for a strong team dynamic. The small competition is a pavilion that is part of an annual architectural festival. The medium level is a small library in a very determinant historical context. The final, large competition is a 40-unit residential building in a heterogeneous neighborhood where students consider social, urban, and architectural parameters.

s > Michael Clifton + Tiger Thepkanjana **i >** Juan Medina Revilla

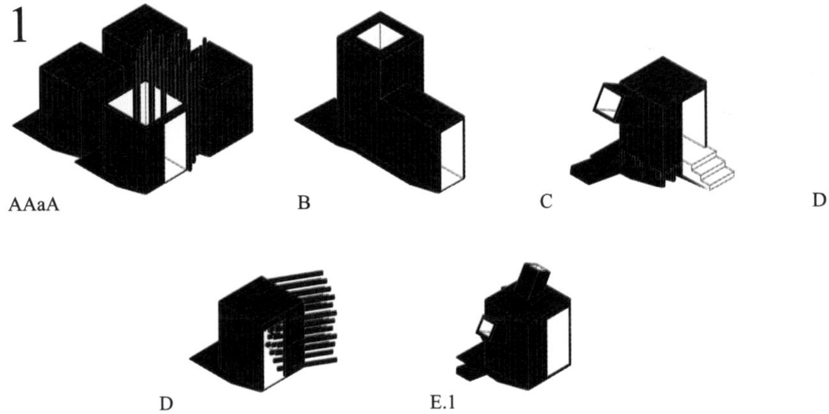
1
AAaA
B
C
D
D
E.1

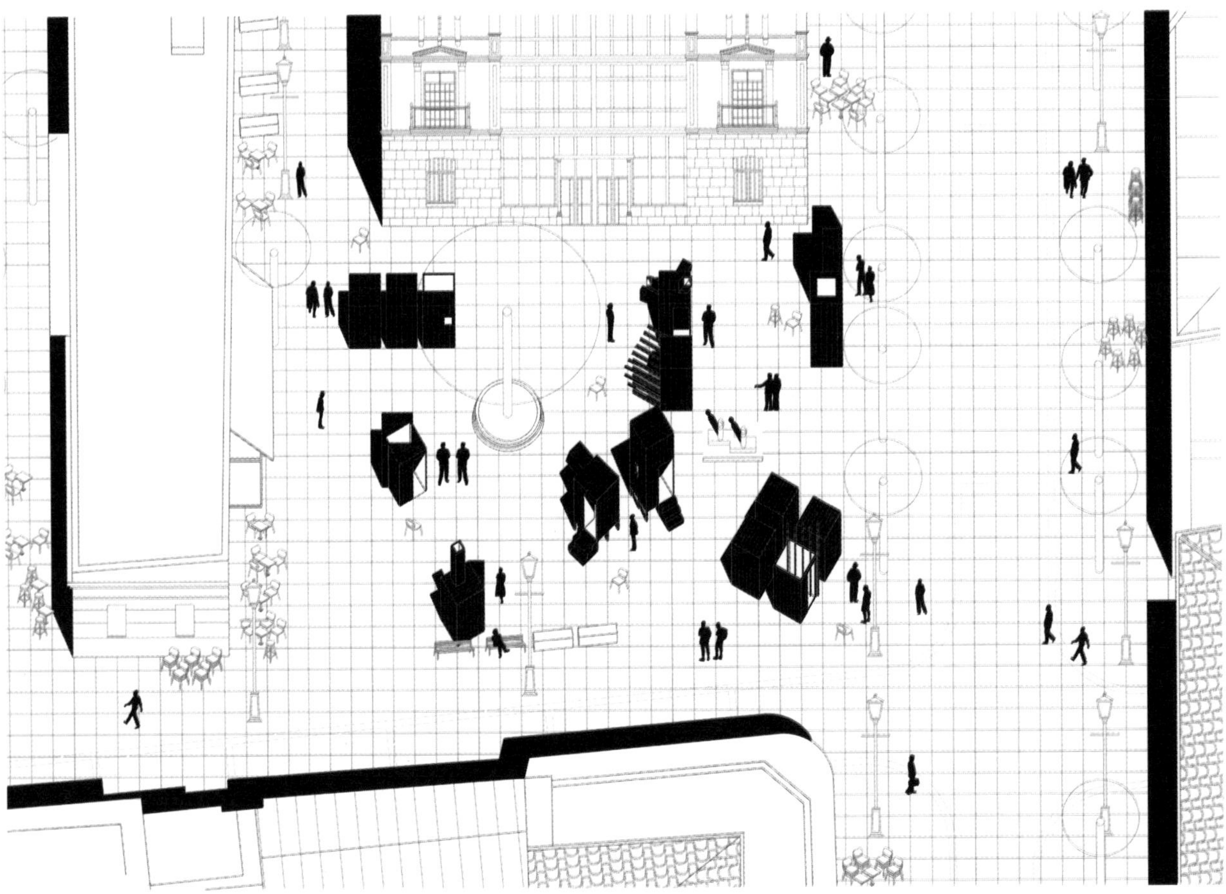

s > Alex DiSimone

i > Juan Medina Revilla

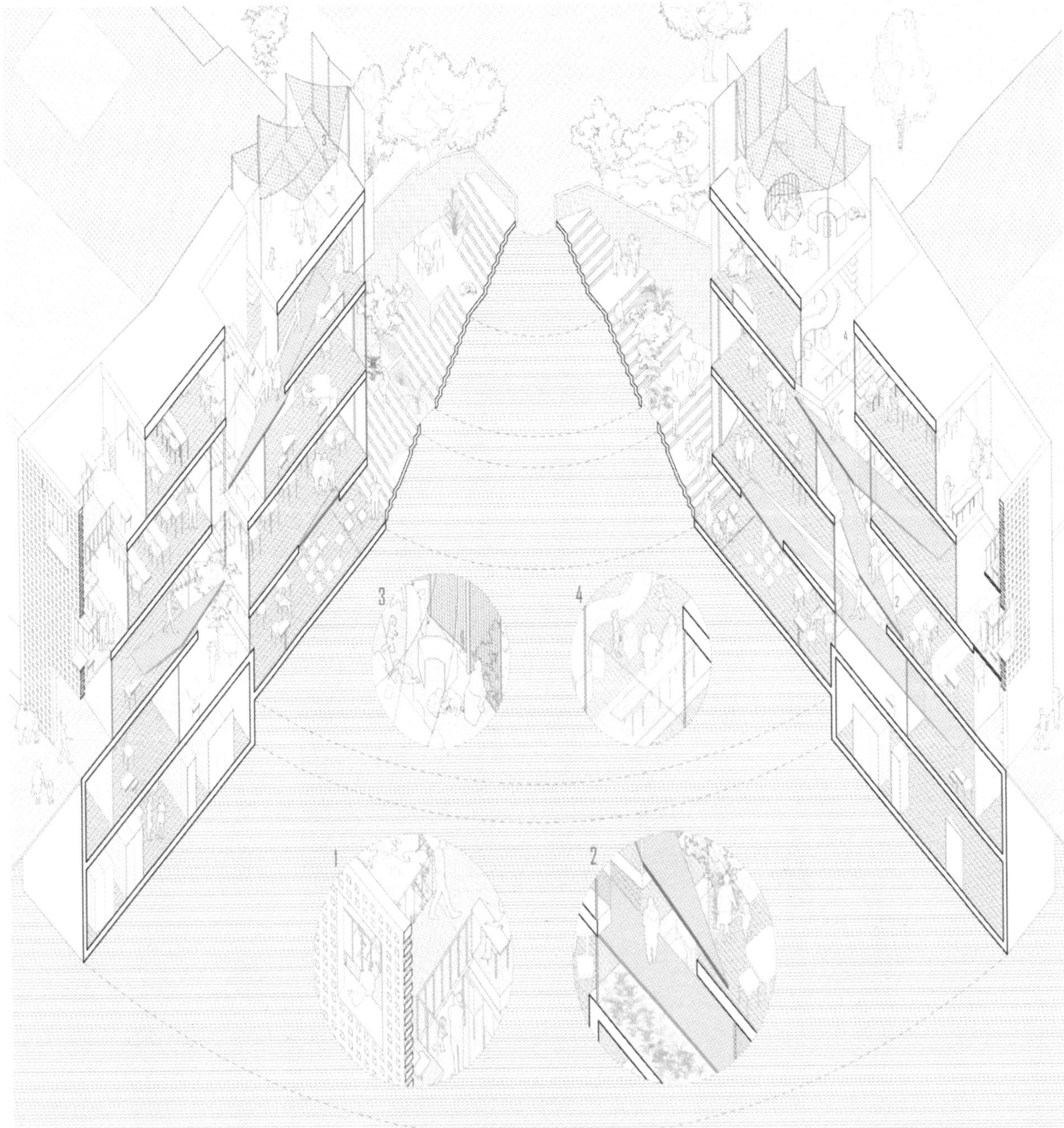

3
4
1
2

s > Alyssa Barber + Andreea Dan **i >** Juan Medina Revilla

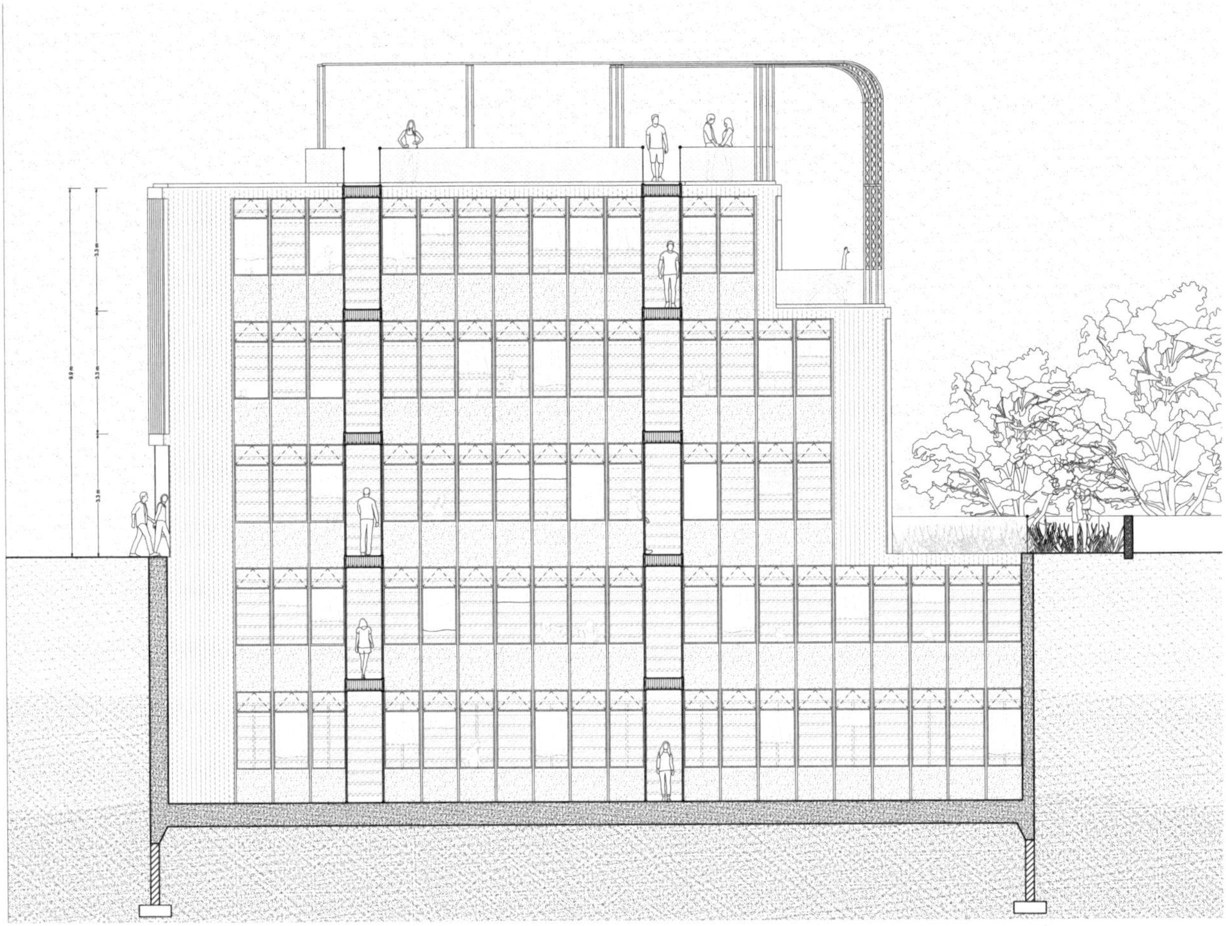

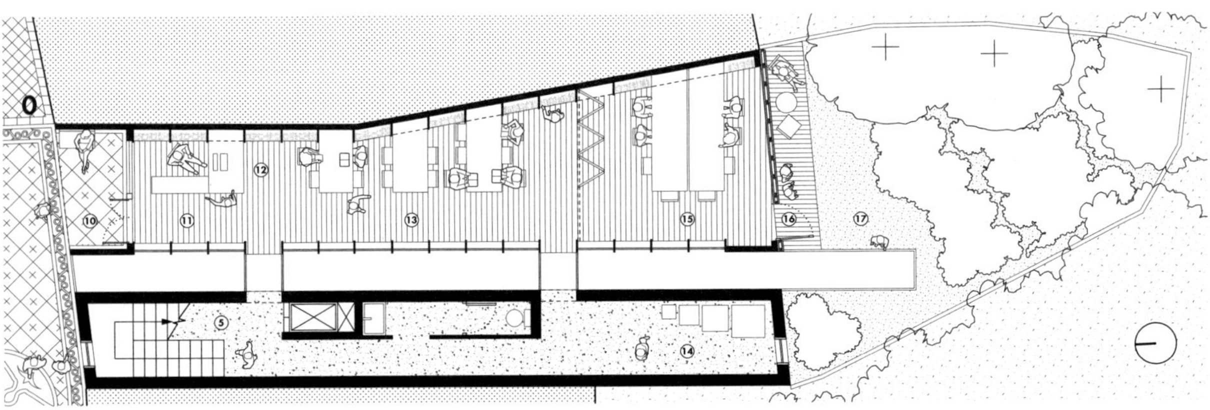

5
10
11
12
13
14
15
16
17

s > Ethan Lewis + Alex DiSimone

i > Juan Medina Revilla

s > Alex Langley + Michael Clifton + Tiger Thepkanjana + Seth Laskin

i > Juan Medina Revilla

ALYSSA BARBER & OLIVIA GEORGAKOPOULOS

LIVING IN DEAD SPACES

Mitigating the Housing Crisis Through the Means of Adaptive Reuse

ARCH 5990 Instructors
Cordula Roser Gray & Ammar Elouini

s> Alyssa Barber & Olivia Georgakopoulos **i>** Cordula Roser Gray & Ammar Elouini

San Francisco's housing crisis is one of the biggest problems that its population faces. Between 2000 and 2018 it was projected that over 1 million housing units would need to be built to accommodate the city's rising population. Only 350,000 of those necessary units were actually built. This lack of available units has had a major impact on the housing market, making San Francisco one of the least affordable cities in the US. By 2070, an additional 2 million housing units will be required. An abundance of vacant buildings in the bay area could be a solution to this problem (California YIMBY, The Bay Area's Housing Deficit – Past and Future). While housing should be viewed as a fundamental human right, with a lack of housing units, this is not the case. In dense urban areas like these, the number of vacant buildings is also on the rise due to either a lack of funding or an obsolete program. Because cities like these have a shortage of developable space, vacant buildings offer an opportunity for real estate that is often overlooked.

Among the abundant dead buildings in the Bay Area, the vast majority of buildings with the same typology and size are religious. With declining membership rates exacerbated by the COVID-19 pandemic and scarce funding, more and more of these once-lively buildings now lie stagnant. These churches pose a new opportunity to utilize buildings that have sat empty for years. The adaptive reuse of abandoned churches presents an ideal means to lessen the affordable housing crisis. Not only do these abandoned churches provide a basis for revitalization (as they have been maintained over the years despite their abandonment), but their transformation into affordable housing to aid the surrounding community aligns perfectly with the overall mission of religious institutions.

Fig. 1
Minimum, Moderate, & Maximum Budget Sections

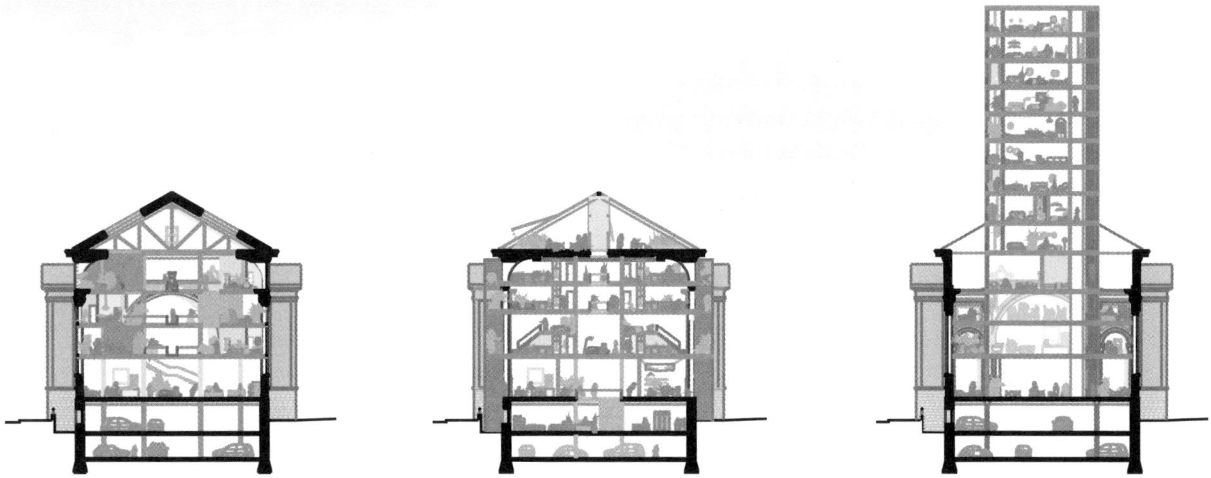

Fig. 2
Moderate Budget Sectional Model

Fig. 3
Minimum, Moderate, & Maximum Sectional Models

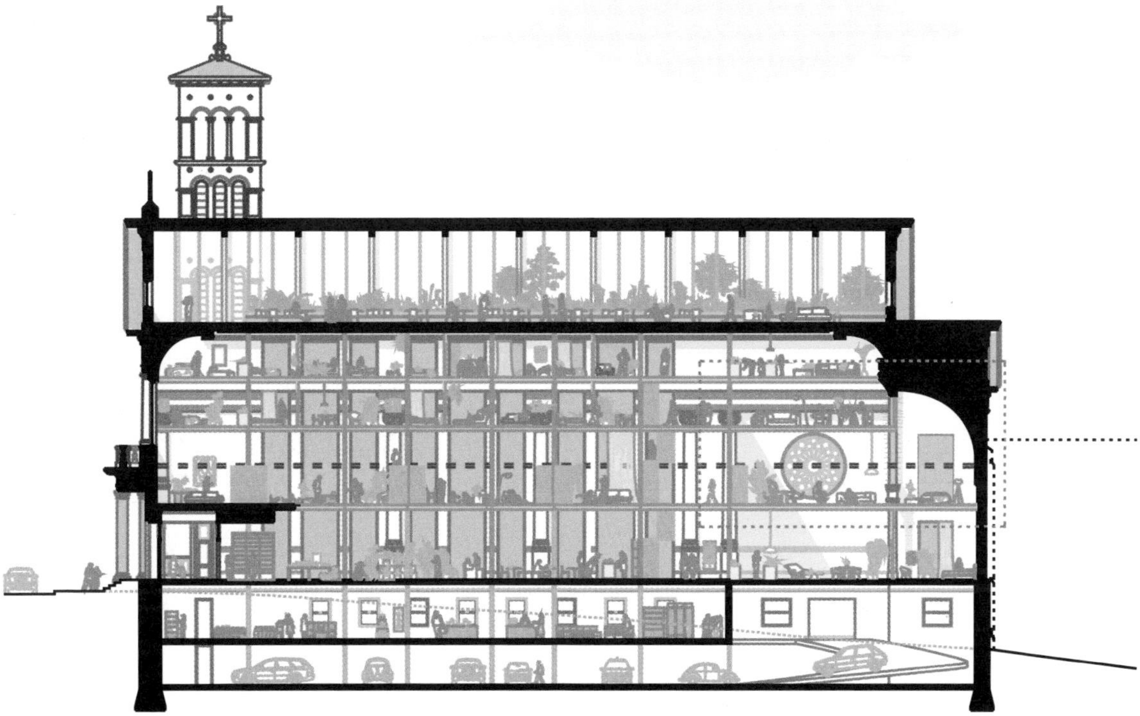

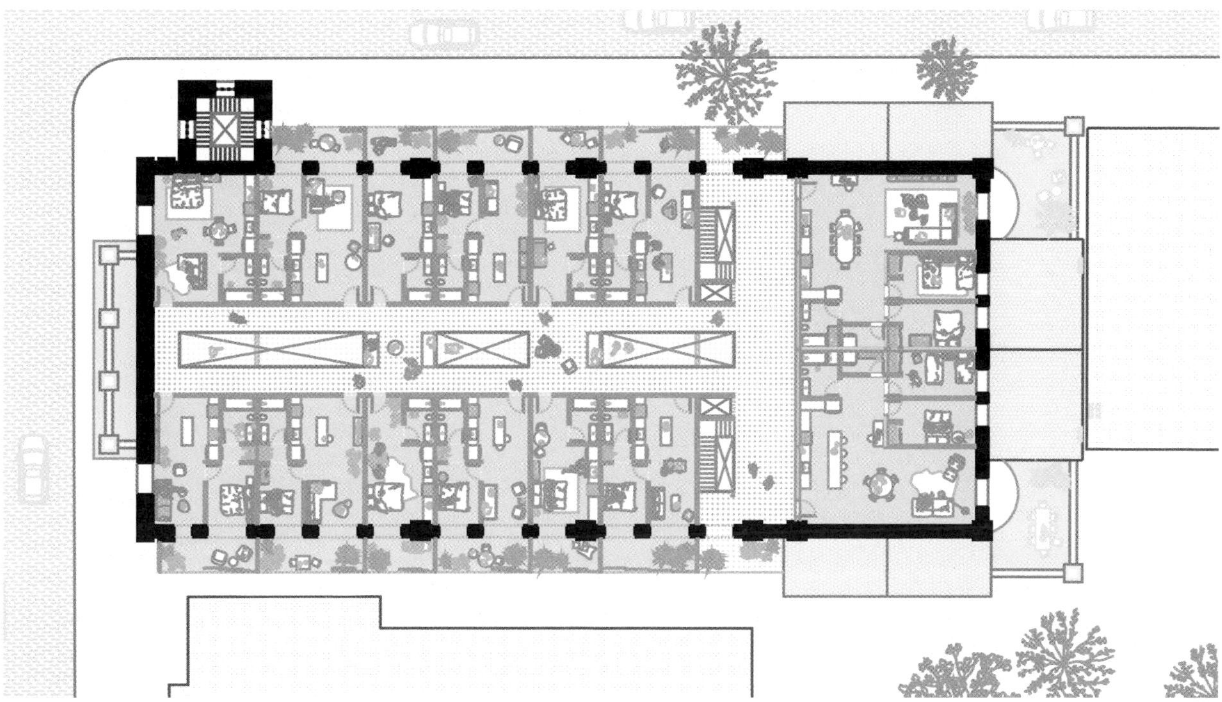

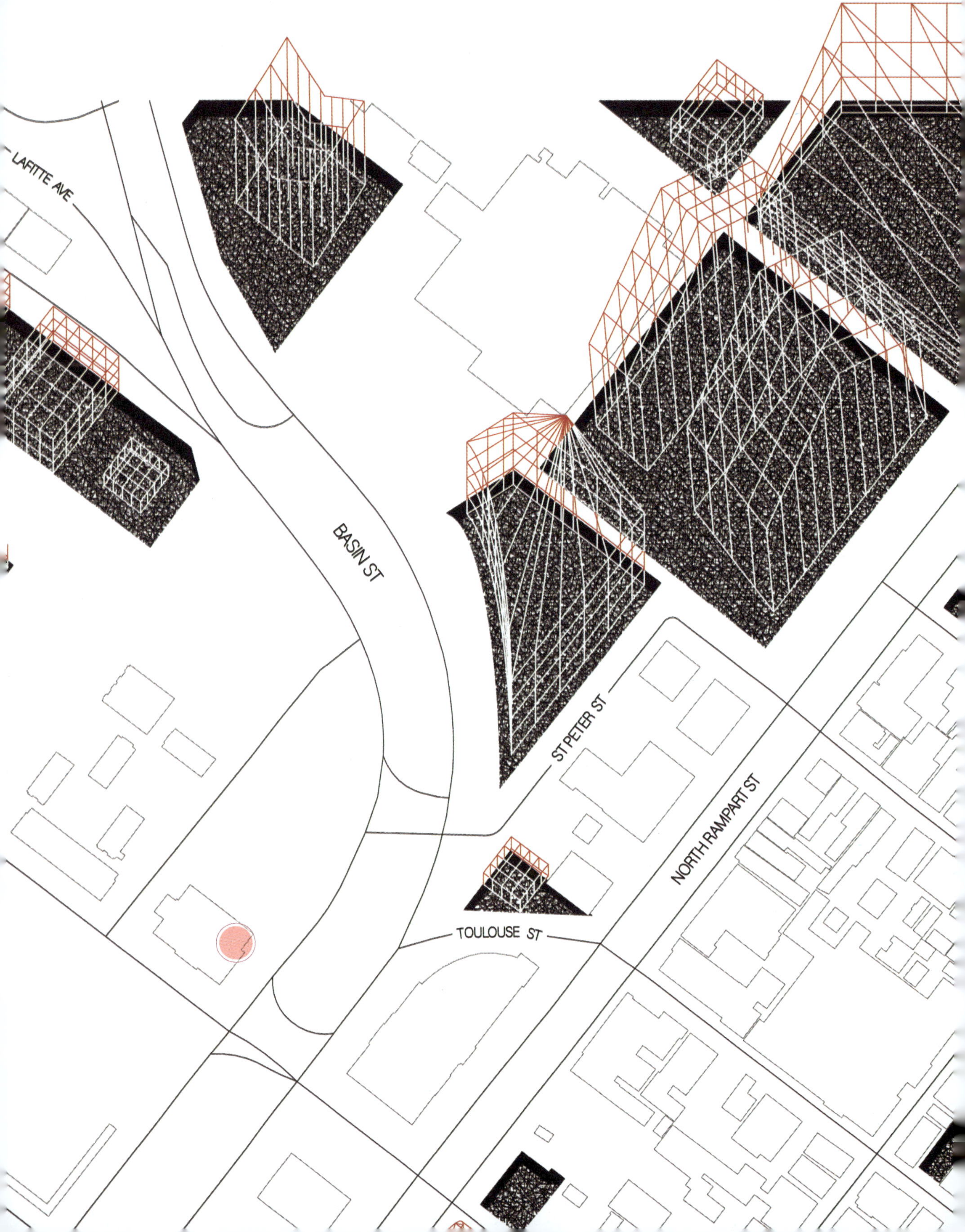
LAFITTE AVE
BASIN ST
ST PETER ST
NORTH RAMPART ST
TOULOUSE ST

SOCIALLY AFFORDABLE
Public Infrastructure

ADAM MARCUS, MOLLY REICHERT,
JOHN KIM & DANIEL DEAN

LINES OF FLIGHT, HUMAN

FACADE DESIGN
Minneapolis, Minnesota 2021

Credits
Adam Marcus, Molly Reichert, John Kim, Daniel Dean, Pete Pham (Design Team.) Alicia Johnson (Data Consultant.) Henning Larsen Architects, MSR Design (Building Design.) Corey Gaffer Photography, Futures North (Photography.) City of Minneapolis (Client.)

Lines of Flight, Human is a large-scale architectural facade design for the new Public Service Building at City Hall Plaza in Minneapolis. One of two projects commissioned by the City of Minneapolis for this new ten-story building, this custom-designed pattern extends along 150'-0" of the building's second floor and 400'-0" of a public skyway bridge that connects to the county office building across the street. The facade is intended to meet bird-safe standards for high-rise glass curtain walls, while also providing a dynamic and engaging experience for both people within the public circulation space of the second floor and pedestrians on the street level.

The project is inspired by the history of human migration to the state of Minnesota. Lines of Flight, Human references immigration's role in the forced dispossession and genocide of Indigenous communities with a visual land acknowledgment of Indigenous place names. The project also redresses human migration's consequences for non-human life by reducing bird mortality with bird-safe design. The pattern spatializes demographic data as a series of nested bands that weave over and under one another to create an interconnected network across the facade. Each band corresponds to a different constituent population of Minnesota and serves as a kind of timeline, mapping the history of that respective population's migration to the state. The variable dash pattern of each band intensifies in density and line weight to indicate moments of greater migration. The resulting pattern produces multiple readings for multiple constituents at multiple scales: as a bold super-graphic from the street, as an intricate, lacy pattern for the thousands of people who walk through the space each day, and as a bird-safe deterrent for migrating avian species.

This project was commissioned by the City of Minneapolis after a national selection process. Futures North worked closely with the building architects Henning Larsen and MSR Design to coordinate the pattern with the curtain wall design.

Fig.1
Inside view
Photo. Corey Gaffer Photography.

Fig. 2 Facade Design.
Axon. Futures North.

Fig.3 Outside view.
Photo. Corey Gaffer
Photography

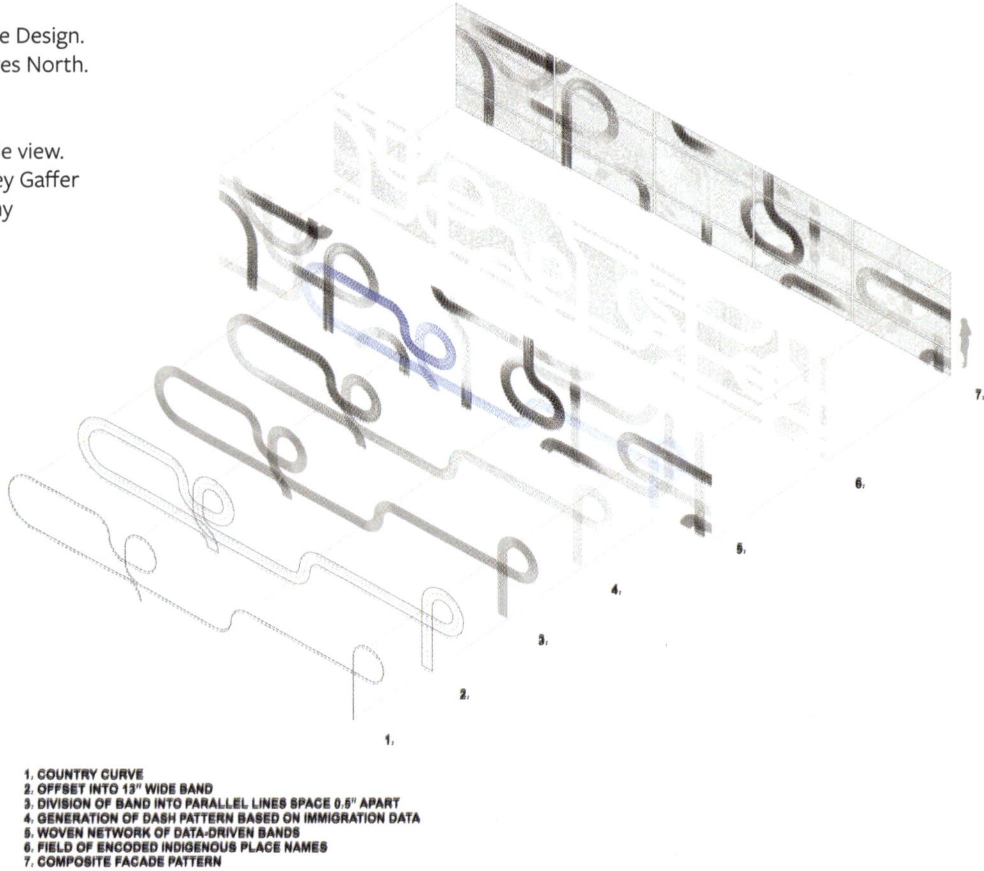

Fig. 4
Outside view.
Futures North.

URBAN RESILIENCE

Ruben Garcia Rubio [C]
Omar Ali
Miwako Hattori
Sonsoles Vela

The core 3031 studio introduces scale in the context of a city. Students had previously studied the urban scale and its integration with the architectural one, as well as the metabolic and territorial scale used by architects and urbanists. This studio reinforces and expands on these lessons, focusing on the integration of urban and architectural design through the understanding of the intermediate scale. The project expands on a previously designed area in Los Angeles.

The studio is divided into three interconnected parts. Students begin by investigating, analyzing, and synthesizing the city's major urban systems transversely and across multiple scales. Next, students identify urban issues and opportunities to propose a resilient urban design for a specific part of the city, focusing on the architectural dimension of urban design. They also analyze significant urban precedents during this phase. Lastly, students develop an architectural design that acts as a catalyst for the previously designed part of a city based on a specific program.

s> Natalie Miller, Naomi Smith, & Joey Tomshe

i> Ruben Garcia Rubio

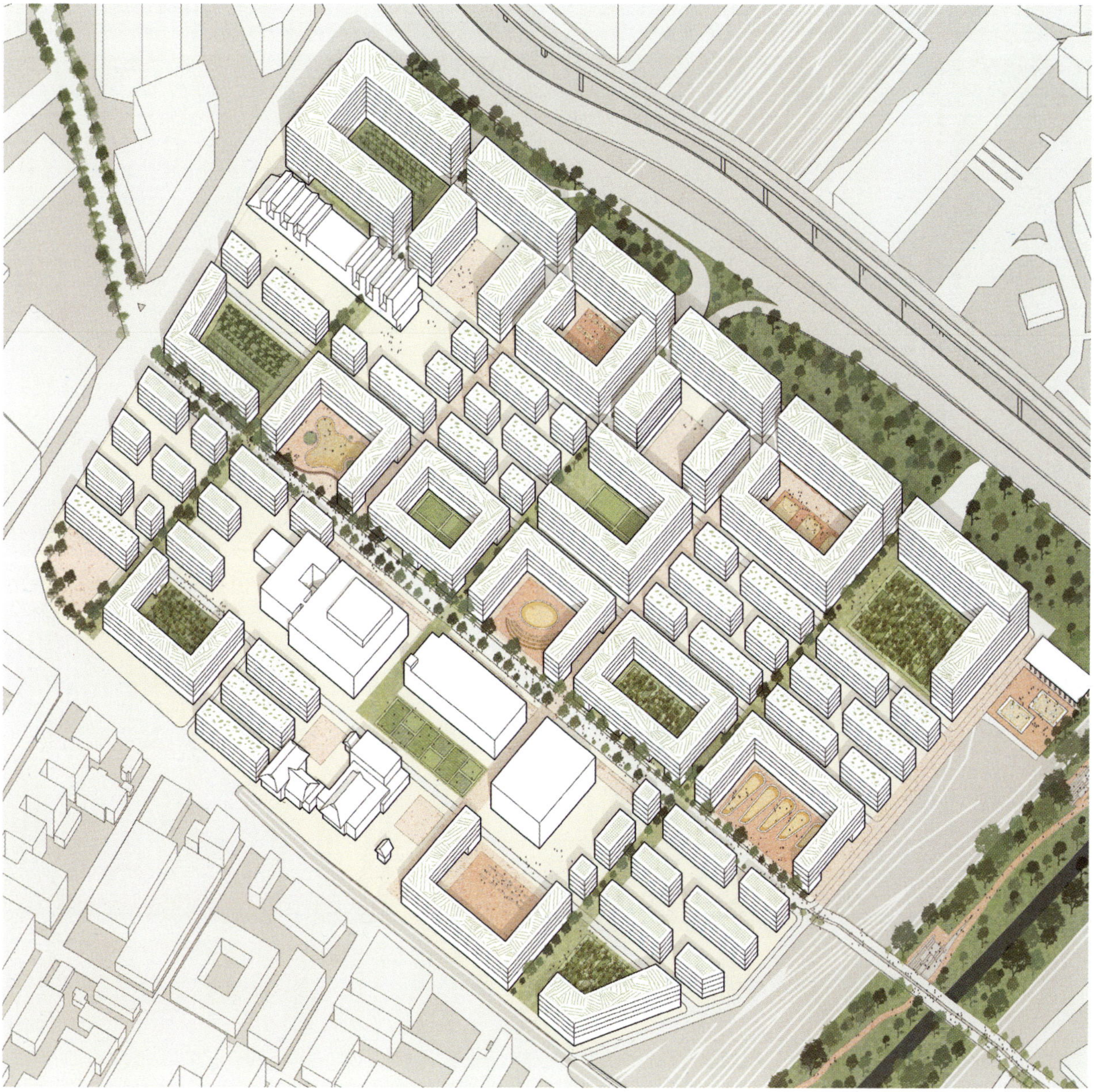

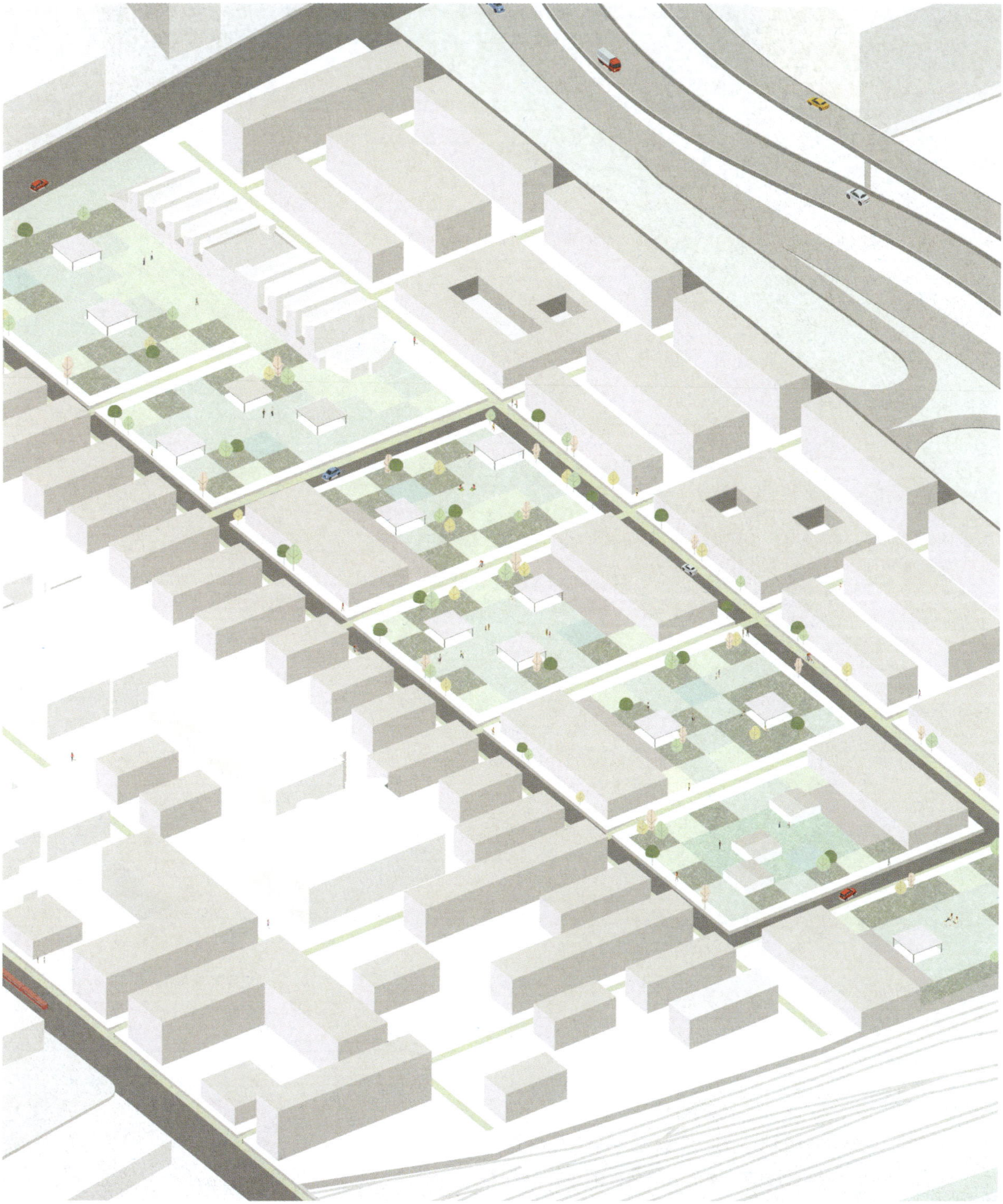

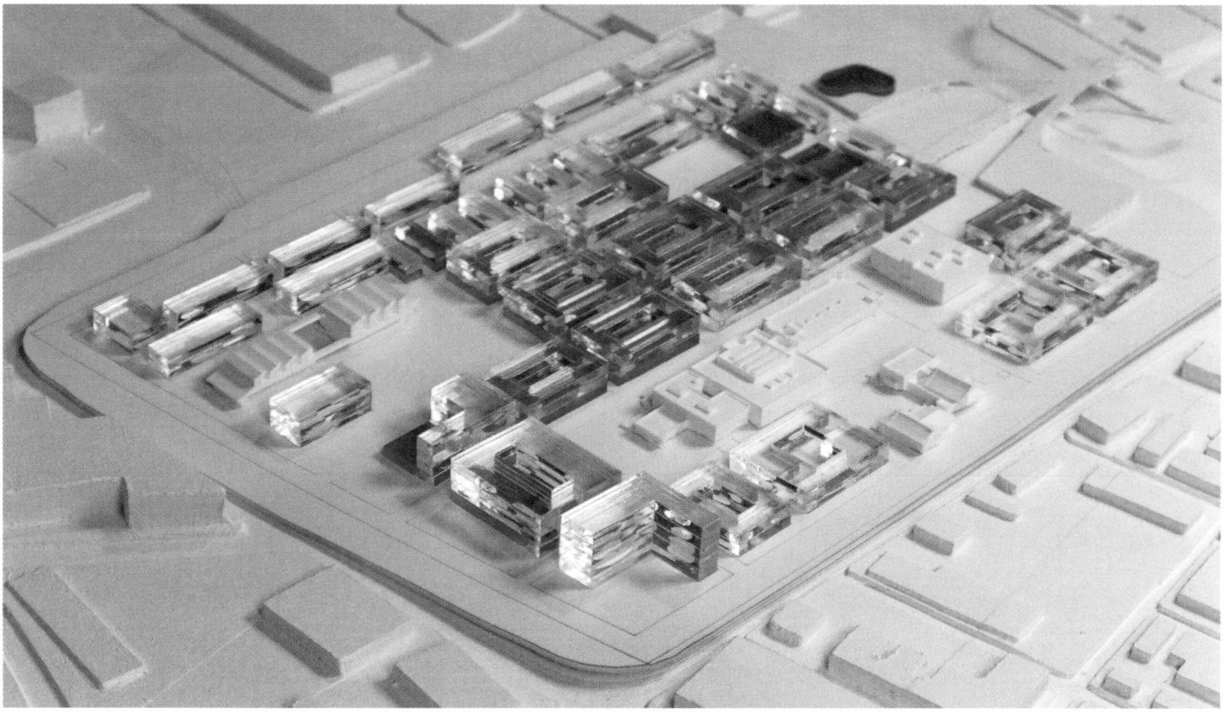

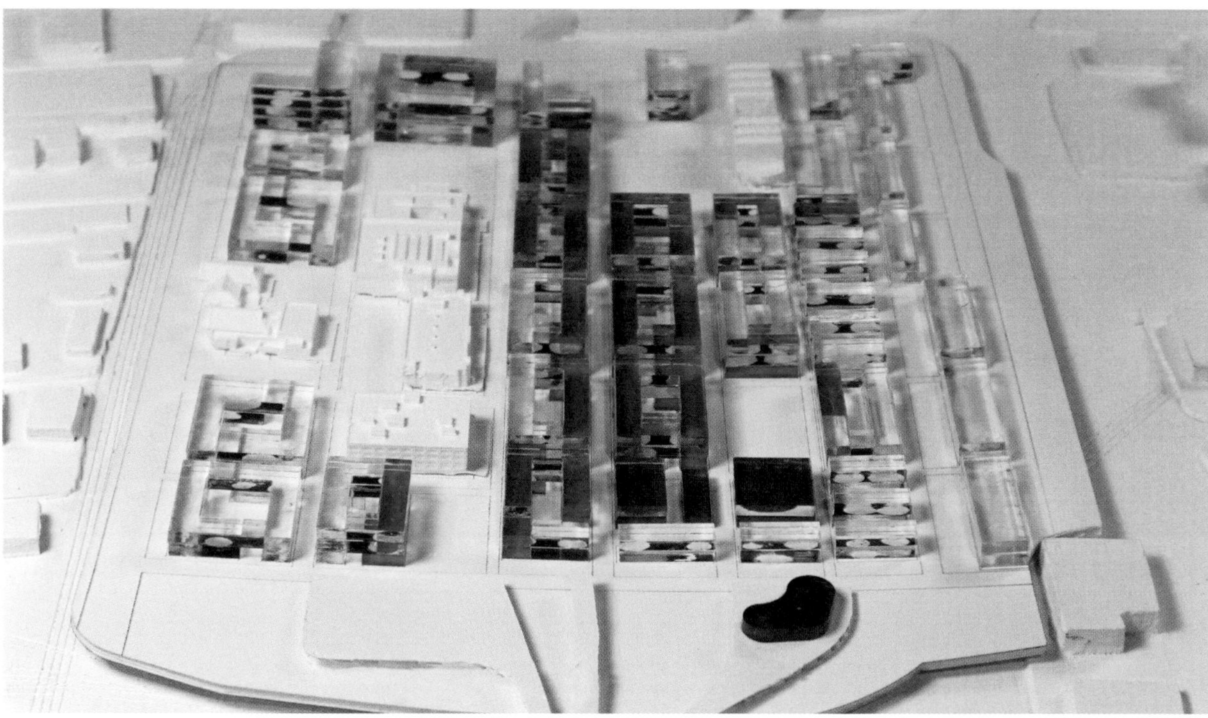

s> Naomi Smith

i> Ruben Garcia Rubio

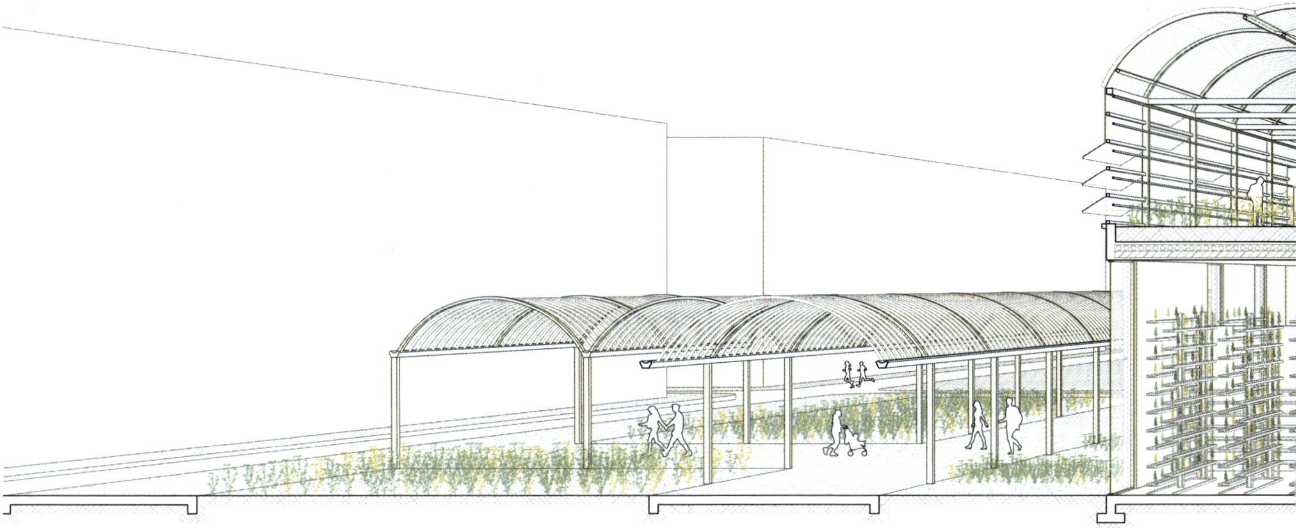

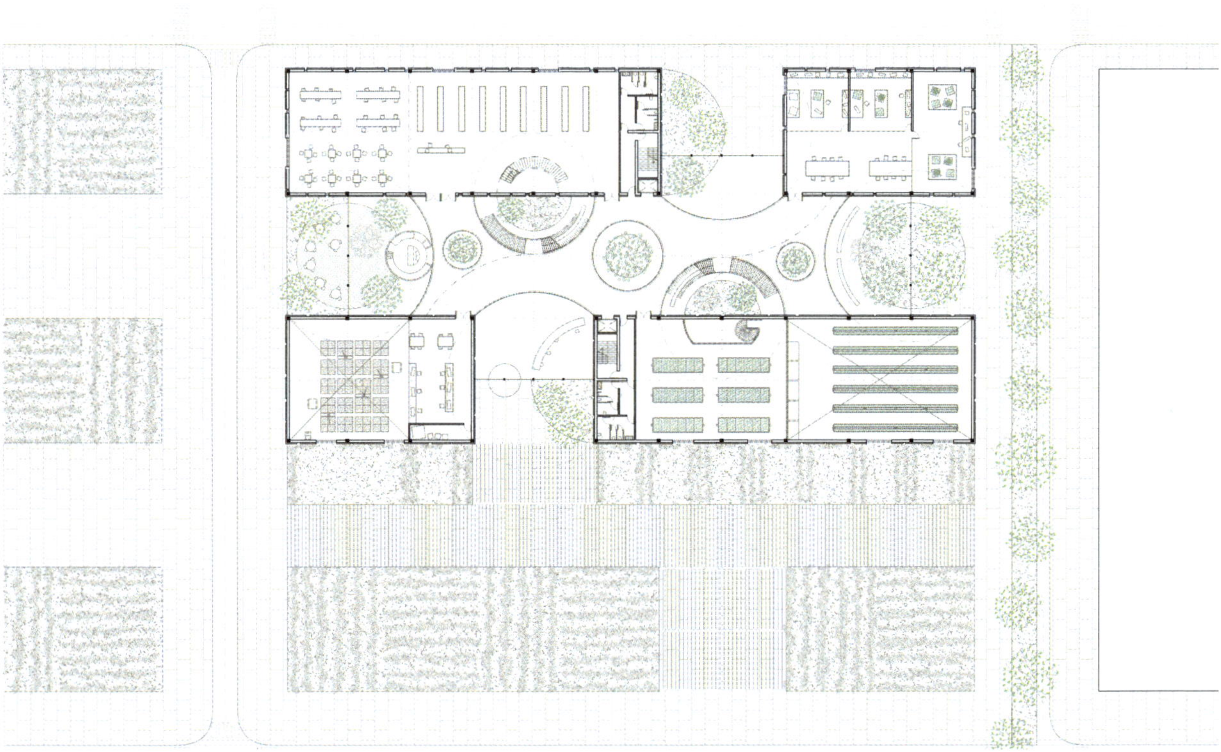

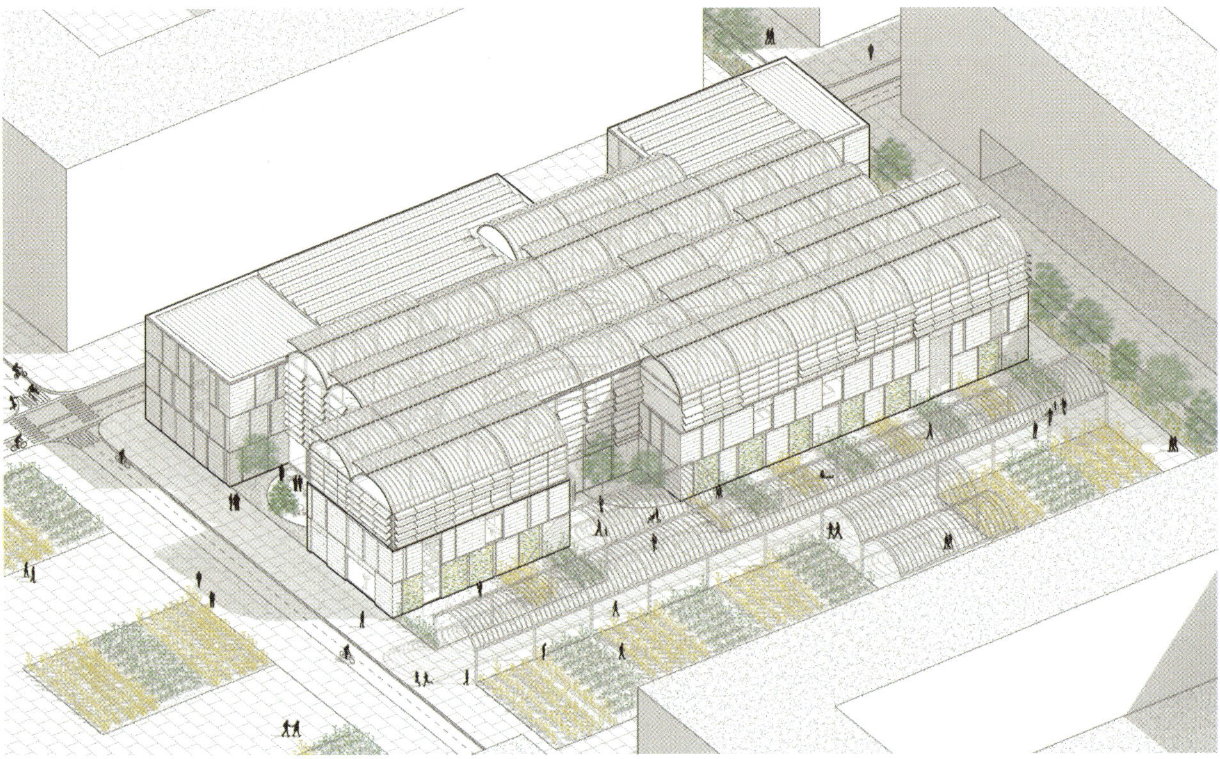

s> Cassie Kadish

i> Omar Ali

STEM LIBRARY LAB

Emilie Taylor Welty
Charles Jones
Jose Cotto

STEM Library Lab (SLL) is a center where teachers from Metro New Orleans schools can find resources to develop STEM curriculum for their students. Started in 2017, SLL has grown to help teachers across the area, offering access to new equipment and a space for collaboration in their 12,000 square foot warehouse facility. SLL reached out to the Small Center to reconceive and redesign their interior warehouse space to support these functions.

Undergraduate and graduate students working on the project began their research by engaging with the center's users. They sent out surveys and held community events where patrons of the center could share their needs, allowing the team to better understand how the space is used. User engagement played an integral role throughout the project's development.

The final design provides a bright and engaging entryway and front desk, expanded event and collaborative work space, and functional spaces including storage and kitchen facilities. Warm and creatively deployed materials create a welcoming environment for staff, teacher clients, and other visitors. To maximize space and impact, the design team divided the interior into three zones — entry, lounge, and print center. Each zone is connected through materials, patterning, and a rope installation designed to guide users through the space.

The entry area is anchored by a large desk that acts as a workspace where teachers can check out items. A double-height storage wall emphasizes the tall ceilings and showcases various STEM items that are available. The lounge consists of a kitchen area with a moveable island and a series of booths that provide individual and group meeting spaces with more privacy. The CNC milling pattern used throughout the space was inspired by sound waves and DNA structures, drawing upon the science-based educational tools in the STEM-oriented environment.

Support:

Ann Yoachim
Nick Jenisch

Students:

Rachel Bennett
Lauren Liroff
Delaney Smith
Sam Spencer
Frank Taylor
William Trotter
Jose Varela Castillo
Antonia Villafranca
Bryce Winter

Belinda Chau
Charles Crawford
Katherine Dunn
Riley Siltler
Maria Arrieta Vallarino

s > STEM Library Lab team

i > Emilie T Welty + Jose Cotto + Charles Jones

s > STEM Library Lab team

i > Emilie T Welty + Jose Cotto
+ Charles Jones

COMMUNITY
AND PR

s > STEM Library Lab team

i > Emilie T Welty + Jose Cotto + Charles Jones

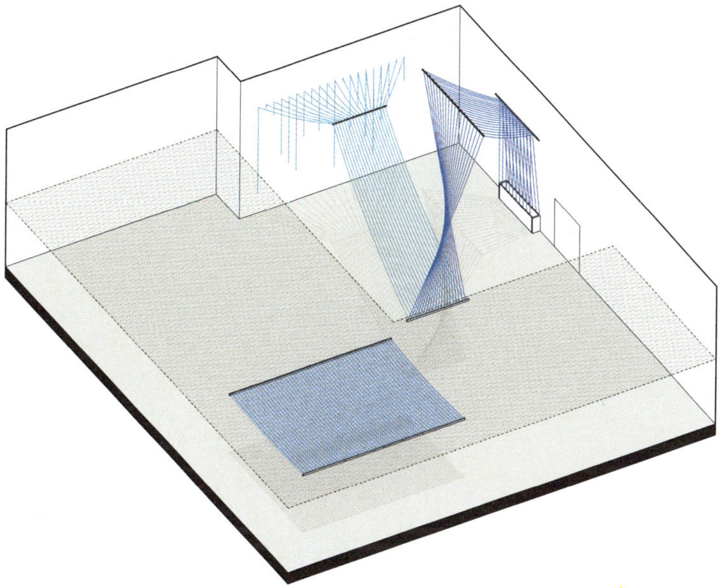

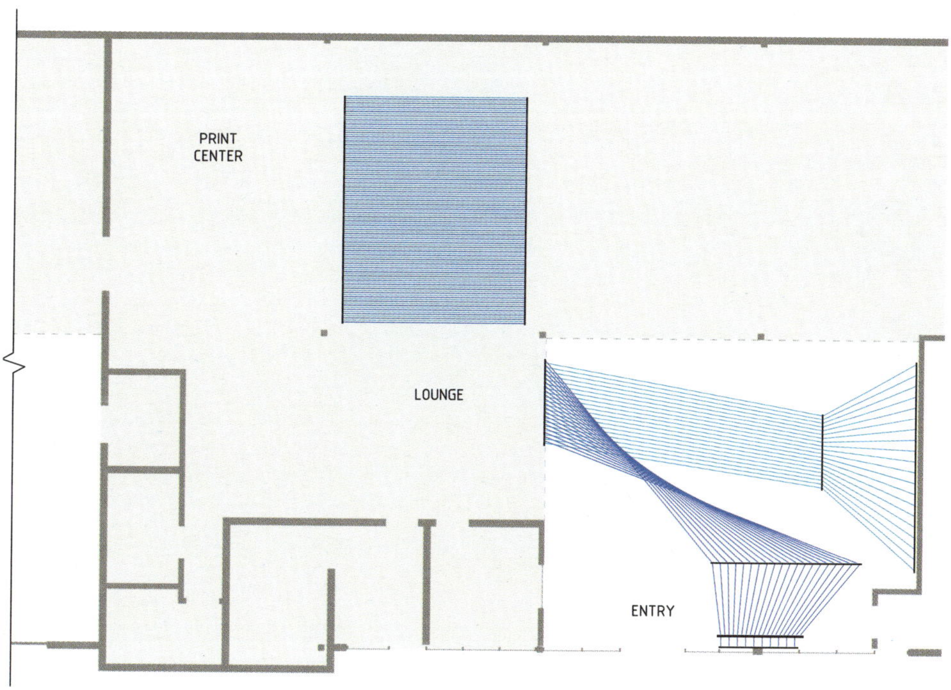

MACHINES

s > STEM Library Lab team

i > Emilie T Welty + Jose Cotto
+ Charles Jones

ARCH **Research** 4041-5051 6041-6051 UG & GR **FA21**

SUGAR ROOTS FARM

Emilie Taylor Welty
Jose Cotto

Sugar Roots Farm, located 8 miles south of the French Quarter in Lower Algiers, invites people from the New Orleans area to learn and practice intentional land stewardship. The farm's mission is to build food sovereignty and community resilience by fostering a connection between local residents and farming in the natural environment. Looking to grow their outreach, the staff at Sugar Roots teamed up with the Small Center to design and build a new outdoor teaching space at the farm, expand educational events and community engagement, and address stormwater challenges at the site.

Students from a Small Center undergraduate and graduate research studio led the design and build of this project. They met with the Sugar Roots team, proposed schematic ideas, and volunteered at the site to understand how the space was used. Sugar Roots' vision served as a filter for design decisions about materials, placement, and overall strategies as the team aimed to make concepts of full cycle farming visible throughout the site.

The design process resulted in an outdoor classroom that sits on a raised concrete slab, and integrated drainage that anchors a large steel and wooden roof structure. The roof provides a large shaded area for classes and directs water to alleviate localized flooding. The structure includes wash stations for the farm's produce and large counters for teaching canning and cooking classes. The space is bracketed by a storage area for kitchen equipment and supplies and an educational wall with fixed signage and a chalkboard for class activities. A rope screen along the street edge provides a surface for vines to grow on over time and leads visitors to an observation deck over a bioswale.

Support:
Ann Yoachim
Nick Jenisch

Students:
Merrie Afseth
Zach Braaten
Kelsie Donovan
Nick George
Connor Little
Valentina Mancera
Johnathan Michka
Malina Pickard
James Poche
Natalie Rendleman
David Rodriguez
Katie Schultz
Brianna Baldwin
Le'Bryant Bell
Kareem Elsandouby
Ellen Feringa
Ella Jacobs
Sam Lindley
Mandii Malhotra
Sofia Perrotta Mensi
Karan Sharma
Bruno Soria
Giuliana Vaccarino
Gearty
Yao Zhang

SUGAR
ROOTS
FARM

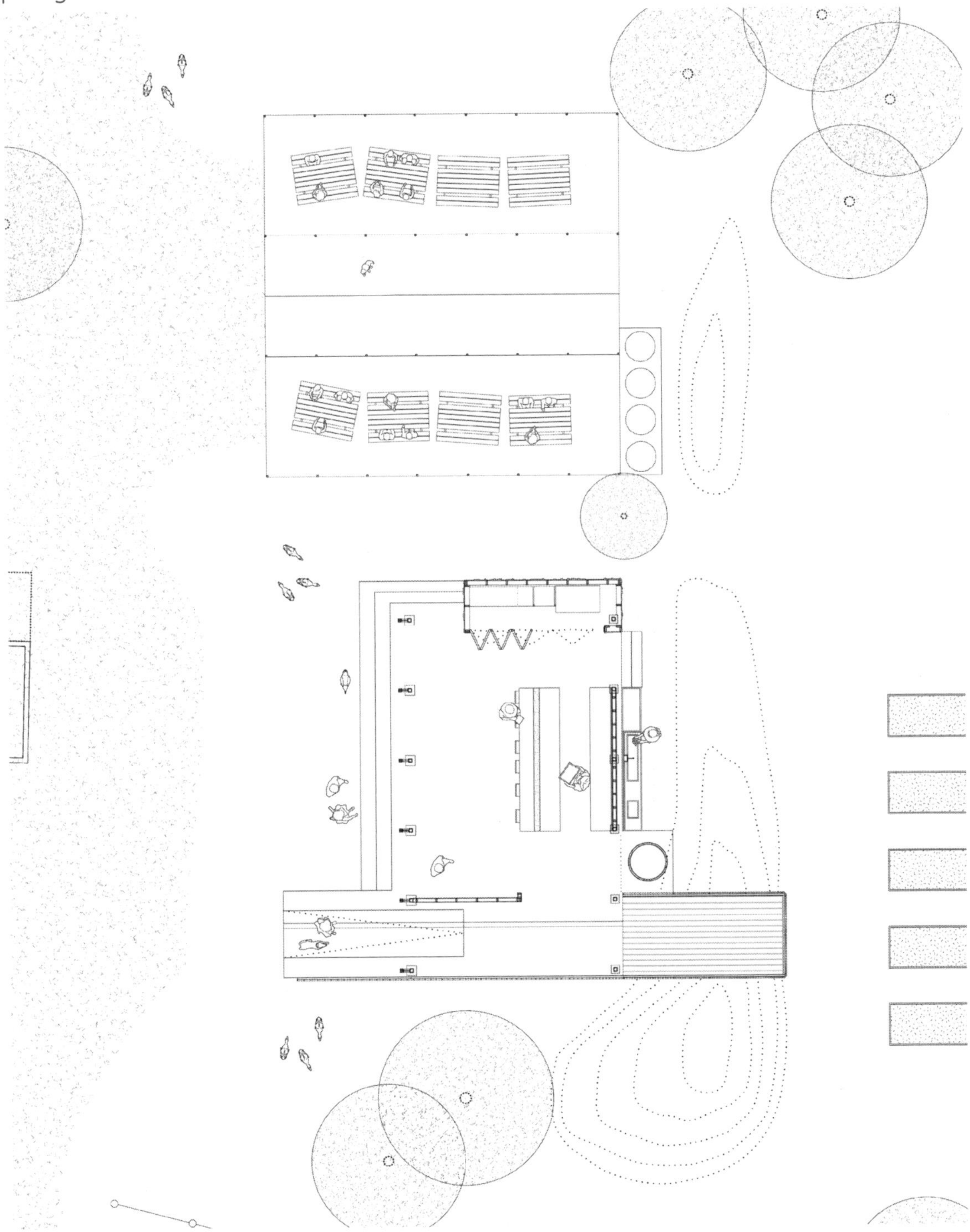

s > Sugar Roots Farm team

i > Emilie T Welty + Jose Cotto

s > Sugar Roots Farm team **i >** Emilie T Welty + Jose Cotto

s > Sugar Roots Farm team

i > Emilie T Welty + Jose Cotto

PUBLIC SPACE AND SCRUTINY

Tiffany Lin
Emilie Taylor Welty

Public Space and Scrutiny invites students to examine existing public spaces, monuments, and memorials--and collaborate on developing a framework for future design. Students learned the basic scientific methodologies and protocols used in social psychological research, including the ethics and compliance training required by the Institutional Review Board for working with human subjects. They also learned about intergroup relations research and applications with a focus on explicit and implicit bias.

Guest speakers were instrumental in contextualizing monuments and memorials, leading class discussions in the context of the national movement. Informed by the distinct perspectives of each invited speaker, the team generated the framework for community discussions and public surveys. In concert with the development of the research materials (e.g., survey questions), seminar students produced a catalog of existing monuments/memorials/ rituals and analyzed each case study in terms of formal and perceptual characteristics, working in groups to diagram and catalog these monuments/memorials in terms of type, subject matter, formal characteristics, and temporality. These studies generated an index of existing monuments and memorials in the Greater region. Existing monuments were exhibited as operable panels, inviting participants to note whether they have seen the monument, whether they knew what it was about, and whether or not they liked it–serving as additional data on perceptions of public spaces, monuments, and memorials.

The studio led students through the process of prototyping a public memorial for a linear park that traverses six urban neighborhoods of central New Orleans. The course collaborated with a non-profit organization to generate a community-engaged process that would inform the design of a new public memorial.

The specific site selected is a public park that used to house a segregated playground. Currently the park is frequented by adults, children, locals and tourists of all races, enjoying the space for recreation and fitness, but there is no recognition of its history in the location's online presence or physical space. The team generated design proposals that acknowledged this divided history, and worked to advance conversations regarding the need to co-design public spaces and memorials that reflect the communities in which they exist.

Working with a community partner with deep-seated knowledge of the site and its conflicted history allowed the studio to test strategies for more inclusive site engagement and designs.

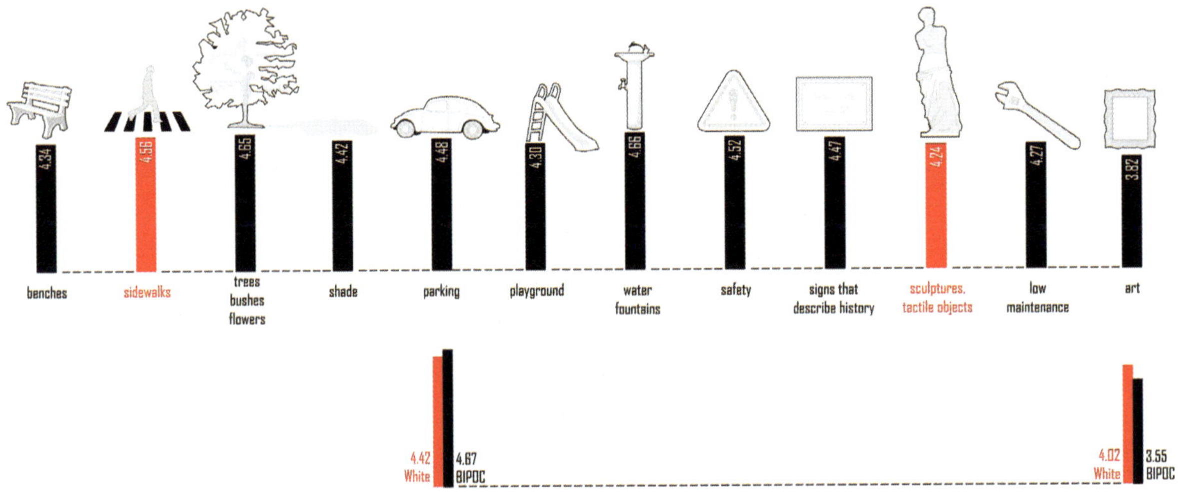

Participants who self-identified as BIPOC rated parking as being more important than those who self-identified as white.
Participants who self-identified as white rated art as being more important than those who self-identified as BIPOC.

In your opinion, what is the role of public spaces, monuments and memorials in our city?

s > Public Space & Scrutiny team

i > Tiffany Lin + Emilie Taylor Welty

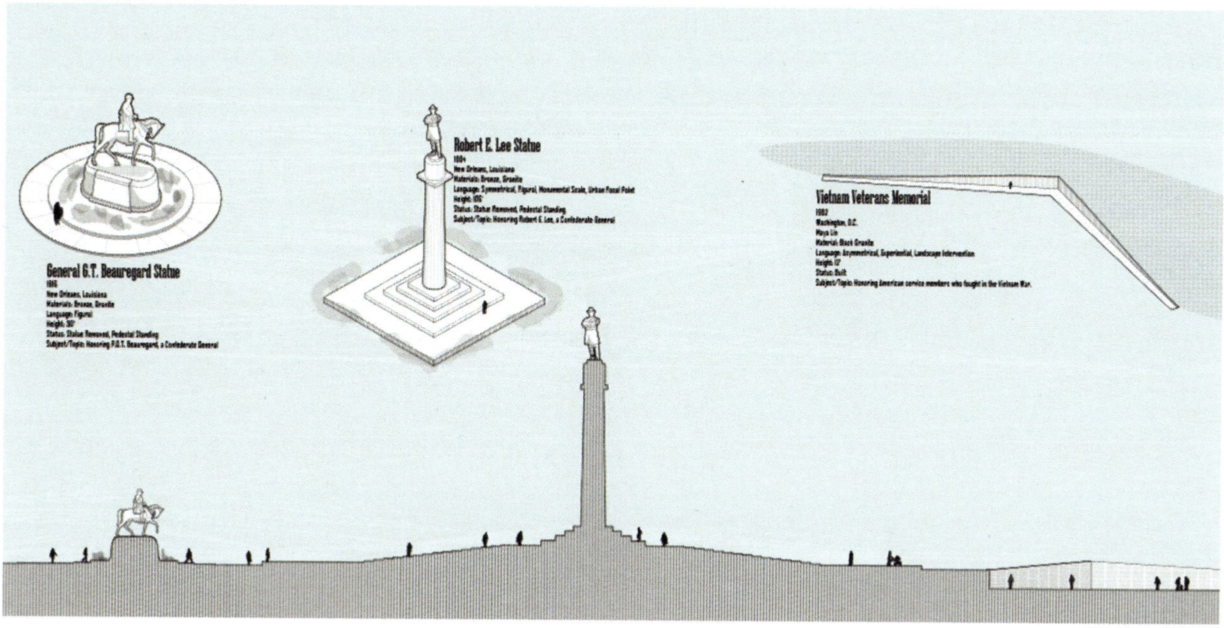
General G.T. Beauregard Statue
New Orleans, Louisiana
Materials: Bronze, Granite
Language: Figural
Status: Statue Removed, Pedestal Standing
Subject/Topic: Honoring P.G.T. Beauregard, a Confederate General
Robert E. Lee Statue
New Orleans, Louisiana
Materials: Bronze, Granite
Language: Symmetrical, Figural, Monumental Scale, Urban Focal Point
Status: Statue Removed, Pedestal Standing
Subject/Topic: Honoring Robert E. Lee, a Confederate General
Vietnam Veterans Memorial
1982
Washington, D.C.
Maya Lin
Material: Black Granite
Language: Asymmetrical, Experiential, Landscape Intervention
Status: Built
Subject/Topic: Honoring American service members who fought in the Vietnam War.

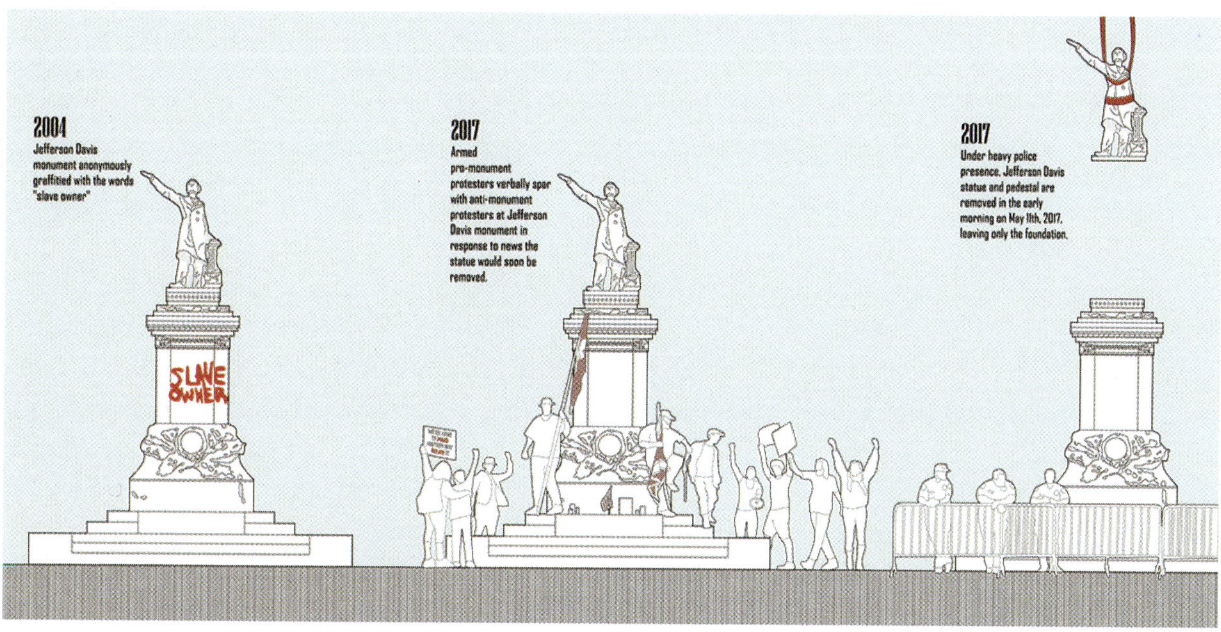
2004
Jefferson Davis monument anonymously graffitied with the words "slave owner"
SLAVE OWNER
2017
Armed pro-monument protesters verbally spar with anti-monument protesters at Jefferson Davis monument in response to news the statue would soon be removed.
2017
Under heavy police presence, Jefferson Davis statue and pedestal are removed in the early morning on May 11th, 2017, leaving only the foundation.

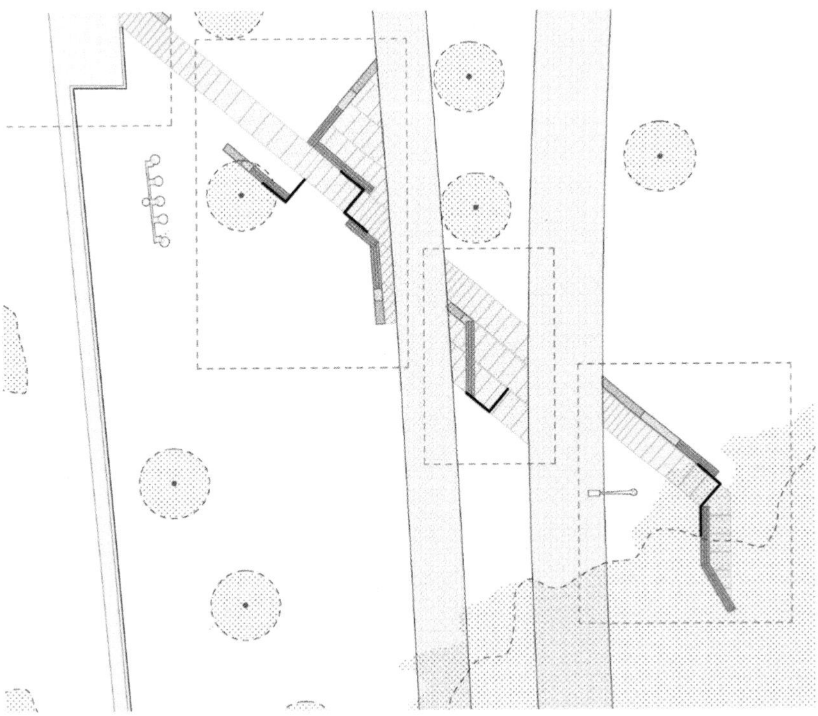

MITCHELL HUBBELL

ADAPTING BOUNDARIES

Maintaining Small Retail Strip Malls While Expanding Affordable Housing

ARCH 6990 Advisor
Judith Kinnard

s> Mitchell Hubbell   i> Judith Kinnard

The strip mall is a ubiquitous building type across North America. These structures function as blank slates for any commercial operation to fill their shells, from the original grocery and retail tenants of the mid-20th century to today's business and healthcare retrofits. In many of these sites, a boundary exists between the large-scale commercial development and the small-scale, single-family residential behind. A result of single-use zoning, this boundary is often resolved with a privacy fence separating grassy lawns from loading docks. This research and design intervention addresses the recurrent boundary condition between a strip mall's rear face and the neighborhood behind, promoting high-density, low-rise housing and encouraging neighborhood connectivity.

The site of investigation is a strip mall in the Los Angeles neighborhood of North Hollywood. Los Angeles was selected for its legacy of post-war urbanism, its history as the birthplace of strip malls, and the recent California legislation promoting commercial to residential site conversions.

This project offers an alternative to the large-scale redevelopment of retail centers with incremental housing that preserves small-scale retail tenants, often vital to communities. Designed to fit on sites with limited depth, the primary living space of the shallow housing units are raised above the ground for visual porosity and the flow of activities beneath. Strategies developed in this project can be distilled and replicated on strip mall sites across the United States to provide housing for communities in need.

Fig. 1
Site Axon

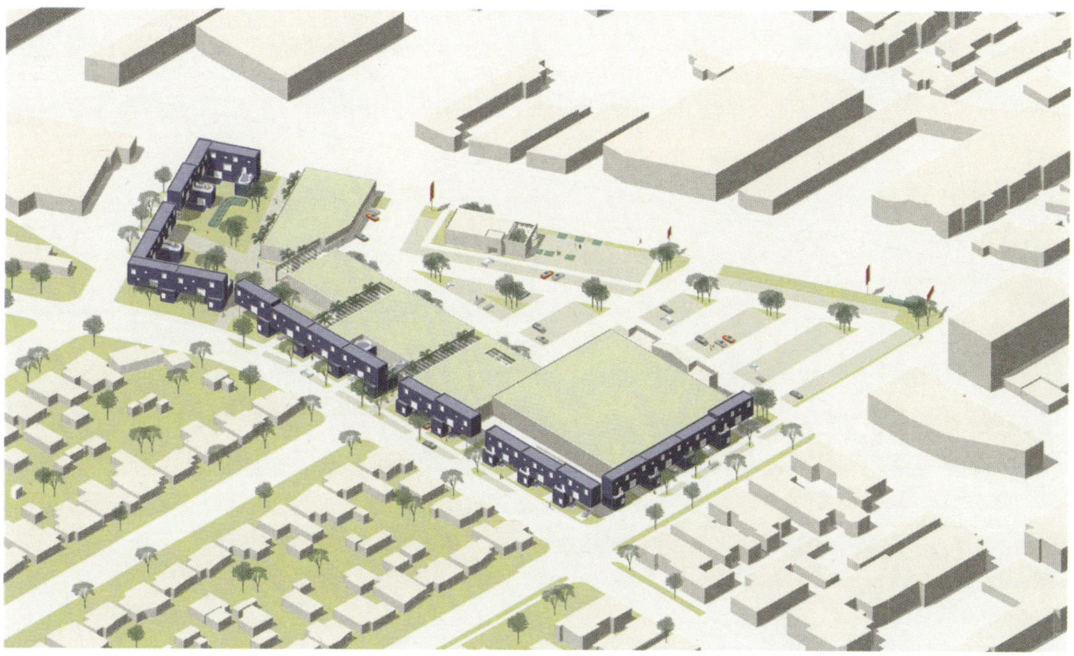

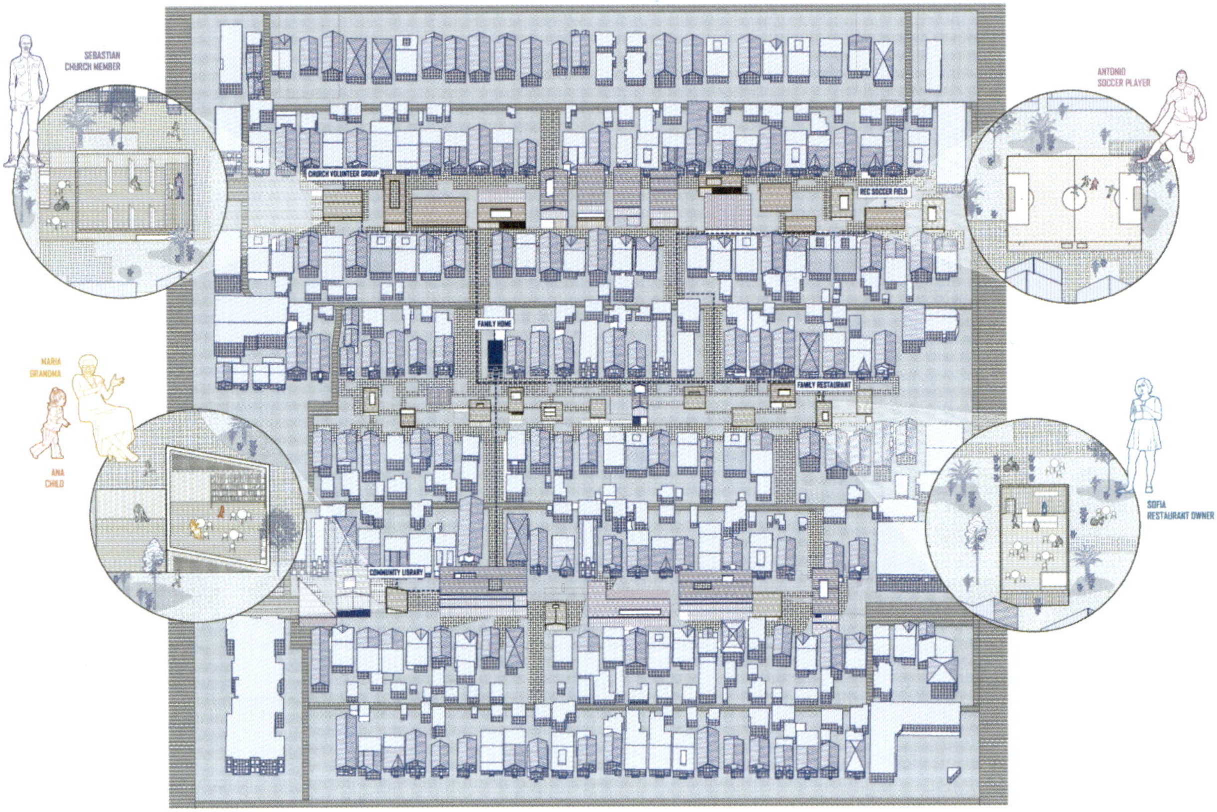

ANDREEA DAN & TESS TEMPLE

AMENDING DEAD SPACES OF AMERICAN DOMESTICITY

Community Vitalization Through the Public Private Intermediary

ARCH 5990 Instructors
Cordula Roser Gray & Ammar Elouini

s> Andreea Dan & Tess Temple

i> Cordula Roser Gray & Ammar Elouini

The standardized suburban typology of the American home populated the nation in the mid 20th century, infiltrating the growing fabric of low-rise city dwellings. This contrast forms a continual disconnect between the realms of private domesticity and public urbanism. A sense of community is lost in this duality, as the inefficient detachment of single-family housing within a contemporary metropolitan creates a divide. It is necessary to re-imagine the variables that define the American home in such a way that encourages continual community growth of neighborhoods, such that they do not decline. This can be possible through the proposal of a new urban corridor introducing the public domain into the previously private residential.

The period between 1940 and 1960 saw unprecedented growth in the urban metropolis of Los Angeles, a city whose version of urbanism has been historically critiqued due to its sprawling neighborhoods, referred to as a "suburban metropolis". South LA, known for its historically African American population, quickly developed in a manner adhering to this suburban home vision, thriving with culture and life specifically in the 1950s. The presence of community in the neighborhood deteriorated quickly as the setup of the individual home proved inefficient, creating dead zones in the areas traditionally labeled as "front yard" and "driveway", and allowing the vehicular street to overtake the neighborhood. As the area was left to decline by the rest of the growing metropolis, violence increased and residents started to under-use these spaces of their homes even more than is seen in other parts of the country. However, the sum of the front spaces between facing residences poses a significant opportunity to re-imagine a new area for urbanism that mediates the scales of the city and individual domesticity.

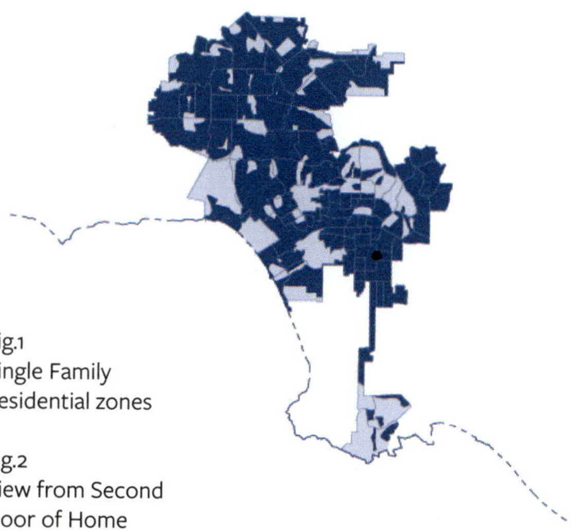

Fig.1
Single Family
Residential zones

Fig.2
View from Second
Floor of Home

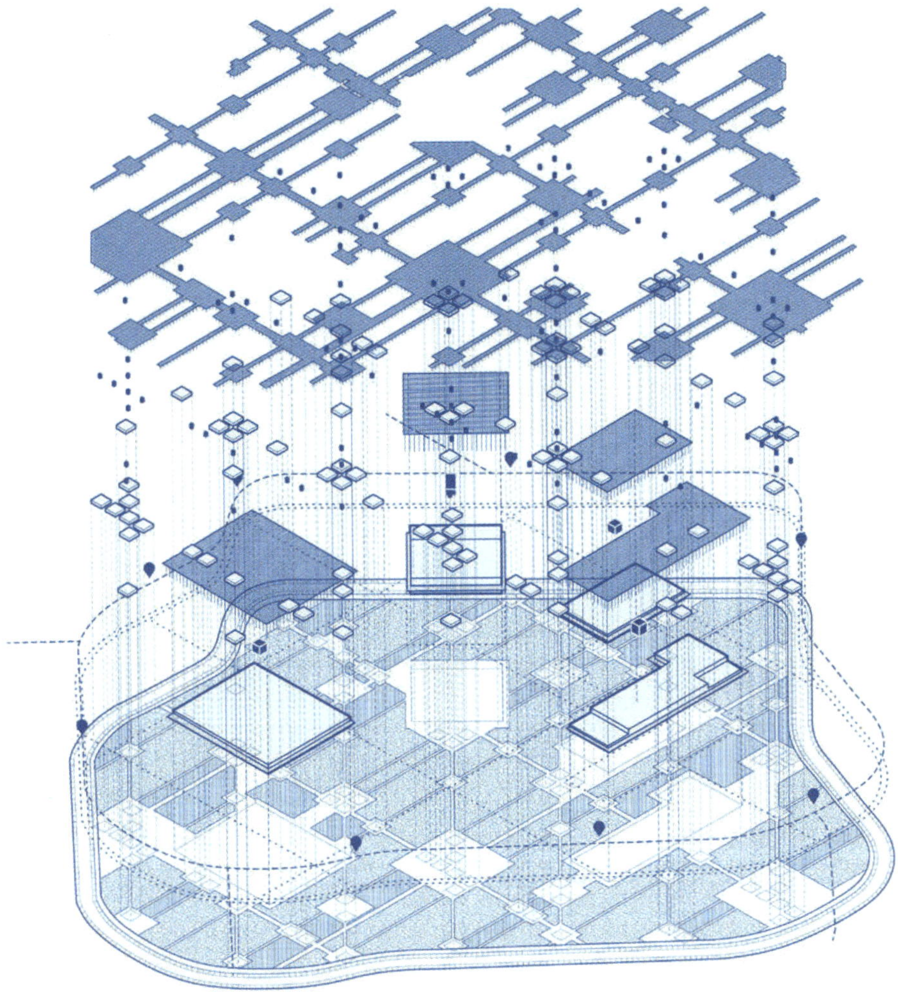

MERRIE AFSETH & CONNOR LITTLE

WHAT TO DO WITH YOUR FAILING MALL

A Guide for Infrastructure Driven Redevelopment

ARCH 5990 Instructors
Cordula Roser Gray & Ammar Elouini

s> Merrie Afseth & Connor Little **i>** Cordula Roser Gray & Ammar Elouini

Once a nexus of the suburban lifestyle, the Acadiana Mall now stands as a shell of its 1979 image, with one of its four box stores six years abandoned, and with eighteen of its smaller stores sitting vacant. The 2004 renovation has contributed to an outdated, dark, and dingy interior, and the site is essentially a pedestrian island only accessible by car: surrounded by a sea of concrete and manicured grass lawns. Conversely, younger generations find walkable communities much more appealing than auto-centric suburbia, opting to move into already-established historic cores. In the Gulf Coast region, sprawling systems of life (primarily energy, water management, single-use zoning, and daily transportation needs) contribute to ineffective disaster management and exacerbate global warming through excessive energy consumption.

Denser communities with more localized daily life patterns and infrastructural systems have proven to be more resilient in the face of disaster. The walkable environment of denser communities can also curb the impacts of climate change; in the U.S, 76% of commuters travel by car to get to work, and the transportation sector contributed to a whopping 1,018 million metric tons of CO_2 emitted into the atmosphere last year alone. Transitioning energy dependency to more localized, greener sources like solar is also paramount in designing for the future. In Louisiana about 86% of energy produced comes from large natural gas facilities, and our energy infrastructure is mismanaged by mega-corporations, with costs associated with damage and outages often falling on the consumers.

These anthropogenic problems are typical of the mall typology regardless of context. Through the development of a field condition, we can organize these demands into a series of general goals for the mall's stakeholders in order to establish an action plan for redevelopment. Through re-imagining the relationships between the box store, the mall, and the community, we can develop a new model for regeneration of the site. Here, the Acadiana Mall serves as a testing ground for a new model of intervention.

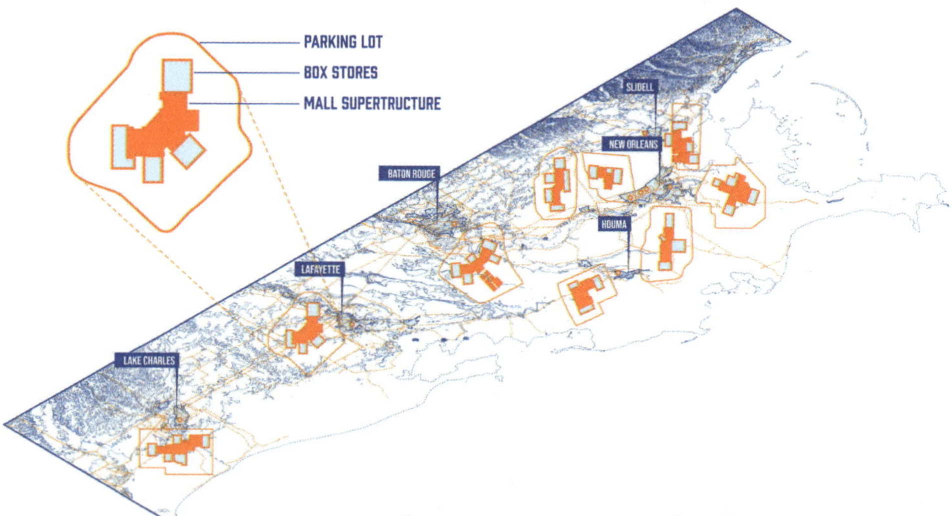

Fig.1
Regional Potential

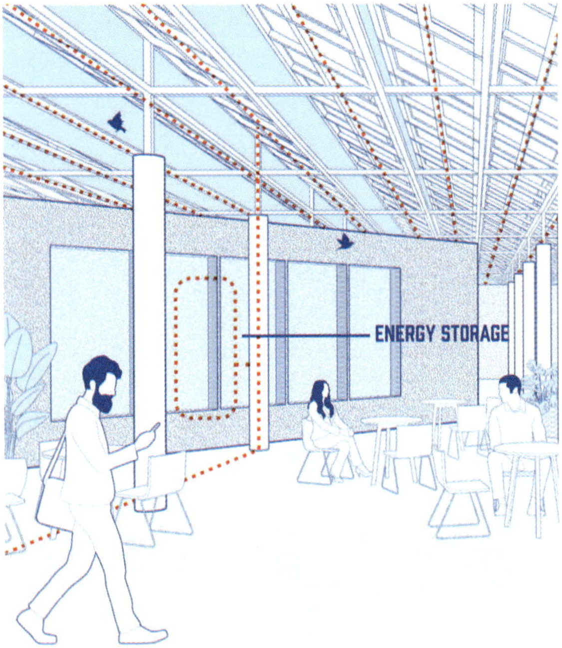

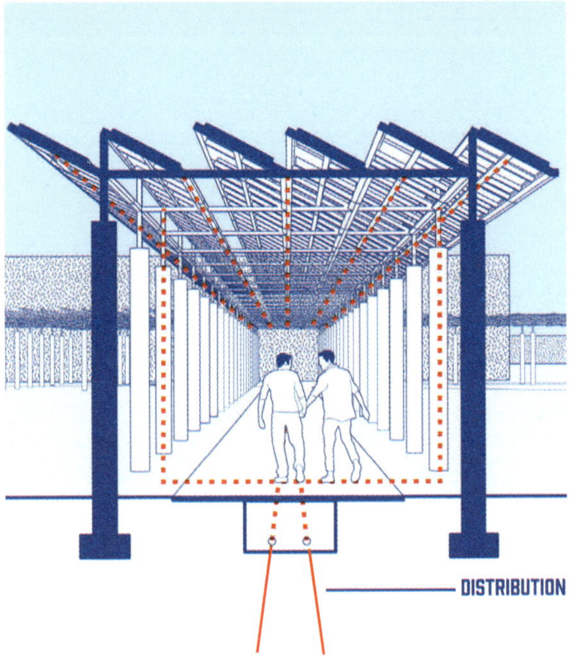

Fig. 2
Energy Park Microgrid Post-Op: Energy Storage

Fig. 3
Energy Park Microgrid Post-Op: Distribution

Fig. 4
Energy Park Microgrid

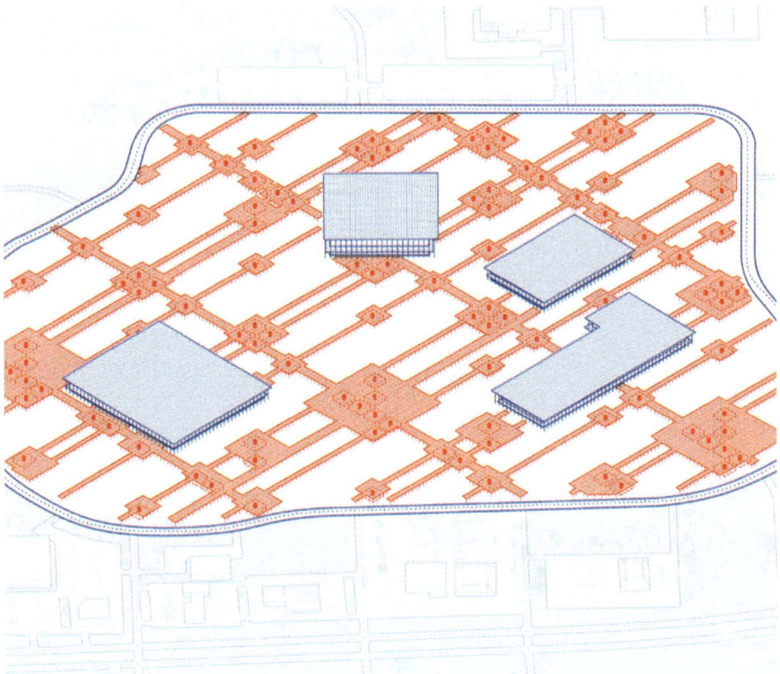

Fig. 5
Facade Face Life Post-Op

Fig. 6
The Prarie

RICHARDS

TULANE SCHOOL OF ARCHITECTURE

2021-2023

LECTURES

Lecture Series 2021-2022
Debate, Delete, Reboot

Rebecca Elliott
October 2021

Rania Ghosn
October 2021

Germaine Barnes
November 2021

Martha Schwartz
November 2021

Washington Fajardo
November 2021

Michael Meredith
November 2021

Hernán Díaz Alonso
December 2021

Lecture Series 2022-2023

Peggy Deamer
September 2022

Shalini Agrawal
September 2022

Kate Wagner
October 2022

Jenny Sabin
November 2022

Ronald Rael
February 2023

Luz Briceño
March 2023

Other lectures

Caribbean Voices: Contemporary Architecture Practice in Puerto Rico
José Coleman-Davis (A '75), FAIA and Luis Ayala Rubio (A '93), AIA
[February 2022]

h2o architectes "Openings"
Jean-Jacques Hubert and Antoine Santiard
[March 2022]

Short Housing Stories
Cynthia Dubberley (A '97), Juan Medina Revilla, and Pia Mendaro
[April 2022]

Historic Preservation Lecture:
Dr. Charlotte Andrews, Bermuda National Trust
[September 2022]

Historic Preservation Lecture: "Cultural Heritage Conservation in the Anthropocene"
Morris "Marty" Hylton III
[March 2023]

Joan Busquets
[April 2023]

EXHIBITIONS, INSTALLATIONS AND DESIGN COMPETITIONS

Small Center Exhibit: The Atmosp(HERE) of Space. 2021-2022
Albert and Tina Small Center for Collaborative Design The center is also hosting a series of public writing workshops called "Scribe Tribe w/ Sharita A. Sims: The Spirituality of Home" Tuesday evenings in November.
[November 2021]

Small Center Exhibit: Legacy of Central City
Legacy of Central City, curated by Civic Studio and featuring work from Polo Silk, Big Chief Beautiful, Sharita Ariel, José Cotto (alum M.Arch *14 & faculty)
[February - May 2022]

New Orleans Architecture + Design Expo
The National Organization for Minority Architects - Louisiana partnered with Lecturer Laura Blokker's 'Identity and Agency in Architecture' class to host the first Architectural and Design Expo in New Orleans. [March 2022]

Newcomb Quad Pavilion Design Competition
Winning Design: MELT by students Jose Varela, Chelsea Kilgore, and James Poche
[Spring 2022]

Exhibit: "Public Space & Scrutiny: How Do We Remember?"
Tiffany Lin, Emilie Taylor Welty and, Students in Public Space & Scrutiny Research Studio (ARCH 4042/6052)
[May-June 2022]

Sukkah 14 Design-Build
Seth Laskin and Ethan Lewis
[October 2022]

Small Center Exhibit: INHABITATION
The Albert and Tina Small Center for Collaborative Design's Fall 2022 DEPART-MENT exhibition and program series.
Kira Akerman, Flora Cabili , Dr. Robin McDowell and Annabelle Pavy
[October - December 2022]

Exhibition of Light and Pattern
Students in Design Studio I (DESG 2005)
[November 2022 and November 2023]

Small Center Exhibit: EXTRACTIVISM
The TuSA Albert and Tina Small Center for Collaborative Design's Spring 2023 DEPART-MENT exhibition and program series. Anya Groner, Jazmin Miller.
[February - April 2023]

Memorial Installation
Emilie Taylor Welty, supported by the Tulane Duren Professorship program, and architecture students in the Duren seminar installed a memorial to ancestors at a local non-profit that works with BIPOC - LGBTQ youth in the city.
[Spring 2023]

EVENTS & WORKSHOPS

ASG Gallery + Auction
Organized by and in support of the Tulane Architecture Student Government
[December 2021 and February 2023]

Design Symposium: Utopia Dystopia
Organized by the Tulane Architecture Student Government. Cruz Garcia (Wai Think Tank), Tomas Musca (Cassius Castings), Bryan Maddock (Dimensions.com)
[April 2022]

Colloquium: Bringing New Life to an Old City
Organized by Tulane School of Architecture Graduate Government
Regina M. Davis, AIA LEED AP BD+C (Trapolin Peer Architects)
Kenner Cardmody, AIA (Waggonner & Ball Architecture and Environment
William H. Riehm (University of Louisiana at Lafayette)
[April 2022]

Data by Community by Design (DXCXD)
Co-organized by Tulane School of Architecture and Tulane School of Science and Engineering
Edson Cabalfin, Associate Dean for Equity, Diversity, and Inclusion, worked with Tulane's Department of Computer Science to host Data by Community by Design (DXCXD).
[April 2022]

Design Symposium: Material World - Futures of Tactile and Digital Expression
[March 2023]

Student Travel Fellowship Presentations
Karan Sharma (M.Arch '23 candidate), Seth Laskin (B.Arch '23 candidate), Chris Kolodey (M.Arch '23 candidate), Ethan Lewis (B.Arch and BSRE '23 candidate), Giuliana Vaccarino Gearty (M.Arch '23)
[March 2023]

Colloquium: The Future of Housing - Community, Commerce & Carbon Neutrality
[March 2023]

Design Showcase + Fashion Show
Students in the Design Program
[May 2023]

Small Center Exhibit: Home, Sweet Home
Emmanuel Osorno, Omar Ali and Nimet Anwar
[May 2023]

FACULTY

2024. Regular and full-time faculty

Fallon Samuels Aidoo
Assistant Professor in Real Estate Development and Historic Preservation

Iñaki Alday
Richard Koch Chair Professor in Architecture
Dean

Scott Bernhard
Associate Professor in Architecture,
Associate Dean for Academics

Hannah Berryhill
Professor of Practice in Design

Will Bradshaw
Professor of Practice in Real Estate Development

Edson Cabalfin
Professor of Practice in Social Innovation Social Entrepreneurship
Associate Dean for Equity, Diversity and Inclusion
Program Director, SISE

Richard Campanella
Mintz Senior Professor of Practice
Associate Dean for Research

Liz Camuti
Assistant Professor in Landscape Architecture

Rebecca Choi
Assistant Professor in Architecture

Ammar Eloueini
Professor in Architecture
Director of Graduate Programs in Architecture

Ruben Garcia Rubio
Assistant Professor in Architecture

Bruce Goodwin
Associate Professor in Architecture

John Huppi
Professor of Practice in Real Estate Development
Associate Program Director, Real Estate Development

Margarita Jover
Professor in Architecture and Landscape Architecture
Program Director, Landscape Architecture and Engineering

Jesse Keenan
Favrot II Associate Professor in Real State Development

Irene Keil
Senior Professor of Practice in Architecture

Judith Kinnard
Professor in Architecture
Harvey-Wadsworth Chair
in Landscape Urbanism

Tiffany Lin
Favrot V Associate Professor in Architecture and Design
Program Director, Design

Adam Marcus
Associate Professor in Architecture

Wes Michaels
Associate Professor in Landscape Architecture

FACULTY

2024. Regular and full-time faculty

Byron Mouton
Michael G. Lacey Senior Professor of Practice in Architecture
Director, URBANbuild

Casius Pealer
W. Henry Shane Professor of Practice in Real Estate Development
Program Director, Real Estate

Wendy Redfield
Associate Professor in Architecture

Carol McMichael Reese
Favrot IV Professor in Architecture

Daniela Rivero-Bryant
Professor of Practice in Real Estate Development

Cordula Roser Gray
Senior Professor of Practice in Architecture

Megan Saas
Profesor of Practice in Design
Associate Director, Design

Kenneth Schwartz
Professor in Architecture
Michael Sacks Chair in Civic Engagement and Social Innovation
Director of Phyllis M. Taylor Center for Social Innovation and Design Thinking

Versé Shom
Professor of Practice in SISE

Jill Stoll
Professor of Practice in Design

Emilie Taylor-Welty
Favrot III Professor of Practice in Architecture
Program Director, Architecture
Design-Build Manager, The Albert & Tina Small Center for Collaborative Design

Kentaro Tsubaki
Favrot Associate Professor in Architecture

Sonsoles Vela Navarro
Assistant Professor in Architecture

Heather Veneziano
Professor of Practice in Historic Preservation
Associate Director, Historic Preservation

Ann Yoachim
Professor of Practice
Director, The Albert & Tina Small Center for Collaborative Design

2024. Visiting faculty

Tyler Antrup
Visiting Assistant Professor in Real State Development

Luz Briceño
Visiting Assistant Professor in Design

Sean Fowler
Research Assistant Professor in Architecture

Patricia Fraile
Visiting Assistant Professor in Architecture

Zaid Kashef Alghata
Favrot Visiting Assistant Professor in Architecture

Hannah Kenyon
Visiting Assistant Professor in Architecture and Design

Austin Lightle
Visiting Assistant Professor in Architecture

Andrew Liles
Visiting Assistant Professor in Architecture and Design

Javier Marcano
Visiting Assistant Professor in Real Estate Development

Jesus Meseguer
Visiting Assistant Professor in Architecture

Cristobal Molina
Favrot Visiting Professor

Nick Perrin
Visiting Assistant Professor in Design

Mark Rabinowitz
Christovich Visiting Professor
Interim Director, Historic Preservation

Maria Redondo Perez
Pre-doctoral Fellow in Architecture

Sofia T Shwayri
Visiting Assistant Professor in Urban Studies

BJ Siegel
Favrot Visiting Professor

Kelly Tierney
Visiting Assistant Professor in Design

FACULTY

2024. Adjunct faculty

Sienna Abdulahad
Adjunct Lecturer in SISE
Dwan Adams
Adjunct Lecturer in SISE
Jacob Alter
Adjunct Lecturer in Architecture
Erinn Banks
Adjunct Lecturer in SISE
Stephanie Barksdale
Adjunct Lecturer in SISE
Alon Barzilay
Adjunct Lecturer in Real Estate Development
Sarah Berger
Adjunct Lecturer in SISE
Jason Blankenship
Adjunct Lecturer in Architecture
Natalie Blaustone-Dye
Adjunct Lecturer in SISE
Erika Bradford
Adjunct Lecturer in SISE
Bryan Bradshaw
Adjunct Lecturer in Design
Patrick Beard
Adjunct Assistant Professor in Real Estate Development
Ben Bolz
Adjunct Lecturer in Real Estate Development
Jada Buckner
Adjunct Lecturer in SISE
ZoAnn Campana
Adjunct Lecturer in Historic Preservation
Robert Cangelosi
Adjunct Lecturer in Historic Preservation
Bear Cheezem
Adjunct Lecturer in Real Estate Development
Gabe Christian-Sola
Adjunct Lecturer in SISE
Annie Clark
Adjunct Lecturer in Real Estate Development
Ira Concepcion
Adjunct Lecturer in Architecture
Jackie Dadakis
Adjunct Lecturer in Real Estate Development
Michael Dalle Molle
Adjunct Lecturer
Woodshop Manager
Cynthia Dubberley
Adjunct Lecturer in Architecture
Michelle Duhon
Adjunct Lecturer in Historic Preservation
Todd Erlandson
Adjunct Lecturer in Architecture
Maille Faughnan
Adjunct Lecturer in SISE
Marion Forbes
Adjunct Lecturer in Design
Vic Franckiewicz
Adjunct Assistant Professor in Real Estate Development
Anne Gibson
Adjunct Lecturer in SISE
Darryl Glade
Adjunct Lecturer in Real Estate Development
Pankaj Vir Gupta
Adjunct Lecturer in Architecture
David Harms
Adjunct Assistant Professor in Real Estate Development
Austin Hogans
Adjunct Lecturer in Architecture
Jonathan House
Adjunct Lecturer in Architecture and Design
Hugh Jackson
Adjunct Lecturer in Architecture
Daniel Jacobs
Adjunct Lecturer in SISE
Dasjon Jordan
Adjunct Lecturer in Architecture

Shallene Joseph
Adjunct Lecturer in SISE
Jillian Keiffer
Adjunct Lecturer in Design
Gina LaMacchia
Adjunct Assistant Professor in Real Estate Development
Julia Lang
Adjunct Lecturer in SISE
Liz Langlois
Adjunct Lecturer in SISE
Joseph Latson
Adjunct Lecturer in Real Estate Development
Wendy LeBlanc
Adjunct Lecturer in SISE
Patrick Long
Adjunct Lecturer in SISE
Alex Lopez
Adjunct Lecturer in Historic Preservation
Abigail Lukens
Adjunct Lecturer in SISE
MC Matucheski
Adjunct Lecturer in Architecture
Andrew (John) Mayronne
Adjunct Lecturer in Real Estate Development
Juan Medina Revilla
Adjunct Lecturer in Architecture
David Merlin
Adjunct Lecturer in Architecture
Anna Monhartova
Adjunct Lecturer in SISE
Adam Newman
Adjunct Lecturer in Design
Becky Otten
Adjunct Lecturer in SISE
Martha Pearson
Adjunct Lecturer in Design
Matt Perrenod
Adjunct Assistant Professor in Real Estate Development
Emily Riemer
Adjunct Lecturer in SISE
James Rolf
Adjunct Lecturer in Historic Preservation
Cassidy Rosen
Adjunct Lecturer in Architecture
Catherine Sckerl
Adjunct Lecturer in Architecture
LeShawn Simplis-Barnes
Adjunct Lecturer in SISE
Megan Slattery
Adjunct Lecturer in Architecture
Z Smith
Adjunct Lecturer in Architecture
Shaina Spector
Adjunct Lecturer in SISE
Jesse Toohey
Adjunct Lecturer in Design
Millhaus Manager
Nichole Valenzuela
Adjunct Lecturer in SISE
Tucker Van Leuwen-Hall
Adjunct Lecturer in Architecture
Valerie Vides
Adjunct Lecturer in Historic Preservation
Bruce White
Adjunct Lecturer in Architecture
Thomas Wimberly
Adjunct Lecturer in Design
Rosie Yates
Adjunct Lecturer in SISE

EMERITI

Errol Barron
Emeritus Professor of Architecture

Geoffrey Baker
Emeritus Professor

Eugene D. Cizek
FAIA, Emeritus Professor of Architecture and Historic Preservation

Ronald Filson
Emeritus Dean

Karen Kingsley
Emerita Professor

John P. Klingman
Emeritus Professor of Architecture

Richard Powell
Emeritus Professor

John Stubbs
Emeritus Sr. Professor of Practice

Ellen Weiss
Emerita Professor

STAFF

Alexandria Andara
Program Manager, Faculty Employment

Savannah Chiasson
Administrative Program Coordinator

Skyler Choice
Financial Services Specialist

José Cotto
Collaborative Design Project Manager, Albert & Tina Small Center for Collaborative Design

Christy Crosby
Assistant Dean for Administration and Operations

Allison Cruz
Assistant Dean for Academics

Michael J. Cusanza
Assistant Director of Admissions

Andrea Elnems
Director of Finance and Budget

Naomi King Englar
Communications and Marketing Manager

Victor Garcia
Technology Manager

Alvin Green, Jr.,
Building Manager

Sean Huff
Executive Administrative Assistant

Nick Jenisch
Project Manager, Albert and Tina Small Center for Collaborative Design

Kristen Jones
Assistant Director, Student Affairs

Nick LiCausi
Director of Fabrication

Andrew Lorelli
Academic Advisor

Angelle Keller
Program Manager of Academics

Hailey Mathieu
Administrative Program Coordinator, Student Affairs

Keyoka Nelson
Program Manager, Office and Student Employment

Abraham Passman
Administrative Program Coordinator

Catherine Restrepo
Visual Design Manager

Jazmine Smith
Web Developer I